BATTLE WINNERS

Australian Artillery in the
Western Desert 1940—1942

Alan H. Smith

ECHO BOOKS

First published in 2014 by Barrallier Books Pty Ltd, trading as Echo Books
Registered Office: 35-37 Gordon Avenue, West Geelong, Victoria 3220, Australia.
www.echobooks.com.au

Copyright ©Alan H. Smith

National Library of Australia Cataloguing-in-Publication entry.
Author: Smith, Alan H., author.

Title: Battle winners : Australian artillery in the Western Desert 1940 - 1942 / Alan H. Smith.

ISBN:9780987586438 (paperback)

Subjects: Australia. Army--Artillery--History. Australia. Army. Division, 6th. Australia. Army. Division, 9th. Tobruk, Battles of, 1941-1942. El Alamein, Battle of, Egypt, 1942. World War, 1939-1945--Campaigns--Egypt--Participation, Australian. World War, 1939-1945--Campaigns--Libya--Participation, Australian. World War, 1939-1945--Campaigns--Africa, North--Participation, Australian. World War, 1939-1945--Regimental histories--Australia.

Dewey Number: 940.5423

Book and cover design by Peter Gamble, Ink Pot Graphic Design, Canberra.
Set in Garamond Premier Pro 12/17 and Minerva

www.echobooks.com.au

OTHER BOOKS BY THE SAME AUTHOR

The Life and Times of Major General Timothy Frederick Cape,
CB,CBE,DSO

Gunners in Borneo,
Artillery During Confrontation 1962-1966

Do Unto Others,
Counter Bombardment Artillery in Australia's Military Campaigns

Contents

Foreword	ix
Foreword	xi
Acknowledgements	xiii
List of Maps and Diagrams	xvii
Introduction	xix
Author's Note	xxiii
1 From Volunteers to Veterans	1
2. Artillery Development Between the Wars	13
3. The Development of Artillery Doctrine	25
4. The BEF Artillery in France—1944	37
5. The Artillery Mobilisation and the Field Regiment	49
6. Palestine to Benghazi	69
7. The 9th Division in the Western Desert	95
8. Cyrenaica and Tobruk	107
9. The 9th Division artillery (2/7 and 2/8 Field Regiments) in Egypt—1941	145
10. The British Experience, 1941–42	159
11. First Alamein and Alam Halfa	175
12. The Other Side of the Hill: Enemy Command, Tactics and Resouces	197
13. The 9th Division Artillery Prepares	211
14. Artillery Command, Resources and Counter Bombardment	223
15. A Regiment Prepares–the 2/7 on 23 October 1942	241
16. Lightfoot and Supercharge: The Anatomy of an Artillery Battle	251
17. Aftermath and Analysis	277
18. The Western Desert 'Balance Sheet', 1941–42	295
19. Accolades	305
20. Epilogue and Valette	313
Notes to Sources	321

Endnotes	325
Appendix 1–Establishment of a Field Regiment, RAA	347
Appendix 2–9th Australian Infantry Division Order of Battle	353
Appendix 3–El Alamein—Characteristics of German, British and Italian Guns	357
Appendix 4–9th Division Operation Order for The Battle of El Alamein	361
Appendix 5–Artillery Honours and Awards	367
Appendix 6–Artillery Casualties in the Western Desert	371
Appendix 7–The Post—War Lives of the Major Figures in the Wetern Desert	383
Glossary	389
Index	395

Dedication

*This book is a tribute to all those officers and other ranks of the 2nd Australian Imperial Force
and supporting British, New Zealand and South African gunners
who participated in battles in the Western Desert of Egypt and Cyrenaica from December
1940 to November 1942.*

Foreword

In Australian Military history the apogee of the artillery as a decisive combat arm arguably was at El Alamein during the period October to November, 1942. During the climactic last 12 days of the battle, Eighth Army artillery fired more than one million rounds of 25 pounder ammunition, the 354 25 pounders in XXX Corps fired 577 rounds per gun on the first night and overall averaged 1,000 rounds. The Australian guns fired almost 50 per cent more than the corps average. The guns that performed this prodigious task came from the field regiments of the famous 9th Division (2/7th, 2/8th and 2/12th). That they were able to perform at this exacting level is a tribute to the adoption of Royal Artillery doctrine by the 6th Division's artillery in early 1941 and its subsequent development throughout the North African campaign.

This book gives a detailed insight into how artillery came to be applied with great success in North Africa especially by the 9th Division. They were to master the frequent need for re-deployments, to develop gun areas against tank attack, to develop kills and discipline to participate in major fire plans, and to handle vast quantities of ammunition. The book covers artillery tactics, fire planning and the intricacies of counter-bombardment. The Royal Artillery units and headquarters that were an integral part of the story have been included. Importantly, this book does not restrict itself just to artillery techniques, tactics and procedures. Some outstanding gunners served in the artillery of the 6th and 9th Divisions: Alan Smith summarizes their various contributions. Among these were Edmund 'Ned' Herring, CRA 6th Division and later to command New Guinea Force, Alan Ramsay, the CRA of 9th Division, who was later to command the 5th and 11th Divisions. It should be a matter of pride for Australian Gunners that Ramsay's operational orders for the battle of El Alamein were afterwards circulated as a model within the British Army. Another to be mentioned is Tom Eastick, initially the CO of the 2/7th Field Regiment and then CRA of both 7th and 9th Divisions. They were all later knighted.

A whole lexicon of gunner terms has all but disappeared since El Alamein. The 'Stonk', a splendidly descriptive word for a divisional defensive fire concentration had, to my knowledge, fallen into disuse by the early 1960s; so have other terms, 'Zero Line' by 'Centre of Arc'; 'check window', a drill between troop and battery command post, is no longer necessary, and the observation post officers and forward observation officers have all been replaced by a succession of other terms. Yet, it is worth remembering in these days of command post computers and other technical aids that, in 1942, a well-trained battery could report 'shot' two minutes after receipt of orders using tabular and graphic data. Of course, manual fire planning (and there was 120 plus serials in some el Alamein fire plans) took considerably longer than now!

In many respects this is a specialist book, but it is not only for Gunners. Anyone interested in the application of combat power on a high intensity, conventional battlefield will profit from reading it. Alan Smith is to be congratulated on producing this record of our artillery at its peak.

Major General Steve Gower, AO, AO (mil) (Retd)
Director, Australian War Memorial
Canberra, 2003

Foreword

This is a book which will appeal both to the serious military historian and to the general reader with an interest in the Western Desert battles of 1940-1942, and the part played by the Australian Army, including the climax at El Alamein While Alan Smith goes into great detail regarding the activities of individual artillery regiments and batteries, and the personalities involved, he also deals with the successes, and, until the arrival of Montgomery, with many failures and defeats sustained by the British Commonwealth forces as a whole.

Although critical of certain senior commanders, he is full of praise for the front line soldier, and quite rightly emphasizes the excellence of the Australian and British artillery which was considered by both Montgomery and Rommel to be the outstanding arm in both defence and attack.

Alan Smith refers frequently to the 7th Medium Regiment of the Royal Artillery. After commissioning in 1943 I was posted to this regiment on its return to England from Sicily as the youngest officer in the resurrected 25/26 Battery. From our landing on the beaches of Normandy in June 1944 to our final 'Empty guns' in North Germany, by way of the Seine, Nijmegen and the Ardennes and the assault across the Rhine, I hope and believe that we maintained the reputation rightly earned with our comrades-in-arms' in the Western Desert.

> Brigadier David Baines, RA, MBE
> (Senior surviving officer of 7th Medium Regiment, RA)
> Surrey, UK, 2003

Acknowledgements

I owe a great deal of gratitude to many people who have helped me compile this record. I would like to thank retired officers Major Generals John Whitelaw, AO, CBE and Steve Gower, AO, AO(Mil); Lieutenant Colonel C. F. Dodds, OAM, RAA; Colonel T. A. Rodriguez, MVO, MBE, MC; and Lieutenant Colonel Roger Fitzhardinge, formerly Command Post Officer with the 2/7th Field Regiment at Alamein for their very helpful comments. I am grateful for other valuable contributions from Mr Max Parsons, historian of the 2/12th Field Regiment, and for his help in other fields, particularly in publishing matters. I am beholden also to Mr W. Stevens (formerly 2/7th Field Regiment Adjutant and later Staff Captain, HQRAA 9th Division), Mr C.C. Bailey, OAM, who generously provided material, as did the late Mr Charles G. McKenzie (per favour of Mr Graeme Petterwood) from the 2/8th Field Regiment Association publication The Tannoy that filled a critical gap. I trust that those readers connected with the 2/8th and 2/12th Field Regiments will forgive my more extensive use of material from the 2/7th Field Regiment sources. Those events recorded by the 2/7th Field Regiment historians in this account are representative of the deeds and exploits of all three regiments in the Battle of El Alamein.

My thanks to Major Alan Sandbach, RAA (Ret'd), a grandson of Major General Alan Ramsay, who provided copies of Ramsay's personal details and papers pertinent to the battle for use in this volume. I would also like to thank the staff of the Research Centre, Australian War Memorial, particularly Margaret Lewis, for their collective help. Finally, I am very grateful to Mr R.C.M. Toplis at the Library/Research Centre of the Royal Australian Artillery Historical Company, National Artillery Museum, at North Fort for his valuable assistance.

Likewise, I am grateful for the help of the Hon. John Smith for his research at the Public Records Office, Kew, and the assistance of the staff; Captain Alan de G. Benson, formerly

Regimental Survey Officer and Intelligence Officer, and Captain Peter Langrishe of the 7th Medium Regiment, RA, whose reminiscences have provided me a source of inspiration; Major Frank Hamer, MC, 4th Survey Regiment, RA (TA), John Bensley, Esq., and Sr Max Mangilli-Climpson, the latter two associated with RA survey regiment histories who generously provided accounts of operations and photos. Major John Timbers, RA, former Editor of the Journal of the Royal Artillery (Woolwich) and Lieutenant Colonel Francis, Executive Officer, Tactical Doctrine Retrieval Centre, Upavon, Wiltshire (UK), have also obliged unhesitatingly and to them I extend my gratitude. The Major Hugh Skillen (Royal Signals) kindly added verbatim to his written record on vital intelligence matters. Mr Ken Jorgenson verified data on the 2/1st Field Regiment and Brigadier Keith Rossi (Ret'd) did likewise in connection with the 2/2nd Field Regiment, for which I thank them. For my coverage of the 2/3rd Field Regiment and its brief role in the affairs of the 6th Division I wish to thank Mr Les Bishop, the 2/3rd's historian for his kindness. Mrs Wendy Salmon graciously saved me from my desperate typing skills in times of trial. My thanks to her and to those others who have helped me and who I hope will recognise their contribution.

Finally, to Mr Roger Lee, Head of the Army History Unit and his expert staff, Dr Andrew Richardson and Lieutenant Colonel Bill Houston (Retd) and to Dr Albert Palazzo, formerly of the Australian Defence Force Academy, I owe a great deal for their critique, comments and support. Likewise, Mr Ric Pelvin tendered expert editorial advice for which I am most grateful. Mr Keith Mitchell created the 'bespoke' maps and diagrams that illustrate the battles and explain the arcane artillery science necessary for excellent gunnery.

I have been saved from including some egregious errors in the text, diagrams etc. by coincidental circumstances, chance discoveries when checking data and, in some cases, intuition. The balance is heavily in favour of enhancing the narrative and illustrations, I'm pleased to say. Any errors remaining are mine alone.

List of Maps and Diagrams

1. 6th Division attack on Bardia, Phase I, showing artillery tasks and gun positions
2. The second 6th Division attack on Bardia, Phase II, to capture the fortress, showing artillery tasks and dispositions.
3. 6th Division attack on Tobruk showing Italian Hostile Batteries, and British/Australian gun regiments dispositions on the evening of 21 January 1941
4. Locations of British and Australian gun regiment areas in the Tobruk salient during the siege in 1941.
5. DAK panzer, artillery and lorried infantry formations A - On the Move; B—Encounter Battle.
6. DAK tactical formations for C - Assault Phase and D, a version of a panzer mobile defence.
7. The Survey Plan of 4th Survey Regiment, RA for the El Alamein front.
8. Flash Spotting and Sound Ranging Bases at El Alamein.
9. 30 Corps and 9th Division artillery communications networks at El Alamein.
10. 30 Corps Front showing divisional objectives, initial gun areas and enemy forces dispositions, including Hostile Batteries.
11. 9th Division gun areas including 7th and 64th Medium Regiments, RA
12. Divisional Fire Plan for 23 October El Alamein from infantry Start Line to infantry objectives.
13. An overview of the brigade/battalion attacks from 23 October to 2 November
15. The detailed fire planning for infantry attacks north and eastwards towards Thompsons Post.
15. Barrage and Defensive Fire Tasks for Operation Supercharge on 1/2 November

Introduction

For me, becoming immersed in writing about the Australian contribution to first, the Battle of El Alamein and second, other battles of the Western Desert of 1940—42 was entirely adventitious. I was researching for articles on the Australian Military Force's experiences with counter-bombardment since the Anglo—Boer War for the Royal Australian Artillery Historical Company's journal *Cannonball* in 1995. As I worked through World War I in all the histories and journal articles there were plenty of references to 'artillery duels' and 'barrages', both terms beloved by most writers and journalists but which are more accurately expressed as 'counter-bombardment or counter-battery fire' and 'concentrations'. Real 'barrages' come in three forms—creeping, rolling and block and expend much artillery and ammunition. This research was highly educational for me because I had only very limited exposure to counter-bombardment as an art/science in my regimental service—it was the purview of another branch of the Regiment that was highly specialised. My only exposure came during command post technical work (exercises) when a member of the Directing Staff person would introduce a piece of paper entitled 'Hostile Battery List'. This list provided coordinates and other details which we plotted studiously and, from time to time, 'fired'. But it was still a 'black art' to me, something to catch up on when I had time.

While I knew and recognised the importance of counter-bombardment in the Battle of El Alamein from my study of military history, I could never find a more detailed record of the contribution of field artillery and other branches of the Regiment to its successful employment in the other battles involving AIF artillery. I felt that its contribution to the overall results achieved by Australian and associated artilleries had somehow been overlooked. Thus I turned from examination of one major part of the battle to the broader view. From the 1950s only David Goodhart's *We of the Turning Tide* and later *The History* of 2/7th Australian Field Regiment described in detail the Regiment's part in the battle providing support for the infantry over the four-month period July to November 1942. Regimental histories of the 2/1st and 2/2nd Field Regiments contained brief references to counter-bombardment, but have much more on supporting infantry with, for example, barrages. Embedded in Barton Maughan's official volume there are several references to artillery's importance, a few to the British, New Zealand and South African field and British medium regiments and none to the XXX Corps survey regiment and counter-bombardment staff.

The Australian artillery worked very closely with three Royal Artillery units, all of which had long associations with the AIF artillery of both the 6th and 7th divisions in the early years of the Western Desert war. There was a dearth of information on these Royal Artillery units, but a request in *Everywhere*, the newsletter of the 94th Observation Regiment, RA, produced four men, two of whom had served with the 7th Medium Regiment, RA (one of whom was a joint author of its history) and two who were RA Survey Regiment historians. Thus I was able to provide much new material to augment RAA artillery history in my other wrtings.

Later I was to discover that there was no history of AIF divisional artilleries in World War II with the possible exception of two chapters of Stuart Sayers' biography of Lieutenant General Sir Edmund Herring (Chapters 14 and 15). Thus this history hopefully fills a gap, admittedly for a relatively short but very intense period of warfare. It is significant because this period covers the first time that the 6th and 9th Division artilleries functioned (mostly) as an organic part of their division. After the 6th was diverted to Greece and the 7th was about to engage in the Syrian campaign, the 9th Division remained in Palestine assembling its units. Its 2/12th Field Regiment was sent to Tobruk in 1941 for the siege while the other two regiments, the 2/7th and 2/8th, served with equal distinction in Cyrenaica with XXX Corps and British formations.

Thus, when Major General Leslie Morshead's 9th Division headed west out of the Levant and Egypt in late June 1942, his division was up to strength, well trained and highly motivated. Its artillery, under Brigadier Alan Ramsay, had become extremely efficient—the best in the Middle East, as will be revealed, and why. Its performance and achievement constituted the highest point achieved by the biggest branch of the Regiment during World War II—hence the title for this history. Ramsay, by virtue of the important role the division was to play, commanded more artillery than any other Australian gunner in World War II and subsequently. In terms of ammunition expenditure, the 9th Division outstripped both the 6th and 7th divisions handsomely. It rightfully earned its high reputation for gunnery and infantry support.

Alan H. Smith

Author's Note

By the time of the 70th anniversary of the Battle of El Alamein hundreds of books covering campaigns in the Western Desert of North Africa had been produced. These ranged from eyewitness accounts written by other ranks and officers—from field marshal to lieutenant—or by others with a point to make about some aspect of the battles. Most of the latter focus on commanders and their decisions and are based on an examination of records and data that the commanders concerned did not have access to at the time. These include wireless and telephone logs, Ultra decrypts, War Cabinet decisions, signals intelligence that emerged after the smoke and dust settled, and so on. Who was it said that post-mortems were seldom wrong? The outbreak of hostilities in the Western Desert in 1940 saw Major General Iven Mackay's 6th Australian Division join Lieutenant General Richard O'Connor's XIII Corps with the British 7th Armoured Division. Within six weeks the Italian Army had suffered debilitating losses in men and materiel at their hands. To prop up Benito Mussolini's corrupt fascist regime, Adolf Hitler ordered *Generalleutnant* Erwin Rommel to North Africa early in 1941 to lead German and later Italian land and air forces against British/Commonwealth formations. The siege of Tobruk that followed from April 1941 to October 1942 was a significant military event. This involved another Australian infantry division, the 9th, its combative and talented commander, and mostly British armour and artillery.

The siege has attracted its share of historians, as has El Alamein. One common thread in the historiography of both these campaigns has been the detailed emphasis on descriptions of armour and infantry encounter battles, from section level to battalion/regiment level and beyond. Man against man, tank against tank and anti-tank gun and gun versus tank. Seldom do general histories dwell on the all-important artillery battle segments. In the front line, scene of the action, artillery personnel are few and far between. The officer (or senior NCO) with his 'OP party' of an assistant and a signaller and/or driver is an adviser to infantry and tank commanders. An axiomatic military truth is that, without field artillery fire in its various forms, battles are lost. When artillery fire is applied to support armour and infantry (the four being known, with engineers, as 'the arms') in accordance with proven, documented principles, it wins battles or avoids defeats. This is generally referred to as 'firepower'—weight of artillery shells at the right time, place and quantity to achieve desired outcomes.

It is also evident that in Australia's military historiography its artillery is mentioned in general terms, a sentence or two, or perhaps a paragraph, describing its contribution to the battle or campaign. This is understandable insofar as the real fighting is what the infantry arm is equipped for and does. The same applies to the Armoured Corps with its armoured cars, light and battle tanks. But they cannot achieve their general's aims without the gunners. Australia's official historians who covered the period of this account have been very mindful in their narratives of the quality of Australian artillery and its vital contribution to the key battles described later. The exception is the siege of Tobruk, where British artillery strength outnumbered Australian by a wide margin.

At Bardia and Tobruk Australian field regiments were supported by Royal Artillery (RA) units as part of corps troops. The AIF had a medium and a survey regiment on its order of battle in the theatre; but for compelling reasons neither were available. A detachment from the latter served in Tobruk, but the remainder of corps support came from RA units. As these units were to support the AIF for almost two years off and on, I make no apologies for including details of their contribution. The key lesson here was that the Royal Regiments of both countries (and later two others) were able to mesh convincingly in a matter of days to produce battle-winning firepower. Gavin Long and Barton Maughan, official historians for the Western Desert battles, are generous compared to other historians in their treatment of Australian field artillery. It is a pity, from a gunner's perspective, that they did not include more descriptive material on the contribution of the RA units and headquarters to the spectacular results they achieved. This is especially so when one understands that almost half the number of shells fired in support of armour or infantry are on counter-bombardment (counter-battery) tasks (against enemy guns) and organised by their higher artillery headquarters. This is the unsung contribution of the RA units in our history and victories. Thus the crump of bursting artillery shell thousands of yards from where the encounter battle rages can be just as crucial to the outcome of a battle as those that fall on or near the combatants. My wish to record this contribution is one underlying reason for this account.

The aims of this history are to record the events of the period December 1940 to November 1942, the active service of the Australian 6th and 9th Infantry Division's field artillery regiments and the application of their firepower. This was a decisive element in defeating their opponents on the field of battle, namely Bardia, Tobruk, the siege of Tobruk, the July 1942 battles (First Alamein) and General Montgomery's set-piece battle of October 1942 (Second Alamein). This account begins with a review of the state of the Regiment following World War I, the hiatus over doctrine development in Britain given the arrival of several new entrants onto the field of battle—the tank, the aeroplane and trench mortar, anti-tank guns and various types of mines, and the subsequent mechanisation of armies. The narrative moves on to describe how the British Army fared facing the German Army (*Wehrmacht*) and Air Force (*Luftwaffe*) blitzkrieg tactics in Belgium and France in 1940, doctrine development and the Australian General Staff's reaction these developments. The story turns then to the arrival of the 2nd AIF's 6th, 7th and 9th divisions, shipped to the Middle East (although elements of the 9th were diverted to the UK but later married up with their ultimate division for operations in Cyrenaica by the 6th Division. The 9th Division's role in the Siege of Tobruk gives way to a time gap from October 1941 to July 1942, when the 9th Division was part of the Ninth Army garrisoning the Levant as part of a defensive deployment against a likely German thrust for Middle East oilfields through Turkey or the Russian Caucasus. It concludes with the July battles at El Alamein and the final, climactic struggle in which the division played a key role in destroying Field Marshal Rommel's forces.

In these actions Australian field regiments and other detachments, officers and men, combined in an association noted for its professionalism. They used common doctrine and shared a unity of purpose in a kind of international freemasonry with the RA—gifted musicians under the baton of different conductors. Their synergy in ultimately what was a team effort to beat Hitler was a strong, underlying motivation. Perhaps the chapter that was the least enjoyable to write concerned the British experience which chronicles, for the sake of contrast, British and Commonwealth artilleries (New Zealand and South African) experiences when they were not employed appropriately in doctrine terms. Mercifully, our gunners avoided their fate.

I have been helped immeasurably in my analysis by the late Brigadier Shelford Bidwell's seminal work, *Gunners at War; A Tactical Study of the Royal Artillery in the 20th Century*. Bidwell was thought to be an Australian serving in the RA because he praised the RAA highly when he wrote extensively on battles in which they participated. Other artillery historians including J.B.A. Bailey, Gudmundsson and Zabecki are more focused on operations, doctrine and the results of actions over the field artilleries of other major powers. Nonetheless, they have been a useful adjunct to Bidwell.

All three recent titles produced for the 60th Anniversary of El Alamein battles approach it from different perspectives—and all for a British readership. Latimer's is the most scholarly and best researched, as is that of Bierman and Smith, both BBC journalists. The latter approach their subject (as one might expect) with a strong emphasis on personal reflections, experiences etc., in a tactical matrix. They include one fanciful episode of a German *kannoner* (gunner) witnessing a 25-strong Australian patrol (smelling of alcohol) being wiped out on a heavy gun position on 27/28 October, presumably several thousand yards from the Australian forward defensive lines. Bungay is quite analytical at the command level and provides thought-provoking comments primarily on command and staff, but less on doctrinal, strategic and tactical issues. There are relatively few references to the Australian 9th Division in these works, even less on the RAA as an entity, although General Morshead is mentioned several times, as is his infantry's fighting prowess. Of Ramsay? Nothing.

Finally, Mark Johnston's own work and his co-authorship title with Peter Stanley, *Alamein*, published in 2003, which also tell the Australian Western Desert story. *Alamein* focuses on the heroic infantry battle in graphic detail and amplifies Maughan's account very well. However, the frequent references to horrific battlefield trauma can be quite distracting for the reader. While my own account boasts few of such scenes, it does include, for the benefit of future artillery or military historians, details of the methods, systems and technology applied by field artillery at the time, albeit in a unique topography against a skilled and worthy enemy.

Chapter 1
From Volunteers to Veterans

A common system of defence can only be carried out by a federation of military forces of the Colonies ... the supply of ammunition for the Field Artillery and a common armament for field batteries ...

Major General J. Bevan Edwards
Report to Colonial Defence Committee, 1889

The history of artillery in Australia dates from the time of Governor Arthur Phillip whose directions to Lieutenant Dawes in 1788 saw the establishment of the first gun battery on the shores of Sydney Harbour to guard the infant colonial settlement at Port Jackson from seaward attack. From these humble beginnings, the Royal Regiment of Australian Artillery (RAA), in its various forms, traces its long and proud history, distinguished by active service overseas from as early as 1885, when its colonial antecedents first embarked for distant shores in the name of patriotic duty.

Apart from their immediate survival in a hostile land, the primary preoccupation of the early colonists lay with their protection from a foreign foe. Thus it was that many of the state capitals and their major seaward approaches came to be protected by coastal artillery batteries. Expatriate Britons commanded these garrison artillery detachments which largely comprised British gunners with trained local recruits to bolster the garrison numbers and which fired ordnance of British manufacture. From the 1850s onwards, artillery-minded enthusiasts in all colonies formed colonial field batteries using horse-drawn guns. These batteries resembled mounted gentlemen's clubs and were based in the colonial capitals and larger towns. Over time, the British battery commanders were replaced by Australians so that, by 1885, when 'A' Battery, the first Australian field artillery unit to serve overseas, sailed for the Sudan, its gunners were predominantly Australian, although the weapons of choice remained the British 9-pounders.[1]

'A' Battery embarked for the Sudan in a blaze of patriotic glory. However, the thrill of the battle remained largely unknown to the gunners, as the battery failed to fire a shot in anger. The gunners trained, ever ready for action, patrolled with the cavalry, groomed their horses, and finally returned to a rousing homecoming, full of praise for their bearing, if not their prowess in combat. Despite the absence of action, the battery did not escape entirely unscathed, as one man died from fever and three from cholera—the first Australian gunner casualties.

In January 1900, 'A' Battery again embarked for service overseas, this time as part of the contingent destined to fight in the Anglo–Boer War (1899–1902). During this campaign, instead of being deployed as a basic fire unit in support of the infantry and cavalry, the gunners were divided into sections, each comprising two guns. They covered vast distances and saw action with both the mounted infantry and cavalry. Again, their casualties were few: to Gunner B. Gowing fell the tragic distinction of being the battery's only loss when he was killed by a stray bullet at Vryburg.[2]

In 1899 Queen Victoria granted the colonial garrison artilleries their Royal Warrant which conferred on Australian gunners a similar distinction to that accorded their Royal Artillery (RA) counterparts. At the same time, the ordnance used by the Australian gunners was upgraded to reflect technological advances in material science and ballistic mathematics. Federation in January 1901 heralded the creation of the new nation of Australia, but made little difference to artillery doctrine and training, which followed the British manuals and continued to do so for almost a hundred years.[3]

In December 1910 Lord Kitchener, then Chief of the Imperial General Staff (CIGS), visited Australia to advise the Australian Government on the optimum composition of Australian artillery units. In his 1911 report, Kitchener recommended a total of 56 field artillery (18-pounder and 5-inch howitzer) batteries for Australia's military forces. The introduction of the Compulsory Military Training Scheme in January 1911 ensured that Kitchener's recommendation would eventually become reality.[4] The scheme lasted until the outbreak of World War I and was largely responsible for the huge influx of volunteers into the 1st Australian Imperial Force (AIF) artillery batteries and brigades from 1914 to 1915.

Recruitment was also assisted enormously by the willingness of colonial, then federal, governments to fund the training, equipment and purchase of expendable stores for the artillery batteries and brigades prior to the scheme's introduction. The batteries and brigades used the 18-pounder gun and, at first, the 5-inch howitzer. Introduced in 1906, the 18-pounder ranged to 6,500 yards and was manned by a crew of 10. The 5-inch howitzer was introduced in 1902, fired a 50-pound shell to 4,900 yards with a much higher trajectory and was also operated by a crew of 10.

The 1st AIF was raised in 1914 and its first two divisions equipped with the 18-pounder field gun. The gunners at Gallipoli soon discovered the disadvantages of deploying their 18-pounders on that narrow salient. The flat trajectory of this particular gun was unsuited to the short ranges typical of the Gallipoli ridges and the gunners were often forced to site their guns in exposed positions in order to provide critical support to the infantry. The howitzer-delivered round, on the other hand, was regarded as far more effective in this type of terrain.

In 1915, two batteries of Australian Permanent Force garrison artillerymen (a total of 450 men), arrived in Britain under secondment to the British Expeditionary Force (BEF). Trained in the use of heavy coastal weapons, this Australian force, the 36th Heavy Artillery Group, was equipped with new types of heavy guns which were commonly known as siege guns. One of the group's batteries, the 54th Battery, fired the 8-inch howitzer, while the 55th Battery was equipped with a 9.2-inch howitzer. Once the gunners had completed their training, they were despatched to France and the Western Front. They became the first Australian gunner units in the BEF.

As the war progressed and the Australian forces in the Dardanelles were moved to France, another three infantry divisions were raised and, from 1916, all five AIF divisions were engaged in the gruelling campaigns on the Western Front. Within each infantry division, Australian Field Artillery (AFA) brigades provided most of the division's offensive power. By this time, the artillery commanders at brigade level were predominantly Australian. As the war progressed and the experience of the gunnery staff grew, they recognised the need for howitzer fire to engage deep entrenchments given the ineffectiveness of 18-pounder fire. Thus, a howitzer battery was added to the divisional field brigades and the organisation of artillery brigades was adjusted to meet a much-expanded regiment.

Within the AIF artillery, a number of field brigades were designated 'army' brigades and tasked to provide fire support to those operations involving corps and armies. They operated more or less independently, unaffiliated with a specific infantry brigade or division. The army brigades became elements of the heavy artillery groups comprising four or six-gun batteries of guns and howitzers heavier than 6-inch calibre, and ranging up to 15-inch calibre. The primary role of the heavy artillery groups was counter-bombardment of enemy batteries and targeting of enemy machine-guns and reserve areas. They operated independent of divisional command and control, usually at corps level. The heavy artillery groups also engaged enemy headquarters, lines of communication, supply dumps and other important targets well behind the front line rather than enemy infantry and machine-guns which were sited forward in either trenches or pillboxes.

During World War I, a total of 60 field batteries, 15 howitzer batteries, two siege batteries and several light and heavy mortar batteries saw active service with the AIF. Of these, 54 survived to see the end of hostilities. The 18-pounder and 4.5-inch howitzer (firing a 35-pound shell) were the backbone of the AFA and blazed a trail for the development of field artillery in Australia during the post-war period. Australian gunners had comparatively few rest periods out of the line compared to infantry battalions. The war took a heavy toll with 10,926 gunner casualties, some 58 per cent of their total number.[5]

Artillery plays a significant role in the planning of any military campaign, a fact not lost on Napoleon, who once wrote that 'you make war with artillery'. However, the advantages of a powerful artillery arsenal must be weighed against the restrictions it imposes on the opportunity for manoeuvre. When evenly matched opponents lose the ability to manoeuvre, the result is attrition warfare, as typified by World War I on the Western Front where trench lines and fortifications ran more or less continuously from the Belgian coast to the Swiss border. Manoeuvre warfare was virtually eliminated, due in no small part to the superiority of concrete and four fabricated steel products—pick and shovel, machine-gun and barbed wire—which both sides possessed and used to great effect. Opposing commanders grappling with the sheer weight of casualties endlessly analysed their tactics for the employment of both infantry and artillery. They looked to artillery to overcome the effectiveness of trenches, redoubts and wire so that attacking infantry soldiers could kill or capture their opponents before they could regroup or recover from bombardment or counter-attack.

Attrition warfare was not particularly novel. Western military analysts studying the Russo–Japanese (1904–5) and Balkan (1912–13) wars noted that the presence of field artillery and high-powered rifles forced defending troops to construct elaborate trench systems to avoid slaughter from shells and machine-guns. In turn, the need to penetrate these well-constructed trenches and fortifications led to the development of 'plunging fire', a technique that enabled shells to be fired at a steep angle of descent to penetrate earth and demolish man or gun. As always, the battlefield created the opportunity for the ingenious adaptation of the technology of the time.

The initial success of artillery also inspired the development of concepts for the use of armour to shield soldiers from shellfire. Lateral thinking then enlarged on these concepts, encasing the soldier in mobile armour for protection and using mechanical power instead of muscle power for movement. As early as 1878, a British colonel envisaged defensive armour protecting guns, stating that 'artillery might calmly await any attack whatever, certain to destroy an enemy long before he reached the guns.' While the Germans were initially interested in this concept, its development eventually lapsed through inaction. The concept of the tank was resurrected prior to World War I, and the British led its development in the hope that the tank would 'neutralise' dug-in troops and defences.

During any World War I offensive, the attacking troops faced the greatest threat, cut down in their hundreds as they attempted to storm enemy trenches. Heavy bombardment of trench lines thus became the primary task of artillery field pieces early in the war. The introduction of the 4.5-inch howitzer and its 35-pound shell in 1916 gave Allied artillery the edge in pulverising earthworks and emplacements, as the howitzer was at least twice as effective as the 18-pounder. Both sides soon discovered, however, that a heavy bombardment churned the ground so badly that it impeded the attacker's rate of advance, particularly when the earth was sodden. In addition, bombardments were often ineffective against well-trained troops who could recover quickly from the effects of the barrage and repel attackers. Soldiers in deep shelters could emerge unscathed once a bombardment ceased and quickly return to their machine-guns and rifles to defeat the attack that followed. The Germans often manned defences in depth, beyond the range of Allied artillery, conceding the ground and blasting advancing troops with artillery and machine-guns. Defence in depth also allowed the movement of reserves beyond the reach of the murderous artillery fire.

The Allies soon learned to be clever in their use of artillery. An important innovation was the 'creeping barrage' which saw the barrage move forward at the infantry pace of advance. The infantry followed as closely as possible, 'leaning' on the barrage—that is, maintaining a minimum safe distance behind the line of bursting shells so that when the fire lifted, the enemy soldiers were at their mercy as they emerged from their trenches or dugouts.

A key element of artillery planning at this time was the effective neutralisation of enemy artillery and machine-guns on the flanks to prevent them firing on advancing infantry in enfilade. The development of an 18-pound high explosive shell with a sensitive fuse to 'cut' barbed wire and enable the infantry to advance was hailed as a major step in breaking the trench warfare stalemate. Likewise, the use of smokescreens was aimed at reducing infantry casualties by screening advancing troops and blinding observers.

The Allied artillery plan also had to take account of assembly areas and start lines from which counter-attacks could be mounted, as well as hostile batteries and machine-guns that would support the counter-attacks. As long-range fire tasks became increasingly vital, more and more artillery was allocated to these tasks. Heavy artillery groups were particularly suited to this role. Later in the war, more than 40 per cent of the artillery employed to support assaulting troops was concentrated on depth fire tasks.

The continual sequence of stalemated attacks involving unsustainable advances followed by withdrawals forced artillery staff to devise new forms of fire concentrations and various types of barrages. Barrages were expensive in ammunition and difficult to modify if the tactical plan went awry in its execution. Given the unstable nature of communication on a shell-torn battlefield, the modification of barrages was next to impossible to achieve. Covering the flanks of the barrage was also expensive in ammunition. Above all, a successful barrage required good intelligence and quality staff work at all levels, including an understanding by the soldiers themselves of the limitations of the barrage's effectiveness against the enemy defenders. Until 1917, during the attack phase of the war, lengthy bombardments were the norm, with preliminary bombardments lasting four days, sometimes more. Later in the war, the length of bombardments decreased to a matter of hours to suit a static front and an increasing use of and reliance on short and intense artillery fire to achieve victory.

In defensive terms, artillery staff focused primarily on employing two types of fire: counter-preparation fire and defensive fire. Counter-preparation fire was aimed predominantly at infantry and artillery assembly areas, supply dumps, headquarters and any target likely to be vital to an enemy plan. Defensive fire covered the most likely lines of enemy approach as assessed by the units to which the artillery was providing support. The employment of fire was prioritised, the most urgent fire missions designated 'SOS' tasks.

Artillery innovation not only involved changes in tactical doctrine for support to infantry and later tanks, but extended to encompass improvements in both guns and ammunition. Innovation in shell type involved the manufacture of a variety of fillings to produce specific effects and the production of new and improved types of shells and fuses. While smoke shell and shrapnel were well known and frequently used, chemical, star, and incendiary shells made their first appearance on the battlefield in large quantities from 1916 onward. Chemical shells were filled with lethal and debilitating gases, while star shells were used for illumination. Incendiary shells contained phosphorous that inflicted horrible burns.

The British used a mix of ordnance—field, medium, siege and heavy—to fire these shells. The timely and accurate delivery of these rounds was vital to achieving the desired result and, ultimately, to final campaign victory. This led to innovation in the use and placement of guns, such as mounting artillery pieces on railways for ease of movement in engaging a range of targets well behind the front line and confusing the enemy as to the location of Allied artillery. Further adaptation followed. The requirement for close support of the infantry gave rise to the creation of light and heavy mortar companies that deployed to the rear of the battalions. The appearance of enemy reconnaissance aircraft heralded the employment of anti-aircraft artillery in the field to supplement the Lewis gunners of field units who had taken on anti-aircraft roles.

However, the primary aim in the advance was to move the guns forward to a position from which their fire could disrupt and defeat counter-attacks from infantry and tanks. The role of artillery was effectively to allow the infantry to consolidate its objective. Direct and indirect fire from enemy guns could be neutralised by counter-battery fire or by overrunning the guns with infantry or tanks. Once terrain was seized from enemy control, it could not be held by artillery alone, or by infantry or tanks in isolation. Together, however, artillery, infantry and tanks could successfully hold ground through coordinating fire and movement and using firepower for protection. The essence of the tank's impact as it lumbered towards enemy infantry was the effect on enemy morale of a seemingly continuous, unstoppable advance. The British were never able to achieve this image of a mass advance with their tanks because their artillery was not equivalently mobile and could not help them hold the territory they won from the Germans. For this they needed their infantry, artillery and engineers to be equally mobile and working in concert.

The Battle of Cambrai on 20 November 1917 proved a milestone. Unlike other battles, there was no preliminary artillery bombardment and the tanks (which moved in threes) became, to all intents and purposes, mobile pillboxes in armoured batteries, armed with 6-pounders and/or Lewis machine-guns. By lowering a curtain of fire ahead of the leading tanks, infantry soldiers advanced and speedily dealt with enemy resistance on their objective. Integrating artillery, tanks and infantry weapons such as mortars proved successful again in 1918 when they combined in John Monash's Australian Corps in the battles of Hamel and Amiens. Artillery smokescreens protected the tanks during their advance while the engines of No. 3 Squadron, Australian Flying Corps (AFC), created a 'sound screen' to disguise the noise of the tanks. At Accroche Wood, platoon commanders from the 14th Battalion indicated strongpoints to tank commanders. Carrier tanks shadowed the advance with supplies of defence stores and ammunition.

The plan Monash devised for Amiens on 8 August 1918 was similar to that executed at Hamel. While corps and divisional artilleries fired a mixture of smoke, chemical and high explosive to deceive and neutralise the defenders, the 24 tanks allotted to each of the advancing infantry divisions crushed the defenders.[6] While Monash had absorbed many of the lessons of British arms at Cambrai, he was also aware of the reasons the German Army had recaptured so much territory. He blamed these losses on a combination of faulty intelligence and planning, and the positioning of the headquarters too far to the rear. He also believed that poor artillery command and control after the long advance had been confounded by the resilience of the German soldier. Monash had made the rapid discovery that 'the gun is the master of the tank'.[7] Ironically, later at Cambrai the Germans also enjoyed success with their use of the tank, employing similar tactics to the British.

Artillery historian Shelford Bidwell suggests that the Army's senior gunner in the field at General Headquarters (GHQ), Lieutenant General Sir Noel Birch, turned Field Marshal Earl Haig from a cavalryman into an 'artillerist' during the war.[8] Birch was a practical gunner and to him fell the enormous responsibility of applying more effective tactics, equipment and science to the battlefield. This he did by overcoming the old-fashioned notions held by most of his peers and subordinates on the capability of modern artillery. He promoted talented and forward-thinking officers such as Herbert Uniacke, Alan Brooke and Henry Tudor. He paid particular attention to analysis by his gunnery staff of the performance of Allied and German artillery, and especially the tactics employed by Colonel Georg Bruchmuller, the German artillerist. He advocated the establishment of schools of instruction covering every facet of gunnery so that gunners moving to the front line would be adequately trained in the latest doctrine and techniques. Despite their opposition, he sought to train commanders at the senior level to use artillery to optimum effect and to persuade infantry commanders to do likewise. He was not always successful in this last endeavour, but he had a powerful ally in Haig.

Thus, by 1917 the senior gunner generals were well aware that expertise acquired and required by the artillery arm signalled the way equipment and tactical doctrine development should proceed. These lessons and those of the future were not lost on Monash and his artillery adviser, Walter Coxen, nor on gunners Talbot Hobbs, Charles Rosenthal, William Sinclair-Burgess, George Johnston and Harold Grimwade. Such recognition also extended to the AIF divisional commanders and Commander Royal Artillery (CRA).

Several significant technical advances by the RA were also adopted by Australian artillerymen. The first was the establishment of a Counter-Bombardment Staff Office in 1916 following the horrendous losses of the Somme campaigns. The Counter-Bombardment Staff Office (at corps level) was formed to collect, evaluate and verify data from observation posts (OPs), flash spotting sections and sound ranging detachments. It also analysed the results of aerial observation by the Royal Flying Corps (RFC) and AFC airmen, including aerial photographs, and examined data from the interrogation of prisoners of war and analysis of captured documents. The integration of battlefield intelligence, based on a practical and efficient methodology, into a single office, and the communication and cooperation of all its component parts proved a war-winning innovation.[9]

The evolution in British doctrine in 1917 and 1918 involved the crucial cooperation of infantry with artillery and was assisted by the RFC, which later became the Royal Air Force (the RAF—in which Australia had three squadrons). The growth of this force from an enthusiastic band of amateur pilots in 1911 to a formidable fighting force is well documented. What is less well known is the commitment of air staff, pilots and observers to reducing casualties among their brethren in the trenches. They spared no effort in their dealings with the Counter-Bombardment Staff Office and higher headquarters to allow the Allies to gain the upper hand in a fairly even contest with their German adversaries. The gunners were the interface between the two services in the forward areas and they developed much expertise together. Yet, however effectively the Army and the airmen worked together, there were still many unanswered questions on the coordination of artillery with armour, infantry and engineers. These concerns would drive future development in which, for the present, Australian gunners would be interested spectators.

One development of critical importance was predicted fire which enabled the artillery to accurately engage targets without using ranging procedures. Ranging had two major disadvantages. It compromised the advantage of surprise and allowed the enemy to take countermeasures and it revealed the location of the ranging gun(s) to the enemy counter-bombardment and locating agencies, thus inviting retaliatory fire. Prediction required an understanding of the way in which the external ballistics of shell flight were affected by variations in the air temperature, humidity, wind speed and direction and barometric pressure. Calculations were made from ground and air observation data several times a day and corrections applied to the guns as 'add' and 'drop' for range and 'left' and 'right' for line.

The collection of accurate survey data which made coordinated gunfire possible in areas such as France and Belgium where several different national bases for mapping existed was another major development of this time. Survey data was also necessary for mapping the front and for the production of maps for artillery and other purposes.

Such data would have been of little use, however, without the development of calibration. When a gun fires, its barrel wears and propellant gases leak past the shell as it moves up the barrel. This results in a reduced muzzle velocity causing shells to fall short of their intended range. Calibration is the process of adjusting gun data for firing in order to incorporate this wear factor. It enables fire units from section level (two guns) to division (24 guns) to produce more accurate fire.

Accurate information and efficient communication were vital at a time when the situation was generally confused on both sides. The advent of wireless communication between headquarters and artillery spotting and reconnaissance aircraft crews opened up the prospect of gunner-to-gunner control of fire using either Morse code from aloft and/or by using gun batteries to lay out patterns of ground signals. Field artillery fire in 1914–15 was 'direct', meaning that it proceeded from gun to target with the observer at or near a battery. From 1916, however, 'indirect' fire, with an observer at some distance from his battery passing fire orders by telephone or Morse code (if the observer were aloft in a balloon or aircraft) became the norm. Indirect fire required the accurate fixing of the position of the gun and the target in relation to the map of the area. With 'fixation' as it was known, came alignment of the gun to grid/map north, an imaginary line shown on a map from which clockwise angles were measured. This was termed 'orientation'. The gun was directed at the target using a theodolite that had been accurately 'fixed' from map data and bearings to reference objects (such as church steeples). Several trigonometrical calculations provided a range to the target. Innovations such as these laid the base for neutralising enemy artillery and machine-guns using predicted fire. This introduced the element of surprise, a feature of Monash's successful operations at Hamel and Amiens in 1918.

In 1918 Monash agreed to the formation of two corps topographical companies which were primarily concerned with activities such as mapping, survey and printing, while flash spotting and sound ranging data was provided by British sections allocated to their corps areas. A strong partnership had been established between the topographical staff and RA gunnery staff, particularly those in the Counter-Bombardment Staff Office.[10] Australian gunners were extensively trained in these techniques and, by 1918, they had earned a reputation as a highly professional body of troops. Monash noted frequently the importance of both training and scientific advancement in the science and art of gunnery. Both the RAA (Permanent Forces) and the AFA (Militia) further developed these techniques during the post-war period, attracting many officers of scientific bent.

As a result, both sides increased the number of guns exponentially and relied on medium and heavy ordnance in order to achieve what manoeuvre could not. The expansion of the artillery regiment in terms of manpower, weaponry and ammunition supplies was nothing short of enormous. The British began the war with 72 field and six heavy/siege batteries.

By November 1918 they had deployed 568 field batteries and 440 batteries of heavy ordnance greater than 6-inch calibre. The Germans increased their total from 838 to 4,460, and the French from 1,040 to 2,286. Expressed on an individual scale, the British and French artillery began the war with 6.3 and 4.0 guns per 1,000 infantrymen respectively, compared with the German artillery's 6.0. By the end of the war the figure for the British and French was 13.0 compared with the Germans' 11.5. Both sides increased the ratio of heavy to light guns, reflecting the increasing importance of the heavier weight of shell to destroy fortifications and longer range for counter-bombardment fire as the war progressed.[11]

Commanders, including Australia's John Monash, sought to break the deadly stalemate that was trench warfare and restore manoeuvre to the battlefield through the clever integration of technology and manpower, both on land and in the air. Battles were won and lost on the strength of elaborate schemes, but hostilities continued unabated until the breakthrough battles from August 1918 finally produced a victor in November, when the Germans sought an armistice. The first 'great war' was over and military analysts and historians recognised that the war had been won with artillery.

CHAPTER 2
ARTILLERY DEVELOPMENT BETWEEN THE WARS

The superior weapon of the future is the gun, the superior soldier is the Gunner and the superior army is a force based on mechanically propelled guns.

Major General J. F. C. Fuller 1926

At the end of World War I, British foreign policy returned to its pre-war military stance which focused on the protection of its colonial empire. The Royal Navy (RN) was dispersed to the seven seas and the RAF deployed to the Middle East to protect British oil interests and the Suez Canal. While Hong Kong, Singapore, West Africa and the Caribbean absorbed various regiments, the largest British military commitment was to the Indian subcontinent to what is now India and Pakistan. Most of the artillery batteries stationed abroad were in north-western India (modern-day Pakistan), which absorbed around a third of all British regular forces.

By 1922 the Imperial War Office had completed its analysis of British wartime operations and now began to consider the future direction of British military development. This was no simple matter, as the War Office believed that military development rested on its ability to meet three main requirements. The first of these was the desperate need for new ordnance—both guns and ammunition—and the question of where this should be obtained. Second, the War Office viewed acquisition of armoured vehicles such as tanks and armoured cars as a high priority. The third requirement concerned communications equipment and other war-like stores. The War Office also recognised a parallel need: that of the enunciation of a tactical doctrine that would structure a force so that, once suitably trained, it could be readily deployed and capable of delivering success. The British Government's foreign policy, of which Europe and its Commonwealth were key components, was to shape the nature and capability of the British military forces.

Within the War Office—which was also the headquarters of the British Army—expert committees of experienced officers from all arms and services pooled their wisdom and experience. Committee members assessed emergent trends in other armies from intelligence supplied by military attachés in embassies worldwide. By 1924 the War Office had become the Ordnance Committee. Its members reached consensus on criteria for the design of weapons systems, defence stores, heavy artillery, tanks and lorries, wireless sets, ration scales and even medical instruments. Expert gunnery staff input to these decisions came from artillery and other directorates within the military hierarchy. The overall direction of the artillery arm was determined, in fact, within three discrete functions and positions.

Pre-eminent was the CIGS Committee—the committee that drafted specifications for ordnance and ammunition based on current or projected Army structure and doctrine. The Committee would decide, for example, on anti-aircraft artillery for the RN. The second determinant of artillery development was the Director General of Artillery whose role was to implement CIGS Committee requirements. The third determinant was the Commandant of the School of Artillery at Larkhill. From an historical perspective, it is difficult to discern which one of these important and influential positions was *primus inter pares*. When development was viewed from the practitioner's perspective, the Commandant and his staff took the lead, particularly in the development of drills in field, anti-tank, anti-aircraft and survey (locating) equipment. The Commandant and his staff were keen to capture improvements in gunnery techniques and use these to update the school's instruction. The 'bright ideas' and 'short cuts' of clever Regimental officers were also eagerly harvested. The gunnery staff at the school wings were selected from the ranks of Regimental officers who were posted into the heady atmosphere of the school and returned to units or other postings to spread the latest doctrine like religious missionaries. Larkhill was commonly referred to as 'the Temple'—a title that aptly described both its role in the development of artillery doctrine and the zeal of its acolytes. The school's overarching purpose was to marry the 'new idea' to the current doctrine, or seek doctrinal change through logical argument, influencing superiors, or by any other means possible. Thus the school served a vital role in the promulgation of tactical doctrine.[12] It was into this heady atmosphere that Australian Staff Corps officers allotted to artillery were regularly thrust, returning to Australia imbued with the latest doctrine.

By 1922 the Regiment's extensive analysis of field artillery effectiveness during the war signalled five areas for improvement. These comprised: excessive wear of guns; insufficient range and shell power; inaccurate time fuses; and structural weaknesses within the guns and their carriages. Modification and improvement in all five areas proceeded in parallel through the interwar years with progress ever-dependent on the crucial availability of funds. The most significant initiative, however, was arguably the development of equipment.

Equipment development required a multi-disciplinary approach as equipment requirements were incorporated into the development of a 'weapons system' and included such considerations as motive power and portability; shell type to suit the destruction or neutralisation of targets of various kinds; muzzle velocity; propellant type; calibre; weight; detachment size to work the gun; and sight system, among others. The gun itself was viewed as merely the means of the shell's delivery to the right place at the right time. A gun requires a calculated type and number of shells to produce the desired result. Thus, within the Directorate of Artillery, areas of development were: propellants; high explosives; ballistics; gun carriage and design; ammunition design including mechanical fuses; and instrumentation.

By far the most important development involved the replacement of the 18-pounder field gun. The gun's replacement was contingent on the defensive configurations of infantry battalions and brigades in operations other than trench warfare. The new weapon would replace the 13 and 18-pounders and the 4.5-inch howitzer. The initial specification drawn up in 1934 involved a piece that weighed up to 30 cwt (hundredweight) or 1.5 tons with a range of at least 12,000 yards with armour-piercing protection. Research into projectile weight led to consideration of shells of 18.5, 22 and 25 pounds. The 25-pound shell was preferred and was adopted in April 1935. The new gun would have to provide neutralising fire at a rate of 8 rounds per minute (rpm) for two to three minutes. Further development and additional doctrinal input increased the range to 15,000 yards. In the end the internal ballistic considerations settled the final design in 1938. Given the proximity of the outbreak of hostilities in World War II, it proved to be none too soon.[13] Other design details were soon finalised including a feature that allowed the gun to be towed so that, if its prime mover broke down, it could be towed by another lorry or even a tractor on a European terrain well endowed with roads, lanes and tracks.

The outcome was a specification for a gun-howitzer with a shell of 25 pounds, calibre 3.45 inches, a range of 12,000 yards minimum and a gross weight of 30 cwt. Its range almost doubled the 6,500 yard maximum of the 18-pounder Mark II. As a large number

The 1920-30s development in UK of the Bedford Lorry that evolved into the famous 'Quad' 25 pounder tractor to modernize its field artillery never made it to Australia.
(AHS 1930-03)

of 18-pounders were still in use, many were converted to 25-pounders with a range of 11,800 yards, known as the Mark I or 18/25-pounder. The new Mark II eventually weighed 1.75 tons, had a maximum planning range of 11,800 yards and a maximum range of 13,400 yards. It sat on a wheel-like platform under a box trail which allowed it to be rotated 360 degrees. This was nominally an advantage when big switches (angular measurements from an imaginary centre line) were ordered, as it gave the gun and crew enormous flexibility to engage tanks and close targets. The Mark II had a hydro-pneumatic buffer and recuperator system for absorbing recoil forces. Other configurations which also absorbed recoil forces were the split trail or spade, both of which required time and muscle power to change a gun's line and were more suited to heavier ordnance. The Mark II entered service in 1940 and was an immediate success.[14] The guns and limbers (for carrying first line ammunition) were towed by a field artillery tractor, the Quad. Production of the new design was slow as, unsurprisingly, the manufacture of heavy anti-aircraft guns had been given priority. As a stop-gap solution, over 1,000 18-pounders were converted to 18/25-pounders. These equipped the BEF in France and possessed a moderate anti-tank capability.[15]

The new gun was envisaged as part of a 12-gun battery supporting a four-company battalion with a frontage of approximately 800 yards, depending on the terrain. This was wider than the frontage of fire that could be laid down by a brigade of 18-pounders, so a heavier shell was required. The shells from a well-dispersed brigade of 24 guns (two batteries) falling in an area of around 300 to 500 yards was regarded as satisfactory. Smoke and illuminating shells now entered the equation, as these were more accurate if their descent was steep to the ground from a high howitzer trajectory. The same principle applied to high explosive shells.

In 1938 there was a significant reorganisation of the field brigade which now became a regiment of 24 guns in two batteries each comprising three Troops. The Troops were controlled by one command post and graphical solutions to gunnery firing data sped up engagement times. Modern warfare would require greater gun dispersion as a means of protection from air attack and this prompted the demise of the regimental survey party in a raft of other changes.

The problems that beset Australian field gunners who found themselves at the mercy of the Commonwealth government's tight purse strings in the 1930s pale into insignificance compared with those of British artillerymen at the time. The British dilemma lay in deciding how to modernise the medium and heavy branch of the Regiment at a time of shortage in industrial capacity, lack of spares, crisis in ammunition supply and other critical shortages. The British experience is illustrative of the conflicting requirements faced by the decision-makers on the General Staff, gunnery staff and in the manufacturing industry, both heavy and light, in the crucial lead-up to 1939.

At the end of World War I the medium and heavy artillery branch of the Regiment, the Royal Garrison Artillery (RGA), boasted a mixed ordnance of no fewer than 14 types, including the medium calibre 60-pounder and 4.7-inch guns, the 5 and 6-inch howitzers, the heavy and super heavy 6-inch, 8-inch, 9.2-inch, 12, 14 and 15-inch guns, and the 8, 9.2, 12 and 15-inch howitzers. A number of these were railway-mounted guns. Some of these guns were 'retired' during the conflict and others went to scrap yards or museums once the war was over. The General Staff had anticipated the possibility of mobile warfare, so manoeuvrability was a prime requirement in any ordnance.[16] By December 1935 one of many specifications included a shell weight of 90—100 pounds using three charges to cover a distance of 16,000 yards with a gun weight of 5.5 tons—a calibre of almost 5.16 inches. Largely as a result of an initiative by the then Director of Artillery, a new design was created for a 4.5-inch gun and 5.5-inch gun howitzer using the same carriage. The 4.5-inch gun was designed to counter the German 105-mm K18 while the 5.5-inch gun provided an answer to the German 150-mm SFH18. The British used heavier shells and the range of the 5.5-

inch gun-howitzer outstripped the German gun by some 1500 yards. In 1939 approval was granted for production to begin and the first regiments received the new guns in 1941 in the Western Desert. The mobility restriction caused by the gun's iron wheels was overcome by pneumatisation, and the 60-pounder calibre was reduced from 5 to 4.5 inches by relining. Unfortunately the guns were considered low priority and this production restriction applied also to the heavy and super heavy branches of the Regiment.[17]

The issue of the mobility of medium artillery was regarded by the Armoured Corps as irrelevant in mobile warfare. The maximum speed of iron-tyred guns drawn by a powerful tractor was around four miles per hour at a time when some light tanks could attain 30 miles per hour. If this was a measure of the Tank Corps' disdain for artillery, particularly medium artillery, it was to cost them dearly in the next three years. Nonetheless, until the latest weapons became available, the Regiment had to make do with the best there was; to its credit, it managed to field three heavy batteries in France in 1939.[18]

The Infantry Corps, on the other hand, championed the development of a close support weapon with a high trajectory (such as a mortar) which would be both mobile and inconspicuous. In 1920 a 3.7-inch howitzer made its appearance in an infantry division and also became an integral part of the Indian Army infantry division mountain artillery long after the 3-inch mortar, organic to a battalion, became the infantry close support weapon. The howitzer also found a role in the first armoured division raised in 1937. However it was considered a poor performer against tanks and, despite its value as a fire support weapon, was retired in a climate that saw the anti-tank capability as of supreme importance.

The Australian experience in ordnance procurement dates from the mid-1920s when mechanisation was on the military planner's agenda, albeit not in annual budgets.[19] Nonetheless progress was incremental. Major General John Lavarack was apprised of the 25-pounder concept during the CIGS conference in 1937 and, soon afterwards, Colonel Vernon Sturdee, at Australian Army Headquarters, recommended that production of the 25-pounder commence as soon as possible. His proposal was well received by the Controller of Munitions Supply but foundered on Treasury opposition.[20] When approval was finally gained, the proposal was again delayed, this time by difficulties in securing working drawings, which finally came from Canada. The first gun was not delivered until 1941. The point is ably made that the procrastinators and decision-makers did not themselves have to confront the shells from a more modern German artillery or attempt to return fire with a World War I gun of lesser range and hitting power. Life would have been even more hazardous had Australian gunners been forced to repel panzers.

In World War I historiography, tank and armoured fighting vehicle (AFV) development has received much more prominence than motor/mechanised transport. The same can be said for the development of guns vis-à-vis their tractors. Lieutenant General Herbert Uniacke (a very capable and influential gunner) was one of the more progressive thinkers and a tireless proponent of mobility for artillery, along with other like-minded senior officers, particularly artillery and infantry 'converts' to mobile warfare.[21] General Birch was also to use his influence to create a mobile artillery concept. It had been demonstrated that tactical gains in operations with or without tanks seldom achieved the potential their planners envisaged because the artillery could not be brought up in a timely manner to help infantry consolidate on the objective before the inevitable German counter-attack.

The General Staff recognised that mechanical rather than animal means (except in mountain warfare) of mobility was the way forward, acknowledging that all methods of moving guns had advantages and disadvantages. Guns could be moved by tractor (wheeled, tracked or half-tracked), replacing horse or mule teams; or they could be mounted on a vehicle as a fixture (self-propelled) or could be demountable (portee).

During the First World War, major industrial nations advanced the technology and design of internal combustion engines, drive trains, suspension and pneumatic tyres. The British were not slow to recognise the importance of these mechanical developments for military purposes (including air and naval uses). The 20-year period that followed boasted some exciting times for those officers and other ranks involved in drawing up specifications, liaising with industry, participating in user trials and witnessing technological improvements that resulted from exercises and demonstrations. The scope was necessarily broad; while development concentrated on employment in a European setting, the needs of the artillery batteries that helped to sustain Britain's empire, including the Suez Canal zone and India, also had to be taken into account. Viewed from today's perspectives, some of their early creations could more readily be described as 'contraptions' rather than vehicles.

In 1922 the first RA unit (the 9th Field Brigade) was mechanised with a collection consisting of a car, a van, motorcycles, a three-ton lorry and a half-tracked vehicle (for battery staff). In April of that year the first tracked tractors (Dragon Mark I) were issued to the brigade, and well-publicised exercises were held the following year at locations such as Petworth in Sussex. Other tractor designs followed and were trialled and, while their mechanical elements scored well on mobility, the same could not be said for their suitability for and reliability in operations. Various British firms using American and French technology assisted the Army in parallel developments over the next three years. One of the results of these technological partnerships was the first self-propelled gun, the 18-pounder Birch gun. The Birch was mounted on a tracked chassis with an 82 horsepower (HP) engine, weighed 12 tons, and could travel at 10 miles per hour cross-country with a range of 80 miles. It could be used in either a field or (light) anti-aircraft role. Over the next two years it would be employed in a direct fire role 'supporting tanks onto the objective'.[22]

Several other modes of transport were also evaluated, including a light artillery transporter (for a 13-pounder gun) by Armstrong Whitworth, demonstrated for Commonwealth Prime Ministers; a Citroen-Kegresse half-track; a Morris-Roadless half-track; a Burford-Cletrac tractor; a Guy 4 x 2 portee (one carrying the gun and towing the limber and another carrying the detachment) and a Cletrac tractor.[23] The biggest rig was a Thorneycroft heavy six-wheeled chassis with a 15 foot 9 inch wheel base, complete with two winches. It carried an 18-pounder and Cletrac tractor, both of which were winched aboard. A Morris D Type commercial lorry was also trialled as were half a dozen others and, by 1929, the Experimental Armoured Force was close to making a decision on which mode of transport was worthy of further development. Guiding the final decision was the need to resolve two pressing issues. First, it appeared that, if weight, height profile and length of design were all

The 1920-30s development in UK of the Birch Self Propelled 18 pounder was never seen in Australia to modernize its field artillery.
(AHS 1930-03)

satisfactory, the detachment—including the driver—would become superfluous. The second problem seemed to present an answer to the first: two vehicles were required instead of one. This was, however, an expensive option and consequently the decision was delayed for another six years.

By 1935 the Dragon [Marks I (Light), II, and III] tracked vehicles which towed an 18-pounder and limber with its detachment mounted had emerged as the design frontrunner. The Dragon's laden weight was 3.25 tons with a range of 100 miles and a handy speed of 20 miles per hour with load. It was designed for use with cavalry brigades. For infantry divisional artillery, Morris Motors had seen the potential for its D Type lorry to capture a commercial niche market and, by 1937, had produced a version for the Army. Its performance bettered that of the Dragon and several batteries were subsequently equipped with these.[24]

A War Office specification ushered in the arrival of the front wheel (forward) drive 4 x 4 field artillery tractor and the requirement (arising from the possible use of toxic gases in the next war) for it to be washed down and therefore enclosed. There were several contenders and the Quad of Morris design was judged the most suitable until it was superseded by an improved Fordson design. The FAT (field artillery tractor) that appeared in 1939 was made from standard Fordson components and Marmon Herrington front drive assembly. It passed into artillery history known to gunners as the C2 Quad. A familiar sight in pictures of the desert war, it is shown towing both gun and limber. In the early days the limber carried the 25-pounder gun platform on top before it was slung beneath the trail. The limber carried first line ammunition (weighing around 800 pounds) as there was insufficient space in the Quad. The ammunition and its carriage were to prove hazardous for the Quad, particularly when engaged by land or air forces. Perhaps, as has been suggested, the preservation of the limber was simply the RA's last tenuous link with its horses.[25]

Hand in hand with these creations came other general purpose vehicles for use in transporting command post staff, sound ranging and survey equipment, wireless sets and other items. Prominent at this time were Bedford 8 and 15-cwt trucks, with the 8-cwt truck the prime mover for the towed 2-pounder anti-tank gun. Motorcycles also appeared on establishments for route discipline and carrying messages. The Bren Gun Carrier was to figure prominently in desert operations and a locally produced version was first introduced into the AIF in May 1940, thanks to the foresight of BHP chief executive Mr Essington Lewis, Director of Munitions.[26]

The development of tractors capable of hauling medium and heavy artillery followed a similar pattern with the Scammel and AEC Matador among the most noteworthy.

The former was a well-engineered 6 x 4 powered by a Gardner 98 HP diesel and could manage a towed load of 36 tons on the level or 15 tons on a slope. It was issued to medium regiments and heavy anti-aircraft regiments whose ordnance weighed five tons or more. The Matador's prototype appeared in 1931 as the AEC Monarch commercial 4 x 2, powered by a 4 cylinder 65 HP engine. Later trials against other commercial types confirmed its suitability as a medium gun tractor.[27]

Australian gunner officers on postings to the United Kingdom (UK) were kept abreast of these developments. All Australian field brigades at this time were fully or partially mounted, and commercial trucks were hired for annual camps to allow practice with mechanised guns. The passing of the horse was not universally welcomed by the gunners but, like the lighthorsemen, they could see the writing on the wall. In the mid-1930s, 'A' Battery, the Regiment's only full-time field sub-unit, trialled the Australian-designed Wilton trailer and, 'in January 1936 a convoy of motor transport departed Melbourne for Sydney, where it arrived on 12 February.' The convoy consisted of one Bedford lorry, one Ford lorry, two Bedford 12-cwt utility vans, two Vauxhall two-seater cars, two Fordson tractors, and two Wilton trailers that were intended to carry (portee) two 18-pounders cross-country at speeds up to 30 miles per hour.[28] However, generally in the post-war period there was comparatively little artillery development in Australia and Australian gunners were relegated to the sidelines, watching events and reading of advances in weapons systems in Britain, Europe and America.

To this end there was a steady trickle of Staff Corps officers to Larkhill and to British/Indian Army Staff Colleges at Camberley (UK) and Quetta (India). In addition, there was a regular conference of the Chiefs of the General Staffs of Britain, Canada, Australia and New Zealand held in London. In 1926 the Prime Ministers of the Commonwealth were invited to attend a demonstration by the Army Mechanical Experimental Establishment of the latest designs in field artillery tractors—tracked, half-tracked and wheeled, doubtless with a view to the export trade. Conferences held in the 1930s discussed the emergent threats of German militarism and Japanese 'expansionism' although, because of budget stringencies attendant on the Great Depression, progress in modernisation was slow. It has been argued that successive CIGS were not in favour of—perhaps even opposed to—the direction mechanisation was taking. Whether they were opponents or simply cautious progressives, the course of events leading up to World War II would determine the wisdom of this approach.

Chapter 3
The Development of Artillery Doctrine

Thus far, this discussion has focused on field and medium artillery development central to its role as prescribed in the British doctrinal publication *Field Service Regulations* (FSR). The first edition of a pocket-book-sized FSR in the late 1800s constituted an early attempt by the British Army to codify and promulgate basic information concerning a wide range of military topics. Germane to this were such subjects as the organisation of cavalry, artillery and infantry, the services, signals by Morse and flags, ration scales for men and horses, etc., all documented in the hope of increasing knowledge in a service that was becoming ever more complex. Periodically updated, it became a 'bible', was used in setting examinations, and was considered an essential component of an officer's kit. FSR contained truisms that guided military thought for the arms; for the artillery this was expressed in basic terms: 'to help infantry gain their objective'. Current drills at the guns and command posts were taught and practised to this end. General Erwin Rommel, a future antagonist of the British, indulged himself with the observation that the British were wedded to manuals. It is certainly true that FSR mothered other regimental and corps training manuals and that artillery had followed this well-trodden path since the nineteenth century.

Of particular importance to this narrative is the first artillery FSR issued in the aftermath of World War I: *Artillery Training*, Vol. III, 1928. Written by a Major Kennedy who was to become Director of Military Operations and Plans at the War Office during World War II (the senior Operations Officer below the CIGS), the artillery FSR was to embody the principles of artillery use in war. Major Kennedy's little pamphlet would become the gunners' bible. It was, as Bidwell notes, 'a timeless exposition of commonsense [sic]'. The artillery FSR's detailed considerations on military operations and lessons to be learned are discussed in later chapters.

During World War I, prior to the employment of tanks, British artillery had often been asked to plug a gap in the line or come to the rescue of an infantry counter-attack with a timely concentration of fire. However, the gunners were frequently unable to comply with such requests, either because of the absence of an efficient communications network or because there were no 'direct support' guns available. At Cambrai and Hamel, once infantry and tanks became committed in an encounter battle, the artillery providing indirect fire could disengage. Without the need to move the guns, fire could be concentrated or distributed if a target was in range. If communication between fire units was adequate to the task, it followed that command and control of artillery should be vested in the highest commander who, given his broader view of the tactical situation, could exercise effective control.

This proposition leads to the doctrinal debate of the essential role of field artillery in a war of manoeuvre. From the 1920s to midway through World War II, the role of field artillery became the intellectual preoccupation of the British and other military thinkers. By this time a modern version of Clausewitz's principles of war had been distilled from the three main theatres of World War I: Gallipoli and Palestine; the Balkans and the Italian Front; and the Western Front. There were several principles based on offensive action: economy of force (a sledgehammer to crack a walnut); concentration of force; coordination and cooperation; surprise; and security. Artillery fire tasks in support of offensive action can be divided into four broad areas: direct and indirect observed shooting; predicted shooting; programmed shooting; and shoots for survey purposes. The first two are generally reasonably straightforward, but the programmed are 'time critical' and usually involve more fire units.

Thus drills and standard procedures are necessary. A divisional barrage is a case in point. It comprises, in effect, successive linear concentrations, usually ordered by the CRA or, at corps level, the Commander Corps Royal Artillery (CCRA), whose headquarters issued traces showing the four corners and number of lines (lifts) in the barrage and other relevant data such as allocation of units, lanes etc. In 1939 the General Staff conducted an exercise based on a simulated break in the front requiring a quick barrage to support a counter-attack. The criterion tested was the length of time from issue of orders and traces to the issue of gun programs (with firing data). This process took an unacceptable length of time and, when a similar exercise was held using a series of concentrations instead of a barrage, timings were much the same. When all other common elements were eliminated, a single problematic factor was identified: the trace and the allocation of tasks to regiments, then to batteries and finally to Troop headquarters before calculation of gun data could begin—known as 'breaking down'. If all went well with the production of gun programs and other vital drills on the gun-line, then the outcome would be a line of gunfire that lifted, for example, every 100 yards every two minutes. This would provide the infantry and tanks with a reference line towards their objective. It also took into account the time it took the infantry to engage the enemy after a lift and fight their way to the next line.

In general, a straight line of fire as a 'standard' concentration from 24 guns or more was little use to infantry in a tricky tactical situation. A fall of shot that was oval or circular in form, however, was far more helpful to the beleaguered infantry. Observers simply had to accurately 'map spot' the centre of the target locality and let the breadth of fall of shot plaster the chosen area. Accurately surveyed guns tended to concentrate fire, whereas some dispersion of shot was preferred by the infantry. Nonetheless there were occasions when a linear form of fire was the answer. The perceived wisdom was to advance a standard regimental concentration as a solution, the basis for which would be one grid reference. Six Troops of four guns would all fire on parallel lines at the central point using one range for each Troop. An exploding 25-pounder round had a 17-yard radius of lethality, and 24 guns delivering 72 rpm could be expected to cover nine 100-yard squares. Thus an oblong area 300 x 200 yards would be sufficient if the rate of fire was normal (3 rpm). The area would be wider if the angular displacement of each battery's line of fire was more obtuse through dispersion of their positions. If the situation became desperate, the rate could be increased.

This method of fire was eventually designated a 'Stonk'—literally defined as a standard concentration (of fire units) on a linear target—and trials in 1940 and 1941 proved its efficacy, although the gunners were less convinced of its worth. The infantry, however, embraced it enthusiastically. The gunners' chief concern was the appearance of a gap in the fall of shot that could be exploited. The Stonk had its full workout in the Western Desert and there the 300 x 200 yard oblong proved ineffective because of much greater dispersion between enemy tanks, armoured cars and troops in the open. It transmogrified to a linear form 525 yards long and was also given a bearing to orient it. Under this convention all Troops in a regiment had a specific location on a template, which simplified the process of generating gun data. Medium regiments followed suit with their 16 guns. In the desert, almost all Stonks were employed defensively on relatively flat ground, as a Stonk was more complicated to produce for undulating terrain and could not be adjusted as easily as a straightforward single grid reference target.[29] The use of Stonks in 1942 will be examined in more detail in later chapters.

Debate amongst field gunners now turned to how best to bring down concentrated fire from a division quickly enough for it to be of use to the infantry. The debate occurred at the time when a reorganisation of field brigades was being considered within overall infantry division and corps structures. One infantry view, expressed by no less a person than Archibald Wavell (then a brigadier) was that, as Bidwell notes, 'he wanted his infantry to be part cat-burglar, gunman and poacher whose touchstone was self-support, not close support.' While there might have been some merit in his view of the necessity for infantry versatility, it flew in the face of established artillery-infantry cooperation which, by 1918, had reached a high plane of practice and acceptance. When experiencing difficulty, for whatever reason, the infantry was encouraged to call for and use artillery. This was certainly the view of the German infantry whose aggression in attack was legendary. The Germans routinely used assault guns and howitzers in their battalions and brigades to help their infantry onto the objective.[30]

What was to become the Armoured Corps saw doctrinal matters *vis à vis* field artillery somewhat differently, since most of their tank casualties at Hamel were a result of direct fire from German batteries. Tank-mounted 6-pounder ordnance was no match for the German field gun, and this became a platform principle. Tanks had to be protected by infantry with anti-tank guns and/or artillery using indirect fire (counter-battery fire) concentrations on anti-tank guns or blinding them with smoke. Guns should be moved forward—guns on tracks were preferred to wheels—and close artillery/armour cooperation engendered. As far back as 1920, pamphlets for training had expounded that the greatest danger tanks faced was direct fire and, by 1927, this noble truism had advanced to become: 'the primary purpose of supporting artillery must be the protection of armoured fighting vehicles from observed fire.' The object embraced concepts of counter-battery tasks, concentrations on anti-tank gun areas, liberal use of smoke shell and intimate, close support. Tank and artillery commands would be co-located, if only to allow the artillery commander the flexibility to order his guns to move at short notice. There was no mention of engaging enemy tanks with concentrations or even barrages to assist the following infantry. It was here that the doctrinal battle-lines were drawn in the British Army.[31] This subject became the most contentious between the wars. Indeed, the events of 1941 and 1942 were to highlight these doctrinal inadequacies— flaws that led to needless loss of men, guns and tanks. For that reason alone, they are worth recording—all the more because Australian infantry, gunners, engineers and others were all to witness these inadequacies, live by their bitter fruits and, in some cases, die.

The British Cavalry and Royal Tank Corps (RTC) prior to 1939 were not an homogenous entity. Given the class-based structure of British society, the former recruited its officer class from the wealthy and the aristocracy. Its cavalry regiments, once horsed, had long traditions of service, sacrifice and snobbery; however, under a *laissez faire* culture of command and control, behaved independently and were inclined to be headstrong rather than circumspect. Their attitude to mechanisation was quite typical given their background. Their Waterloo, Balaclava or Omdurman mindset was akin to: 'You charge on a horse, why not in an armoured car or a tank?' Enter the internal combustion engine.

Made of steel, noisy and smelly and not very manoeuvrable, the tank (more so than the armoured car) gave birth to a new corps, the RTC, and its cavalry regiments (lancers, hussars etc.) were relegated to a reconnaissance role. Britain's cavalry (and Royal Horse Artillery) regiments reluctantly gave up their love affair with horses (except for ceremonial occasions). Even after they had been mechanised, the cavalry continued to look down their noses at the newcomer 'tankies'. Amalgamation was finally effected in 1939 when the RTC became a regiment with tank battalions, and with cavalry regiments formed the Royal Armoured Corps (RAC). The dust had barely settled on this reorganisation when World War II erupted. Nonetheless, mechanisation had been completed and the smaller Regular Army formations established. These were not exercised particularly rigorously as was to become obvious in the near future.

In the 1920s, senior Army officers and military theorists considered that the future development of arms would revolve around the use of the tank and the aeroplane. The tank inspired an acerbic debate about the real role of armour which J.F.C. Fuller has covered comprehensively; suffice to say that he and Basil Liddell Hart were the principal public protagonists of the use of the tank's characteristics to create a moral ascendancy on the battlefield. The key influential figure was Hart who remarked acidly that he educated more German generals in tank warfare than British.[32] The main players in the debate were Field Marshal Milne, Generals Martel, Maxse, Birch, Hobart and Broad. It is beyond the scope of this narrative to recount in detail the various approaches adopted by the British General Staff but, insofar as artillery employment and doctrine are concerned, the principal approach was based on the 18-pounder.

Unfortunately, the 18-pounder used a single charge and had insufficient range for mobile operations; indeed, when mounted on a tank chassis (the Birch gun), it adopted something of a hybrid appearance, failed to impress, and was consequently scrapped in 1930. This same year also marked increasing mechanisation and the phasing-out of horse-drawn artillery.[33] The mechanisation of artillery was a complex and time-consuming affair, although not as expensive as the development of AFV. Suffice to say that experience gained with the Experimental Armoured Force and documented in 1929 prompted the establishment of a light (two or three tank battalions) and medium (one medium and two light tank battalions) armoured brigade. The former was equipped with a 'close support tank battery' and the latter two batteries, and both had an anti-aircraft armoured battery. It was clear from exercises that these batteries would have to have a much greater range than the 18-pounder, and a new gun was thus essential.

Armoured warfare historians Harris and Toase suggest that the British lost their technical ascendancy over the next seven years primarily because of financial cutbacks. By 1935 a mobile division concept envisaged a reconnaissance element, two mechanised cavalry and one tank brigade, supported by two artillery brigades and a field engineer squadron, and administrative troops. Some of the arrangements for mounted infantry, if implemented, would have led to drastic role changes for cavalry regiments, and were thus abandoned. Rearmament initiatives from 1936 onward led to the development of the Matilda I or infantry tank (Mark II) and further increases in tank numbers. The creation of a Mobile Division raised the issue of whether it should include the tank brigade (and, in fact, in the 1940 campaign, it performed creditably in a counter-attack at Arras without the brigade). Another issue in the cavalry-RTC rivalry was the RTC 'tank alone' versus the 'armour in cavalry' debate. There was no shortage of views on that subject.

And so, eventually, the Mobile Division became a reality, albeit with one major disadvantage. In 1938 'training pamphlets were issued for various components of [the Division] but apparently [the General Staff] produced no doctrinal document on the handling of the division as a whole.'[34] Thus, when the first armoured division was formed there were two regiments of artillery: one field regiment of 24 guns; and another comprising a battery (two Troops, 12 guns) of anti-tank guns and a battery (two Troops) of light anti-aircraft guns. Given that the tank-mounted and portee 2-pounder was now obsolescent, the General Staff envisaged the 25-pounder in an anti-tank role as a rallying point if the enemy broke through the main line. There were two battalions of motorised infantry in the division's support group and divisional troops including engineers, signals, service and medical corps. Its teeth were one light and one heavy armoured brigade.

The British did not invest the same time and effort in anti-tank artillery as the German Army. At Larkhill in 1930, a demonstration of the 18-pounder versus a (then) modern tank drew the scorn of Liddell Hart. Any re-engineering of the 18-pounder had only served to increase its weight and make it more susceptible to malfunction, resulting in the development of the 2-pounder anti-tank gun. The inadequacies of the 2-pounder then forced an anti-tank role on the 25-pounder. Somewhat surprisingly, there was no serious study by the British

Army of the effect of (solid) shot on armour or the threat of concealed anti-tank guns. Their neutralisation would require field artillery because the weaponry of the light and medium tanks lacked hitting power.

Thus the field gunners found themselves in an anti-tank role on a scale some might never have thought possible in the desert—and in France. There had been no real development of doctrine for the combined use of armour and artillery (field and anti-tank), although there was a discrete FSR covering infantry and tanks. The armour had its purple manuals, the infantry, khaki and the artillery, blue. Missing from their intellectual approach was the notion that the employment of armour was an 'all arms affair'; the assumption was that artillery and engineers, even signals, were on hand and need only be requested. This was the result of each of the arms guarding its institutional verities rather than seeking an all-arms solution. The difference in approach of their former—and future—adversary could not have been more dramatic.

Even while peace talks following the Armistice of November 1918 were taking place, and before the ink was dry on the Peace Treaty, the German General Staff were busy studying their Great War operations (using some 70 committees). Amidst all the civil turmoil of the interwar years, they remained focused on how to fight the next war. The Allies might have forced disarmament on their beaten foe, but they could not force their 'intellectual disarmament'. The German Defence Ministry (the *Truppenamt*) became the centre of military planning and strategy under the redoubtable Generaloberst Hans von Seeckt, a believer in 'combined arms warfare'. Briefly stated, the German military planners prepared to fight a delaying action on one front while inflicting a heavy blow on the other. Three years later, the generals, aided by their politicians, made their first move. Historian Richard Ovary notes that,

> In 1922 Germany and the Soviet Union signed a pact at Rapallo renouncing their mutual war claims. In 1926 the Hindenburg Treaty of Berlin with Bolshevik Russia promised mutual assistance, included in which was an opportunity for German forces to develop prohibited weapons and train German soldiers and airmen on Soviet soil. At the Kama River in Kazan [600 miles east of Moscow] whilst armistice talks were being carried on from 1918 to beyond its signing, a specialist tank unit was set up where German firms could experiment with up to date equipment. In so doing they trained the Red Army and later evolved mutually agreed doctrine relating to tanks and aircraft cooperation. Fighters and bombers were also tested, and hundreds of young Germans gained technical and flying experience.[35]

Heinz Guderian, the architect of German armoured warfare, was handed responsibility for developing panzer tactics. He was a great admirer of Liddell Hart and analysed every aspect of employment doctrine and equipment. Guderian was somewhat unimpressed with the technology being trialled at Kazan, and was rather more interested in what the British were doing. By 1930 the Germans had created an all-arms balanced force (a panzer division) that had refined Guderian's initial concepts.[36]

For the benefit of foreign military attachés in the 1930s, the Germans staged demonstrations of panzer tactics with motor cars, soldiers running over the field with cardboard replicas of trucks and other mobile equipment, and myriad other ruses. If it was intended to mislead, it succeeded admirably. In 1934 the first panzer divisions were formed and Germany rearmed and exercised its formations. Only the command vehicles were fitted with wireless transmitters and their utility was soon recognised. In 1937 an exercise conducted by Guderian (as chief umpire) did not make much of an impression on the visiting British General Staff and its CIGS, Field Marshal Sir Cyril Deverell. In late autumn that year, Deverell and a number of other officers were invited to watch troops of the *3rd Panzer Division* in manoeuvres with their latest equipment. General Guderian noted several deficiencies that needed rectifying immediately. At a later function in Berlin, Deverell asked the German general's opinion on motorisation. Guderian later wrote: 'Younger British officers were interested in the problem of whether it was possible to employ as many tanks on a battlefield in wartime as were accustomed to appear before Mussolini on manoeuvres. They seemed reluctant to believe that it was so and appeared to tend towards the theory of the tank as an infantry support weapon.'[37]

The panzer divisional structure of a reconnaissance battalion with motorcycles and offensive/defensive artillery power was a concept that never seemed to have entered the British mind, much less the notion of combat engineers working with tanks. The panzer division Guderian had helped create was a balanced all-arms structure primarily transported in non-armoured vehicles. They, like the British, were short of AFVs and so infantry and artillery were carried in soft-skinned vehicles: 'Therefore unarmoured motorised infantry and conventional towed artillery had to be trained to work together with tanks as an integrated whole.'[38] What was also important was that the German Army exercised at division/corps level. If Deverell came away from the demonstration with one good idea, it was the formation of the first RA anti-tank regiments.

One episode in artillery development would cost the British Commonwealth dearly. The first British anti-tank gun was the 2-pounder, designed in 1934 and entering service with the Infantry Corps in 1938. A 2-pounder shot could penetrate the armour of any tank of the 1920s or 30s. The 2-pounder was towable, had a low profile, and a high rate of fire using solid shot. Initially placed in a battalion's support company, the infanteers were no happier with this concept than the artillery, and anti-tank guns soon passed to the Regiment as 'their' weapon. The 2-pounder was also adopted as a tank weapon and turret mounted on the close support infantry tanks.

In anti-tank artillery, the effectiveness of a weapon is measured by its penetration power (measured in inches of metal) at a given range at a right angle and an angle of 30 degrees. Gunnery staff were aware of German (and French) tank developments and, in 1938, design work began on a 6-pounder gun which could penetrate heavier armour. The design took a year to finalise and, at the end of 1940, those involved (and it must be remembered the same people were juggling priorities regarding long-term effects over a whole range of ordnance and defence equipment) had to make a fateful choice: produce more 2-pounders now or 6-pounders later. At this time, the threat of German invasion was palpable and real, and to change would have meant that, for a period, no anti-tank guns would have been produced. In the event, they decided on the 2-pounder and that, according to Bidwell, was a decision 'which nearly lost the war in Africa.'[39] The Australian Military Forces (AMF) purchased several in 1940 and, later that year, established the first Anti-tank Artillery School at Puckapunyal.

Anti-aircraft artillery, including locating and searchlight elements, were not forgotten in the British Army's development and the 30 cwt 3-inch anti-aircraft gun remained in service until superseded by the 3.7-inch heavy anti-aircraft gun. Many of these anti-aircraft regiments were sent to France with the BEF and many more deployed around the country defending vulnerable points. For lower flying aircraft, the British chose the Swedish-designed Bofors 40-mm light anti-aircraft gun which also saw service in France. Both of these guns were introduced into the British Army in 1936. Australia subsequently acquired both and manufactured the 2-pounder, 3-inch and 3.7-inch guns.

The aircraft-artillery cooperation that had proven so successful during World War I had also progressed. RAF Artillery Reconnaissance Squadrons had been formed and now regularly exercised with the Army. New aircraft such as the Hawker Hind and Westland Lysander had been produced and the latter aircraft went to France to support the BEF. Little effort seems to have been devoted to tactical air support procedures and communication, let alone on the scale practised by the Germans in the Spanish Civil War. There, tactical and strategic bombing by the *Luftwaffe Condor Legion* (of which Guernica is the best remembered for its barbarity) helped the Nationalist cause and delivered General Franco's eventual victory.

2nd Survey Company Militia surveyors training 'on the cheap' in Victoria during the 1930s, using their private vehicles to do so.
(Private Collection)

Artillery headquarters for British regular divisions and corps were staffed by officers, many of whom had seen active service. While they exercised from time to time, these activities were not conducted on the same scale as those of the Germans. During the French campaign both levels of command and control were left floundering by the rapidity with which the German invasion unfolded. Counter-bombardment and observation assets with marginally improved technology were deployed, but they were quickly overwhelmed and contributed little to the artillery battle. It would take a gruelling large-scale exercise (Exercise Bumper) in south-east England, staged later that year in the aftermath of the evacuation, to expose flawed decision-making at both levels of command.

One piece of technological equipment was particularly important in World War I and was to be so again in World War II—sound-ranging microphones and recorders. Very few people are aware that this sophisticated method of locating enemy guns was developed by a team of scientists working on the Western Front under Australian-born Major Lawrence Bragg, MC (Bragg and his father won the 1915 Nobel Prize for Physics). He was unimpressed with the design of more modern versions of his sound-ranging recorder which produced a trace on film of the sound waves made by guns and howitzers. Nonetheless, they were deployed and made a minor contribution to the counter-battery responses of the field and heavy artillery. Australia procured two sound-ranging recorders from Britain in 1935 for the 2nd Survey Company, RAA, in Victoria. The company exercised with the recorders, which produced good results considering the minimal Staff Corps support and the lack of opportunities to join with field brigades at annual camps where a smaller number of rounds could be used for training.[40]

A field service wireless set was also a necessity if a force of all arms was to engage in battle. Continuous wave Morse Code was extensively used in World War I and, with the invention of single sideband super heterodyne sets in the 1930s using radiotelephony (voice), the future seemed bright. It was not the ultimate in communications technology as telephone lines that offered greater security were still essential, a fact supported by the number of line signallers, as distinct from wireless operators, retained on regimental establishments. The greatest technical hurdle was to make a set sufficiently robust to be 'soldier and vehicle proof'. This hurdle was eventually surmounted, but range was sometimes a problem in mobile desert operations, although wireless sets proved satisfactory in siege (Tobruk) and static positions (El Alamein).

It is difficult to escape the conclusion that British Army thought concentrated on the doctrinal development of an infantry division at the expense of armour, although development of the latter was certainly marked by more passion. The gunners maintained a stringent focus on artillery development programs with the result that, by 1939, they were on the threshold of commissioning the best field and medium artillery (and anti-aircraft) of the war. However, as previously explained, the same cannot be said for anti-tank weapons. This was a function of industrial capacity and perceived priorities at the time which, fortunately, would gradually improve. In late 1942, the 25-pounder was married to a Canadian Ram tank chassis to become the RA's first British self-propelled field artillery, the Priest. Fuller's axiom had come to pass some 17 years from its issue. Initially, there was an imbalance in the heavy/medium to field gun ratio that prevented the amassing of sufficient effective counter-bombardment efforts, but this too was later rectified and followed by the development of field radar, an offshoot of the air defence instruments already in use.

Those responsible for making decisions that influenced the British Army's performance in two critical campaigns have been subject to heavy criticism. There is scant mention in the literature of training, exercises and the like, which begs the question of how fit the British Army was for operations. The critic's safe generalisation was that the Army was well equipped for a war of 1919 but not as well trained. Their putative opponents, with the exception of panzer development, were not as well endowed as many have supposed. In fact, horses were still the main mode of transport for infantry within panzer divisions, whereas the British at that time had the only mechanised army in the world. One German general complained in 1940 that his army was the worst equipped ever sent into battle! The German standard of training was high across all levels of command, and the Germans were confident of their ability to achieve their commanders' aims.

In Commonwealth countries military development was closely monitored and defence requirements were forced to fit the political agenda of the respective governments. To the population at large, it was post-Depression employment and social concerns that mattered most. The AMF experience up until mobilisation was to be primarily as a spectator of overseas events. Given a parsimonious treasury for almost 20 years, it is surprising how well full and part-time soldiers coped with shortages of equipment, a miserly ammunitions allotment at annual camps, and the use of the transport industry's prime movers to tow guns because the Army was underfunded. If there was a bright side to this state of affairs it was that it conditioned the soldier to improvise and do more with less. This experience was to benefit the men who later fought in desert operations as Australian soldiers used captured equipment at every opportunity. The digger's reputation as a resourceful and determined fighter was to be well founded.

Chapter 4
The BEF Artillery in France–1940

Nothing happens for the worst that does not happen for the best.

Old German proverb

As the British military's mind was being exercised by the latest developments in weaponry and doctrine, Germany continued to build its military might and, on 1 September 1939, Hitler's armed juggernaut rolled into Poland. Seventeen days later, attacked also by the Soviet Union, Poland surrendered. With the declaration of war, the British Army commenced a build-up of its forces in France. The BEF crossed the Channel landing at Cherbourg and Le Havre and eventually concentrating around Lille. The build-up continued until, by May 1940, a BEF of six divisions, soon to total 10, was sandwiched between the Belgian Army on its left flank and the First French Army on its right. Commanded by General Lord Gort, VC, the BEF consisted primarily of British regular formations augmented by Territorial forces.

Its frantic assembly complete, the BEF and its allies then sat in static positions for six months during the so-called 'Phoney War'. The effect of this military stagnation was to make the BEF progressively more defensive in its mindset, particularly as the assembled forces spent little time manoeuvring, honing their communications or exercising with their allies. The staff at General Headquarters (GHQ) and the War Office, who had studied German methods to date, suggested that the future war in Western Europe would be 'a more ponderous war of masses rather than a highly mobile type'. This view, supported by a joint French-British political (and High Command) directive not to antagonise the Germans, was duly promulgated in an Army Training Memorandum.[41] It reinforced the rationale behind British planning—that the Army would literally and metaphorically stick to its guns and the German armoured forces would have to respond. The French Army assured its allies that it would do likewise.

While training was intended to focus on producing quick fire plans for mobile warfare, a key aspect of regimental operations, a GHQ order that units severely restrict the use of the wireless, meant that communications were essentially limited to telephone lines and dispatch riders. Accordingly, a division took three hours to produce the most complex plan on the artillery agenda—gun programs for a barrage beginning with written orders and traces. It was almost impossible to quickly modify a barrage once ordered, a situation that had remained unchanged since 1918. The problem was that accelerating the process compromised accuracy, and this dilemma divided the gunner community into the pedants versus the pragmatists. A key side issue was the safety of the infantry soldiers following the barrage.[42]

By 10 May, when the *Wehrmacht* opened the battle of France with 134 divisions, the BEF could only muster a mere 16 infantry divisions, one army tank brigade and five mechanised cavalry regiments equipped with light tanks, including corps and army troops. The artillery component was organised at three levels: GHQ Troops (Army) reserve under a Major General, RA; Corps Troops under a CCRA and a Corps Commander Medium Artillery (CCMA); and Divisional Artillery under a CRA.

The order of battle comprised 46 field regiments armed with 18/25-pounders, 18-pounders and 4.5-inch howitzers; 18 medium regiments equipped with 6-inch howitzers, 5-inch (60-pounder) guns; three heavy batteries using 6-inch and 8-inch guns and 9.2-inch howitzers; and three super heavy batteries of 9.2-inch guns and 12-inch howitzers. At corps level, a survey regiment and two counter-battery organisations using flash spotting and sound ranging made up the field force. There were also substantial anti-aircraft defences represented by 18 light (40-mm) and 42 heavy (3.7-inch) anti-aircraft batteries and a searchlight brigade. These assets were grouped in brigades and overall command was vested in the Major General, Anti-Aircraft at GHQ level.[43]

The campaign in its entirety has been well analysed by historians. German *blitzkrieg* methods of integrated armour, artillery and infantry, supported by dive-bombers and a flexible fighter control system that neutralised Allied air power, paralleled the results achieved in Poland a brief six months before. The French Army, some elements of which fought bravely, was outclassed in materiel, tactics and moral fibre. The Dutch lasted a mere five days. The British staged an ad hoc evacuation at Dunkirk, although reinforcements continued to be brought in through Cherbourg as the entire western front was crumbling.

From the artillery perspective, fire plans generally were no sooner made than German advances and attacks rendered them superfluous—a point discussed by the British Official Historian, Major L.F. Ellis, in his analysis of the campaign.[44] Yet, when the front momentarily stabilised and a division could concentrate its fire on a narrow front, it was accurate and timely, inflicting substantial casualties on the German infantry. However, the campaign narrative is mostly one of batteries and Troops being forced back from position to position as the lightning mobility of the Germans brought rapid changes in the tactical situation and withdrawal became the most prudent course of action. In the 'fog of war' that enveloped the BEF as it retreated, observed defensive fire tasks chosen from the map proved simply a waste of ammunition.

It quickly became obvious that the new organisation of the field regiment into two 12-gun batteries did not work, particularly during the withdrawal. One of the rationales for the 12-gun battery was that, in mobile offensive operations, the simplest and fastest technical control of 12 guns could be effected by one command post. Second, the new organisation was primarily configured for the attack phase of war and for periods of inactivity at the front, but was far less effective in defence and withdrawal. Formations retiring and hopeful of keeping their flanks intact tended to concentrate field regiments in range of their dedicated brigades and allocate observers to units most in need. In the fluid campaign that developed rapidly amid crowded roads and lanes packed with refugees, the divisional CRA's life quickly became a nightmare. The campaign was noted for the prevalence of observed shooting on 'targets of opportunity' and it was recognised in later analysis that the field artillery had not performed well. While there were many reasons proffered for this, the overwhelming factor was the devastating element of surprise, part and parcel of *Blitzkrieg*.[45]

The essence of the 'Flanders debacle', however, was the divergent views within the British Army on the nature of the 'next war'—static front versus mobile operations. There was no shortage of ideas for restructuring the division/brigade/battalion and on how the arms would coordinate their firepower and manoeuvre in open country. The problem was to distil this into doctrine, and that is where some of the worst features of the entrenched regimental system of the Army exerted undue influence on the resulting organisation and equipment. Inevitably a compromise of sorts was reached. Leaving aside the issue of artillery support for armour, the genesis of the problem was the imbalance in the provision of supporting fire between the requirements of the infantry—now a three-battalion brigade structure—and the minimum number of field guns per division (72) in three regiments. At the heart of the problem was the 12-gun battery in two Troops, each battery providing an observation post officer (OPO) who would be co-located with the battalion commander; ergo, only two battery command posts would be required, a saving of manpower. By way of concession, it was appreciated that if one battery command post was crippled, half the regiment would be likewise.

The survey function at Regimental Headquarters (RHQ) was also abolished. Lines of six guns were too dense a concentration to survive aerial attack; but if that were to be reduced, the battery frontage would be bigger and more difficult to control. A change to two 4-gun 'Troops' in a battery meant more manpower and a one-third reduction in guns, an unacceptable alternative. A sensible compromise was finally achieved in three 8-gun batteries of two 4-gun Troops. This fitted with the triangular brigade in which a four-company battalion with two companies 'up' would each have an OPO as his artillery adviser, thus continuing the liaison at three levels of command.[46] It was noteworthy that the Royal Horse Artillery never adopted the larger battery model.

Something of the way ahead was shown by a rising star in the Regiment, Lieutenant Colonel John Parham, Commanding Officer (CO) of the 10th Field Regiment. In France, Parham turned a 'Nelsonian Eye' to the restrictive communications practices and, with a characteristically enthusiastic approach, drilled his regiment until his men boasted an extraordinary level of radio skill and, more importantly, were not afraid to use their wireless. Early in the campaign, when Parham's regiment was positioned on the River Escaut, a panzer regiment pulled into a wood to refuel. The panzers were spotted by an alert OPO who map-spotted its location. He engaged it with his regiment without preliminary ranging, firing 250 rounds of high explosive and causing a huge conflagration. In so doing, Parham made history with the first fully predicted radio-controlled concentration at an opportunity target.[47]

Some of the more senior regular and Territorial regimental survivors of World War I were no strangers to the multifarious arts of counter-bombardment. Its importance was reflected in the 336 medium and heavier guns that comprised the BEF corps artillery, plus some of the Army field regiments with counter-bombardment roles. This was roughly equivalent to the proportion of field to medium/heavy guns that saw action with the RA in 1918, although the variety of medium and heavy calibres in service in World War II never reached that of World War I. As it was, 272 of the medium guns never made it back to Britain. The RA was never to reach that height in the field army again, as longer range fire support was handed to the RAF.[48]

The war of movement soon demonstrated that the centralised counter-battery organisation, the showpiece of British artillery in 1918, was now outmoded. Flash spotters and sound rangers, which still depended on line communications that took too long to lay or reel in, were reduced in their flexibility and therefore their efficiency.[49] When the Army re-equipped in the 1930s, survey regiments were not accorded a high priority for the issue of wireless sets. When deployed on the River Dyle in 1940, they located several hostile batteries. However, as locating assets had been removed during the precipitous withdrawal, the counter-battery organisations had to rely on the OPOs' hostile battery reports. In addition, five RAF Lysander Army Cooperation Squadrons, sent to be the 'eyes of the army' were driven from the skies in short order with insufficient RAF fighter aircraft to protect

Brigadier Parham, RA had the distinction of firing the first regimental target of WW2 in France, where his 25 pounder guns destroyed an armoured force refueling.
(Private Collection)

them as they attempted to conduct their reconnaissance and Army cooperation duties.

As Pemberton notes, the BEF campaign in France saw the emergence of a new battlefield phenomenon—casualties caused by the German 80-mm mortar. The mortar demonstrated the need for a quick response, but neither sound location nor flash spotting could generate a location report in a timely manner. A skilled crew could bring a mortar into action, fire 20 aimed bombs, pack up and be gone in five minutes. This practice was to be a constant concern to the Allies over the next five years. The initial response was to increase the number of mortars in an infantry battalion's support company from two to six.

Some actions achieved success of a kind: K Battery, Royal Horse Artillery, at Hondeghem and St Sylvestre, featured in a battery versus tank fight characterised by boldness and offensive spirit. K Battery's weapon was the 18/25-pounder which had far more hitting power than the 2-pounder and demonstrated what could be done with a field gun. This was to be the precursor of thousands of engagements between British guns and German and Italian tanks over the next two years.

A key element of German tactics was the use of the dive-bomber as a substitute for fire in depth beyond the range of field and medium guns. Some observers maintained that it was the equal of the tank in its effect on morale, particularly against inexperienced troops. However, the dive-bombers were less accurate than heavy calibre weapons, particularly when the target was defended by anti-aircraft artillery. They were relatively easy to shoot down compared to high flying aircraft, as experience later in the war was to demonstrate.[50]

During the BEF withdrawal to the Channel, command and control broke down at several key points. In some areas, the gunners fought as infantry, or single guns fired until subdued. Medium and heavy artillery lacked mobility and range and so no effective counter-bombardment fire could be mounted, despite considerable staff effort, and most of these guns were lost.

Following the evacuation from Dunkirk, the Regiment faced the problem of replacing the vast quantity of equipment left in France.[51] At the end of the short campaign the loss of guns and howitzers represented 60 per cent of the total stock of those weapons held by Britain and the Commonwealth countries (see Table 1).

Artillery piece	with units	reserves	total
field	944	72	1016
medium	224	48	272
heavy	48	2	50
super heavy	6	3	9
Total	1222	125	1347
anti-tank			
2-pounder	480	77	557
25- mm		50	50
guns lost in action or accidents	518	127	645
total for field guns			1347
anti-aircraft			260
Grand total			1607

Table 1. Artillery losses in France.

This was an enormous quantity of material, particularly with the inclusion of the small arms, wirelesses, trucks, ammunition and other equipment of a standing army that was also lost in the evacuation. The extent of the loss was compounded by the fact that it included the most modern guns and communication, ranging and detection equipment. Only 322 guns were returned to Britain.[52] High casualties meant that a significant fund of gunners' expertise was also lost.

Fortunately, Britain was able to call on the forces of India and the Commonwealth. In 1940 these included 72 field force batteries of mountain, field and medium guns. That experienced force was to be the backbone of the Middle East campaigns from 1940 to 1943.[53]

The British infantry division, its organisation and command, shouldered most of the blame—it was too clumsy, too slow and too unwieldy. A brigade was the answer, said one school of thought, being 'handier and mobile'; but this meant that the position of CRA would be left with only the relatively minor functions of training and personnel. It would not have escaped the attention of the military analysts, particularly Liddell Hart, that German armoured and infantry divisions of the *First Army* reached the Swiss frontier in 17 days. French tactics were also partially to blame as their units formed 'hedgehogs' which were not mutually supporting. The panzers bypassed them and left their reduction to the following infantry divisions.

The vital lesson for the British doctrine developers was the way the German artillery and armour cooperated closely with their infantry. They also demonstrated refined procedures that resulted in timely tactical air support that materially affected results on the ground. Even before the Flanders campaign, cooperation between the British artillery and infantry (and armour) was not a strong cultural element of the British Army. Each arm kept assiduously to its own regimental world.

The doctrinal issues that crystallised in the aftermath of the failed French campaign, once all the operations reports were completed and analysed, were several in number. The first concerned fire support for mobile operations. Improved communications and simplified procedures would mean increased fire support and increasingly *flexible* fire support and also addressed the related issue of massing firepower for maximum effect. The Germans were able to mass fire at critical points along the front. British interest in self-propelled artillery had, according to Bailey, all but disappeared after initial enthusiasm waned in the early 1930s. He notes that 'the Birch gun's chief offence was to be 15 years ahead of its time.'[54] Self-propelled equipment was now back on the agenda.

The telling weight of German artillery fire derived from the fact that a 12-gun battery under single command and control worked better for the Germans than two lots of six. This contrasted with the British analysis of the failure of their 12-gun battery organisation. Somehow the German gunners, equipped with superior wireless communication, could deliver larger quantities and more powerful fire by virtue of their mobility, and concentrate it at critical points. The British had to adopt similar methods and design better equipment if they wanted to compete. They had not learned the lessons from Poland. The employment of anti-tank artillery and its coordination with field artillery led to a deduction that one way to counter mobility was by placing anti-tank artillery well forward with the infantry. There were not enough anti-tank guns forward and there was a need for a heavier gun, perhaps even a self-propelled tank destroyer at divisional level. This would take three years to eventuate.[55]

Those senior gunner officers who straggled home with the remnants of the BEF were faced with enormous problems, some internal and one external. The internal dilemmas related to the reorganisation of a field regiment and the design of fire control methods to cope with the German tactics as well as the revival and refinement of the art of fire planning, alongside many tangential issues. The external dilemma was less easily remedied and concerned the education of armour and infantry commanders in the use of field artillery. These issues were addressed as a matter of urgency, although General Alan Brooke first had to attract the attention of his senior commanders; in this he was helped by two other senior figures—Bernard Montgomery and Sidney Kirkman.[56] Brooke, Montgomery and Kirkman were to have an extraordinary influence on the war and the way it was fought from June 1940.

Lieutenant General Alan Brooke was commanding II Corps in France when, on 25 May, he was promoted general and appointed CIGS. The General Officer Commanding (GOC) the 3rd Division, Bernard Montgomery, took over II Corps and returned to Britain to be promoted and appointed GOC Southern Command. His command was the immediate object of Hitler's Operation Sea Lion—the invasion of Britain. Kirkman was CRA of a Territorial Division, ex-Royal Horse Artillery and a Staff College graduate. In an endeavour to come to grips with the field artillery's failure to live up to its promise, the CIGS ordered the conduct of Exercise Bumper, a mock battle between four divisions in two corps, one fighting as friendly troops, the other playing the role of the enemy. Enemy troops represented a German force landed over the beaches, built up and then launched towards London. General Brooke directed the exercise, while Montgomery was Chief

Umpire. Quite by chance, Montgomery had attended an earlier exercise Kirkman had conducted. At the end of the exercise Montgomery had commented that 'The business of Gunner COs is to train their regiments, and then train the infantry brigadiers to use them properly.'⁵⁷

Army training exercises are not written by the Directing Staff to reveal how clever they are, but with the aim of producing hard data on performance against several criteria. Exercise Bumper was noted as the birth of new doctrine and practice that was to last through the war with some minor modifications. Both the BEF campaign in France and Exercise Bumper highlighted errors of judgement and lack of application or misinterpretation of doctrine, many of which were listed by Sidney Kirkman in a post-exercise critique. Kirkman noted that:

- moving, deploying and re-grouping artillery was an endemic problem—particularly the decentralisation of artillery from higher to lower formations—thus weakening the offensive power of the division/brigade/battalion (one CCRA reorganised himself out of a job);
- a brigade had four gun regiments under command with one CO trying to control the lot;
- a division took its artillery with it when transferred to another corps;
- another division left its artillery behind when transferred to another corps;
- formations moved into reserve and took their artillery where it sat idle;
- two regiments never saw action;
- the artillery in an armoured formation was sent into reserve when the battle started;
- in one corps there was no cooperation between artillery and infantry; and fire planning was almost totally neglected.

Perhaps the most important outcome of the exercise was the discussion of these and many other observations with a group of younger officers who were to be the future regimental commanders, CRAs and CCRAs. They listened in rapt silence as Montgomery reinforced the doctrinal message. One deficiency of a systemic kind was the production of a heavy volume of fire from a corps of three divisions. In 1941, the livewire Parham, now CRA 38th Division, demonstrated that the fire of 72 guns from three regiments of a division could be arranged in five minutes, slightly longer for a corps. This, of course, depended on quality wireless communications and, as a result of Exercise Bumper, authority was given for specific officers on the battlefield using an appropriate wireless network to call for fire from a specified number of fire units. An artillery arm with a reliable and comprehensive intelligence network/system on the battlefield was the key to the timely provision of such firepower. A corollary of using the data available was to plan (program) its application. In France there had been a tendency for the process to be too slow or the result too complicated so that, by the time all the data was checked, the battle had moved on. The same occurred during Exercise Bumper.[58]

Yet another revelation related to the use of field, medium and heavier calibres in the counter-bombardment role. Nominally this was a corps, indeed a CRA responsibility. However, the CRA's problem was that, if he delegated control of his guns to divisions, how could he properly advise the Corps Commander from so far to the rear? This problem was solved in the reorganisation which saw RA groups of army and corps artillery become Army Groups, Royal Artillery—AGRAs. These usually comprised one or two field, two or three medium and a heavy regiment/battery depending on the circumstances, under the command of a Commander AGRA (CAGRA).[59]

In World War I, the Regiment had amassed an enormous amount of expertise in survey and post-war doctrine had reinforced its importance. However, survey accuracy *per se* did not sit comfortably with the notion of artillery as an area weapon in mobile operations. In practical terms it boiled down to how accurately and/or how quickly guns could be put on a single grid. In well-mapped Europe, this could be achieved with comparative ease, but in the Western Desert of Cyrenaica which had been mapped by the Italians (the British mapped Egypt) it was a different matter altogether, particularly given the featureless terrain. The problems this presented are the subject of later chapters.

Survey regiments that had been deemed unnecessary in mobile operations and threatened with the axe also received a reprieve, although this had more to do with the need for mortar-locating techniques than regimental officers' maintaining a 'survey obsessed' orientation for regimental and higher level employment, at least in Europe. Survey remained vital for counter-bombardment work. Gunners at that time did not realise how long it took to become highly proficient in locating skills. The expertise amassed by 1918 was incrementally acquired over three years, while the latter-day gunners had been in action for a month. Development work on field radar location for mortars was on the drawing board and trials at Larkhill a year or so later on were encouraging. Meanwhile, wireless solutions were applied in the hope of resolving the time-consuming problem of laying sound ranging line and, by autumn the following year, the use of the wireless had become widespread.

Any examination of the effectiveness of the British artillery from 1941 must begin with the shortcomings revealed by Exercise Bumper and an analysis of the inability of the British Army (of which the artillery arm was a major component) to embrace the doctrine of all-arms cooperation. Exercise Bumper was a rare historical event, and its lessons were never forgotten by General Montgomery. In the long term, in fact, it proved more important even than the brutal lesson imparted by the efficiency of the *Wehrmacht*. When Montgomery was posted to the Middle East in September 1942, he took Sidney Kirkman with him as his CRA. Thus began a doctrinal conversion similar to that worked by Birch on Haig—Kirkman made an artillerist of Montgomery. All his set-piece battles thereafter were opened with massive artillery preparations.

The dismal performance of the RA (field and observation branches) in France and Flanders formed the basis for the problems it faced in the summer of 1941. When Italy joined the Axis cause on 10 June 1940, the Mediterranean and Africa became the new battleground after Norway and France. The Australian Army entered the fray in time to join the burgeoning fight in the Middle East. Unlike its British counterparts, however, the Australian doctrinal wisdom in artillery matters was still tied firmly to its tried and tested procedures and priorities of the earlier war. The Middle East campaign was to prove the ultimate learning experience for the RAA.

CHAPTER 5
THE ARTILLERY MOBILISATION AND THE FIELD REGIMENT

Not once or twice in our rough island story
The path of duty was the way to glory.

Alfred, Lord Tennyson

The lengthening shadows cast by the rise of Nazi Germany in the 1930s saw a belated, uneven, but nonetheless increasing level of preparedness in Britain and the Commonwealth. Compulsory Military Training had been abolished in Australia in 1929 and Australia's Militia units were soon too under-resourced to maintain their previous regime of realistic camps and tough, warlike training. Shortages of equipment and a miserly allocation of live ammunition saw most Militia units resort to night and weekend parades of repetitive drill which led just as quickly to a high turnover in personnel as enthusiasm waned.

As Craig Wilcox suggests, the decline in Militia strength from 34,000 to 24,000 between 1929 and 1935 was the result of a combination of low pay (four shillings a day), unimaginative drill and unsympathetic employers. On the positive side, some of the younger members 'had imbibed tales of military glory at school or, perhaps had a father, older brother or school teacher who had served in the AIF, and many had been Compulsory Trainees.'[60] Artillery and 'technical' units had better retention rates than infantry and other corps and had benefited from a dribble of new or second-hand equipment from Britain. Medium and anti-aircraft guns arrived in 1926 and in 1929 four light tanks arrived to establish the nucleus of a mechanised force. Training with horses retained its appeal and helped sustain many batteries at a reasonable strength. Unit commanders also did as much as they could to retain their men through sheer enthusiasm and leadership, often reaching into their own pockets to fund extras that made training more interesting.

Militia officers and senior non-commissioned officers (NCOs) in the Citizen Military Forces (CMF) continued to qualify by examination under Section 22(a) of the *Defence Act* which required them to demonstrate written and practical competence in the application of current doctrine at the same level as Permanent Force officers for first appointment to lieutenant or promotion to the next higher rank. Senior officers and Staff Corps officers (who were not promoted or promoted very slowly) wrote tactical exercises for promotion purposes from first appointment to lieutenant colonel—also a requirement of the *Defence Act*. Staff Corps officers were also required to sit promotion examinations.

These were the men who would ultimately lead the AIF divisions in 1939–40. They were a mix of mostly CMF in regimental appointments with Staff Corps taking an active part in staff/headquarters roles. An examination of Army List records reveals that Militia officers who were commissioned between 1924 and 1926 were captains six or seven years later. There were exceptions, such as Lieutenant Colonel Thomas Eastick, CO of the 13th Field Brigade in Adelaide. Veterans of World War I rose to lieutenant colonel in the early 1930s and were colonels or brigadiers in 1939. This cadre of enthusiasts followed developments overseas as well as they could. Often the forums of the (Royal) United Services Institute, an all-service organisation devoted to military matters with branches in all capitals, served as useful venues for the exchange of opinions. Brigadier Herring's biographer records that TEWTs (tactical exercises without troops) were held in the Naval and Military Club of Melbourne under the guise of 'conferences of persons of similar interests'.

The focus of artillery thought and development in its broadest sense reposed in three separate branches of the AMF hierarchy: the School of Artillery at South Head (Sydney); the artillery branch of the General Staff, and the General Staff Officer Grade 1 (GSO1); and the Establishments Committee. The GSO1 Artillery interfaced and cooperated with his other two colleagues in preparations, plans and staff duties in a position that was also privy to the latest developments. The Establishments Committee had the last word on how, for example, a regiment was manned, armed, transported and equipped for operations (and for peacetime training purposes). These were based on compatibility with British establishments.

The repository for technical artillery matters [weapons (ordnance), ammunition and mechanical repair and maintenance] was the Master General of the Ordnance (MGO). This was a quaintly titled office of great tradition in which the Director of Artillery handled such items as equipment, postings and matters involving the many facets and traditions of the RAA. He also interfaced with other directors in the branch on associated matters. Colonel Ted Milford, a decorated World War I veteran and renowned graduate of Camberley Staff College, was Assistant Director of Artillery and Director Regimental Artillery (DRA) from 1938 to 1940. He was replaced by Lieutenant Colonel G.P.W. Meredith, a coast artilleryman, who was succeeded by Brigadier Lewis Barker in 1941. After a short term as Assistant DRA in 1940, Barker commanded the 2/1st Field Regiment in the Cyrenaican campaigns for a few months.

Brigadier John Whitelaw was GSO1 Artillery from 1935 to 1939. In June 1940 he commanded the coastal defences and pushed the development of radar. Then, in April 1942, he was promoted Major General Royal Artillery (MGRA), taking the artillery portfolio from the MGO Branch with him. Whitelaw was the only Australian Army officer ever to hold that appointment. MGO Branch was responsible for ammunition, repair, recovery and maintenance, the latter belonging to the Electrical and Mechanical Engineers (EME) Branch. This reflected the increasingly technical nature of weapons, vehicles, searchlights, radar and other warlike stores.

Lieutenant Colonel Cyril Clowes attended Larkhill from 1936 to 1938 and returned with the latest doctrine and artillery developments from Britain. Unfortunately for the RAA, Clowes was posted to the 6th Military District soon after his arrival in Australia. At that time, doctrine and drills were disseminated through Artillery Training Pamphlets and equipment handbooks. Pamphlets of all British arms and services made their way into unit libraries and field brigades. Amendments to doctrine, procedures and drills were received on a frustratingly regular basis.

With the increased possibility of war, Militia recruitment was boosted by Billy Hughes, a former Prime Minister and now Minister for External Affairs. Funds became available for pay and stores began to flow into training depots to facilitate instruction. 'A drought became a flood', was one gunner's observation. Heavy calibre weapons and wirelesses did not flood into units, however, and the fledgling AIF and Militia were going to need much more than what was then available in almost all categories of equipment.

When Army Headquarters issued mobilisation instructions in 1939–40, the AMF order of battle included the following artillery commands and Military Districts:

> Qld: (1 MD) 5th, 11th Field Brigades
> NSW: (2 MD) 1st Cavalry, 1st, 2nd Infantry Divisions six field brigades—21st, 1st, 7th, 9th, 14th, 18th 1st Medium Brigade, 2 MD Survey Company
> Vic: (3 MD) 2nd Cavalry, 3rd, 4th Infantry Divisions five field brigades—2nd, 4th, 8th, 10th, 15th, 2nd Medium Brigade, 3 MD Survey Company
> SA: (4 MD) 13th Field Brigade
> WA: (5 MD) 3rd Field Brigade
> Tas: (6 MD) 6th Field Brigade
> Ten of these brigades were mechanised.

The raw material for field regiments also came from the many coastal artillery heavy batteries and brigades from as far afield as Port Moresby and Northern, Eastern, Southern and Western Commands, as the former Military Districts were now known. The artillery headquarters personnel of the six divisions were also regarded as a suitable recruiting source. Men from the units listed above had already enlisted in the 2nd AIF—for example, the 1st Field Brigade (Militia) 'co-existed' with the 2/1st Field Regiment, RAA (AIF). Major R.O. Cherry, Officer Commanding (OC) 3 MD's 2nd Survey Company, was interviewing potential recruits in Sydney and Melbourne for his 2/1st Survey Regiment at this time.

A field brigade was organised into two 18-pounder and one 4.5-inch howitzer batteries, each of six pieces. Thus Australia had available within its Militia and 'A' Battery more than two hundred 18-pounders and around 100 howitzers. Field brigades were gradually renamed field regiments and the raising of these regiments continued apace. These now comprised 24 pieces, initially two batteries each of three Troops of four guns. In the medium brigades there were 60-pounders and 6-inch howitzers totalling 16 guns. This gave an overall strength on the ground ratio of field to medium/heavy guns of about 12:1, a stark contrast to the situation at the end of World War I on the Western Front when the ratio was 3:2 over all British forces. The AMF was also reassured by Imperial defence arrangements agreed at this time which would see Britain provide sufficient artillery for AMF needs in any expedition involving both Australian and British artillery.

The British, however, were busy with a drastic cull of their medium and heavy guns to suit their new-found reliance on tactical aerial bombing rather than heavy artillery. This must have sat rather uncomfortably with Australian defence planners and artillerymen whose air force had nothing heavier in 1939 than Hawker Demons and Avro Ansons. The ramifications of this have been covered by other historians and its reasons are outside the scope of this work. Nonetheless, this equipment situation was to bedevil the balance of Australian field force artillery units for three years. In the event, two medium regiments (the 2/1st and 2/2nd) were raised as Corps Troops. Excluding anti-aircraft, searchlight and anti-tank artillery units, the AMF raised two complete infantry divisions and ancillary/corps troops in an extremely short period from Sydney to Perth and Cairns to Hobart with a very small Staff Corps and Australian Instructional Corps (AIC), a feat worthy of admiration.

Lieutenant General Thomas Blamey, then GOC, selected the former CRA of the 3rd Division (Militia) as his artillery adviser when he raised the 6th Division in September 1939. Newly promoted Brigadier Edmund Herring brought impeccable credentials to his role. He was one of the few officers who had served in regimental and staff appointments in France and Macedonia, where he won the Military Cross (MC) and Distinguished Service Order (DSO) with the 99th Field Artillery Brigade, RA, fighting the Bulgarians. His entry into the Regiment was unusual in that, as a Rhodes Scholar for Victoria (at Oxford University) in 1912, he was a member of the Officer Training Corps and was commissioned into the Royal Field Artillery (RFA). After the war he returned to Oxford and completed a law degree.

Herring was born in Maryborough (Victoria) on 2 September 1892 and was a stocky, sturdy bespectacled figure. Between wars he was active in both his legal practice and the Militia, where he rose to the rank of colonel and Commander of the 2nd Cavalry Division's artillery. He was also active in 'clubland' and sought political office. Herring concealed a determined focus under an urbane exterior and, for a senior officer, was quite popular with his gunners. He established his headquarters in St Kilda Road, Melbourne, and his first Brigade Major was a Western Australian Duntroon graduate, Bruce Klein, who was transferred to Corps Artillery Headquarters, his place taken by George O'Brien, another Staff Corps officer. His Intelligence Officer was Lieutenant Desmond Cox, and his Staff Captain, Lieutenant 'Spud' Murphy. Units of the division left for the Middle East in two convoys, Herring's headquarters sailing in the second convoy with the 17th Brigade.[62]

Herring's two senior field regiments, the 2/1st (1st and 2nd batteries) and 2/2nd (3rd and 4th batteries), arrived in Palestine on 12 February and 17 May 1940 respectively. His 2/3rd (5th and 6th batteries) would not arrive until the end of January 1941 as it had been diverted to Britain, together with Brigadier Leslie Morshead's 18th Brigade. The 2/1st was raised in Sydney by Lieutenant Colonel Leo Kelly, a school teacher, its ranks filled with men from the country, Newcastle and Sydney. The regiment trained at Ingleburn and embarked in the first convoy to leave Australia in January 1940. When assembled, the 2/1st and 2/2nd were co-located with their respective 6th Division infantry brigades, the 16th and 17th. The 2/1st brought its 18-pounders and 4.5-inch howitzers with it and shared them with the new arrivals so that training could continue. The junior Troops of both batteries were disbanded and the remaining Troop's guns increased from four to six. This was the first of several organisational changes to battery and Troop structure and also encompassed changes in role, particularly during the volatile days of Italy's entry into the war with Germany.[63]

In September 1940, Herring appointed Victorian-born Lewis Barker CO of the 2/1st Field Regiment. In 1939 Barker had been OC of the 3rd Heavy Battery, a regular unit of coastal (garrison) artillery and part of the Sydney Harbour defences. In short order he had filled the positions of Assistant Director of Artillery at Army Headquarters (AHQ) and CO of the 2/4th Field Regiment before taking the 2/1st into battle after training in Palestine.

Lewis Barker was a six-footer and built in solid proportion. His arrival in Palestine coincided with air raids which his unblooded gunners had learned to take in their stride. The regiment's historian, Mick Hayward, an 'original' in the regiment, noted, 'the men may have been used to air raids, but they were not accustomed to men like Barker.' He was a decorated veteran, having won the MC in World War I, and his keen, blue eyes were set in an unblinking gaze, his sharp mind issuing crisp, clear orders and his large, expressive hands generating flourishing gestures. His nickname was 'Doag' and his style initially incurred 'dislike that changed to reluctant admiration and undisguised admiration. He was their dread, their inspiration and acknowledged leader.'[64]

The 2/2nd Field Regiment was raised in Victoria and trained at the Melbourne Showground, Puckapunyal and Seymour before embarking on the second convoy with the remainder of the division. Lieutenant Colonel Bill Cremor, its second commander, was a former teacher who had served in the 3rd Field Brigade in France in 1917–18 and soldiered on in the Militia, commanding the 10th Field Brigade for three years. Between the wars he completed an Arts Degree (law, arts and education) and was a great fighter for the underdog, always cognisant of his own humble beginnings. He was also active in union affairs and politics and in 1934 resigned from teaching to become the Secretary of the Victorian Dried Fruits Board. Cremor was quite a character. He sported a walking stick for effect, 'used direct, picturesque language and if confronted with an unpleasant or awkward situation it never worried him.' He had a quick temper and was impatient with humbug and hypocrisy. On enlistment he became second-in-command to Lieutenant Colonel Alan Ramsay, having dropped rank, like Ramsay, to join him in the regiment. Cremor, a bachelor, was to show outstanding leadership and apply lateral thinking to the tactical problems he faced in this and later campaigns.[65]

Initially, while the 2/2nd was camped at Dier Suneid near Gaza, an acute shortage of guns and stores inhibited effective training. Both field regiments, trained on 18-pounders and 4.5-inch howitzers prior to the war, had expected to be equipped with the new British 25-pounder. In the interim, they trained as two batteries of 12 guns, eight 18-pounders and four 4.5-inch howitzers.

The 2/3rd had been diverted to Britain. The unit was originally raised by Lieutenant Colonel Athol Hobbs, an architect (as was his gunner father, the renowned Major General Sir J. Talbot Hobbs of World War I fame), who recruited men the length and breadth of the country: from NSW, Victoria, South Australia, Western Australia, Tasmania and the Northern Territory. The 2/3rd trained at Ingleburn, Woodside and Northam. While based in

Britain, the unit assisted in the defence of the south-east and was stationed in various parts of the country, including the historic town of Colchester. The gunners trained on 25-pounders and the 2/3rd also acted for a time as Depot Battery for the School of Artillery at Larkhill. The regiment was inspected by King George VI and Lord Birdwood, a singular honour for its CO, Lieutenant Colonel Horace Strutt. Strutt had been commissioned in 1924 and spent his entire service with Tasmania's 6th Field Brigade, CMF, save for a brief period as aide-de-camp to the Governor. The 2/3rd was to experience several changes in appointments and structure before finally arriving in the Middle East on 31 January 1941.[66]

The third part of this chapter describes the way a regiment works on operations. This discussion necessarily focuses on the 25-pounder regiment of field artillery, although some of the regiments whose active service is described were first equipped with 18-pounders Mark V and 4.5-inch howitzers at Bardia, Tobruk and Benghazi. These regiments' total strength in personnel and equipment, weapons and transport allocations, their organisation for manoeuvre and their support from signals, ordnance and EME, would not have materially differed from those of a 25-pounder regiment. Technical command post work and ballistic details, however, differed quite markedly. Within command posts, line, range and angle of sight derived from several arithmetical and trigonometrical calculations that then had to be applied to the gun sights and/or entered on gun programs for detachment commanders and gun layers to follow. Organisation for manoeuvre was similar and gunners training with their new 25-pounders learned new skills quickly. The establishment of a field regiment is described in Appendix 1.

A field (or medium) regiment's size depends on many factors and criteria. Based on experience, the British Army arrived at a structure that meshed with that of the infantry division which it supported in the major phases of war—advance, attack, defence and withdrawal. The field regiment also had to fulfil its basic doctrinal function of 'supporting the infantry/armour onto its objective', to which one might add 'efficiently'. Thus, in a division of three infantry brigades, most units were structured from the top in 'threes' for another two levels. The AMF adopted British structures in World War I and so, with the invention of the 25-pounder, it was envisaged that a field regiment with its 24 guns would support or be affiliated with a brigade. Given the brief but brutal campaign experience in France and Flanders in 1940, the adoption of a configuration of two batteries each of three Troops appeared consistent with that tenet. This lasted only 18 months and, because it proved far from efficient during operations, a restructure saw a configuration of three 8-gun batteries, each of two Troops. For both brigade and battalion organisations a battery supporting a battalion (with two companies 'up') led to a Troop supporting a forward infantry company.

The Troops fulfilled this role using a variety of shell types to suit the tactical situation. The Troop was the basic fire unit for artillery planning.

In deployment orders a regiment could have several roles, depending on the tactical situation. The most binding order was 'in direct support', meaning that the regiment, battery or Troop's fire was guaranteed to the supported unit. Affiliations between batteries and battalions, and regiments with brigades were taken seriously through joint training and in operations. 'In support' introduced more flexibility as to which unit could be summoned to provide fire if not engaged on another task (counter battery, for example). A fire unit could also be put 'under command for movement' or could be 'detached' to another battery/regiment/division etc. In these circumstances the commander at whatever level was co-located with his appropriate headquarters.

By working from top down and bottom up, the staff created a document known as an 'establishment' for each regiment. This highly detailed document covered every conceivable aspect of equipment, vehicles, stores and 'first line' consumables, weapons allocation, personnel rank and skill groupings. A field regiment's establishment listed 39 officers, 53 warrant officers and sergeants (senior NCOs), 80 bombardiers (corporals) and 504 gunner other ranks, a total of 683 souls. Officers and NCOs were paid by rank and the latter, along with other ranks, were grouped into trade groups to reflect skill and (broadly) education levels. Trade groups were divided into I, II, III and Non-Specialists. To assist it to operate efficiently, the regiment was divided into two echelons, fighting and B.[67]

Responsibility for fighting and manoeuvring the regiment, for command and control and the maintenance of high standards of training in all ranks, rested with the CO, usually a lieutenant colonel. His place was nominally at the supported brigade headquarters. The functional responsibilities of the Intelligence Officer meant that, when the regiment deployed, he was best co-located with the CO, as were the Liaison Officers (LO). At RHQ, the second-in-command's role was to 'mother' the regiment and act for the CO in his absence. The principal operations staff officer, a captain adjutant, monitored the tactical battle and, when required, controlled the fire of his regiment for regimental and divisional/corps targets. He particularised operation orders and instructions received from Headquarters RAA for his regiment for each battery (and any attachments). A survey officer understudied the adjutant and established a regimental (map) grid to coordinate fire from the regiment. Also attached was a Signal Corps officer whose troop of wireless signallers established the link between RHQ and higher headquarters and advised the CO on technical matters. For 24-hour operations supernumerary officers or 'Learners' were attached. The Regimental

Quartermaster (nominally an RHQ posting) managed rationing, supplies, transport and logistical requirements with his staff, and the Regimental Sergeant Major (RSM), the senior NCO, arranged ammunition resupply to Troop positions, attended to the administration of personnel, discipline and such matters. A sergeant clerk ran the 'office'.

In the B Echelon, where the Quartermaster spent most of his time, significant personnel included the cooks assigned to each battery and RHQ in their own B Echelons, a butcher, water and hygiene dutymen, motor mechanics and a gun fitter (the only Group I man on the establishment) and equipment repairers. Each regiment had a Medical Officer and staff who established a Regimental Aid Post (RAP) for initial treatment of casualties, injuries, and illnesses, and a Padre for the troops' spiritual needs and welfare. The six most common positions were drivers (128), signallers (120), gun numbers (72), gun layers (48), driver-batmen (41) and driver-mechanics (25). Every person had a designated vehicle in which to travel, each with a distinguishing alpha/numeral.[68]

The attached Ordnance Corps EME Light Aid Detachment (LAD) was commanded by a senior NCO (later an officer) who was the CO's adviser on these practical matters. A Pay Corps and Postal sergeant and shoemaker at B Echelon were vital personnel in any regiment.

Each battery was commanded by a Battery Commander (BC), a major who had a driver-batman, signaller and assistant who went everywhere with him in truck, tank or armoured car. The BC was located at Battalion Headquarters. At Battery Headquarters, a command post officer was the senior subaltern at the gun position, as each Troop commander (a captain) was located with his observing party at the headquarters of the infantry company he was directly supporting several thousand yards from his guns. The command post officer laid out his battery position of two Troops and Battery Headquarters in accordance with his orders and mission. This was usually expressed as an area allotted by RHQ from which the battery's guns could cover the zone allocated to them and in which Troop commanders would observe and engage targets. The area was also designated by either the scope of a topographical feature, a map easting or northing 'to shoot down to,' or a man-made feature such as a road or track. Usually a bearing known as the zero line originated from the centre of the battery's area and guns were laid for line by angular measurements left or right of this. The command post officer, who had responsibility for battery and Troop technical gunnery matters, was accompanied by his assistant. A Wagon Lines Officer attended to the defence of his battery's vehicles and general maintenance. The assistant command post officer's responsibilities included signallers (telephone line) and communications and relief of the command post officer. Battery Headquarters was connected to each Troop by telephone line through an exchange and each Troop headquarters was set up similarly to Battery Headquarters. Here a gun position officer (one of the best subaltern's jobs in the Army) commanded and controlled his Troop of four gun detachments of six men each. He used laid down drills and 'fire discipline' procedures for his fire unit to respond to orders received from his troop commander, other OPOs or forward observation officers (FOO), or simply adhered to his gun programs. His assistant and relief was a junior subaltern, a Troop leader.[69]

An OPO/FOO was responsible for coaching his infantry counterpart in the capabilities of his troop or battery—or even the regiment—as they affected infantry company and battalion dispositions on the ground. The infantry needed to understand the length zone of the gun, the best shell type for a particular mission, the limits of range and OPO zones, 'dead' ground (depressions, wadis, etc.) and how it could be engaged, quick fire plans (known as 'Little Tiddlers'), timings for movement, command and control, communications and other such aspects. One common responsibility of an observer was to correctly map spot his position/locality for the infantry who were, at times, well astray in their plotting.

Artillery regiments of various branches (field, medium, anti-tank and anti-aircraft) comprised about one-fifth of a division's strength in numerical terms. However, the main offensive power of the division was the 25-pounder gun howitzer, of which each field regiment had three 8-gun batteries. The 25-pounder was a British design and represented a rare triumph for something 'designed by a committee'—the War Office Directorate of Artillery and the General Staff. Its development was based on the successful warhorse of World War I, the 18-pounder, combined with the recognition that a heavier shell was required for supporting infantry.[70] It was trialled in 1937 and came into Australian service in 1940. It had a maximum range of 13,400 yards with supercharge; its calibre was 3.45 inches; shell and propellant (8 pounds for charge 3) were separate and loaded and sealed in a breech with a vertically sliding wedge as breechblock. On firing, the gun's recoil forces were absorbed by a hydro-pneumatic variable buffer and recuperator system. Its weight in action was one and a quarter tons, and it was trailed behind a tractor with limber on two wheels by its towing eye. In action, the gun pivoted around a circular wheel-like platform, the circumference of which had scalloped tines to dig into the earth. The platform was carried under the trail. The gun detachment comprised its commander, also known as the No.1, usually a sergeant. The gun layer was the next senior and was usually a bombardier. There were four ammunition numbers, three were gunners and one was a lance bombardier. The gun was manhandled by its crew using dragropes and handspikes. The gun tractor driver often 'took post' as an ammunition number when lengthy gun programs were fired.

A gun layer aimed the gun for line (direction) using a No.7C Dial Sight placed above a range cone on which was engraved ranges for the four separate charges used. A cursor on the 'muzzle velocity' bar could be set at the appropriate muzzle velocity for the charge being used based on calibration shoots that took into account such factors as barrel wear. With the cursor set at the given range (in yards) and charge, the gun was laid correctly through a parallel motion link connected to the trunnion with the elevating handwheel when the sight was levelled by reference to the angle of sight set on an adjustable clinometer. A traversing handwheel gave four degrees left and right movement and the gun could be easily moved about its platform using a handspike for bigger 'switches'.[71]

The gun fired a variety of ammunition: high explosive (HE), smoke (white and coloured), star (illuminating), chemical, propaganda and armour piercing (AP). Four weights of propellant gave artillery planners flexibility in deciding which one to use: charge 1 (red bag, fixed in the brass cartridge case); charge 2 (white bag); charge 3 (blue bag) and supercharge (brown bag). Thus, for charge 3, the most commonly used for accurate gunnery, the cartridge would contain red, white and blue bags. Supercharge was devised for use with solid shot AP round in a direct fire role, usually anti-tank (although it could be used with HE to reach area targets). The gun layer used a calibrated telescopic sight when engaging these targets.

Gun crews could fire at five different speeds: intense—five rpm; rapid—four rpm; normal—three rpm; slow—two rpm; very slow—one rpm. Three main types of fuse were used on 25-pounder and medium projectiles: 117, a direct action impact fuse; 119 fuse with fuse cap for 'graze burst' (a low-level airburst effect obtained on hard ground or a delayed burst in soft ground); and 210, a mechanical clockwork time fuse. The fuse was screwed into the body of the projectile and its 'gain'—a small quantity of HE—was detonated by the fuse on impact (or by time) which detonated the TNT (trinitrotoluene) HE filling. A smoke pellet was included in the base of the projectile to provide observers an indication of its burst point.

The projectile steel case was made from shell steel which was higher in manganese and sulphur than mild steel. When HE burst in the air or on the ground it produced 'man-killing fragments', some pieces weighing around two ounces and about the size of a finger. These spread in a fan-shaped pattern at 3000 feet per second. Troops dug in and under cover had a much better chance of surviving shellfire than those in the open.[72]

The 25-pounder was justifiably renowned as the best field gun of World War II. It was very accurate and increased the confidence of the infantry in tricky situations, allowing them to advance or withdraw under covering fire, a fact seldom acknowledged in battalion histories. The 25-pounder required very few modifications to improve its basic design. Like any other piece of mechanical equipment it needed regular maintenance and, after a move or a period of firing, the detachment commander would order a 'quick sight test' to be conducted. A more thorough full sight test would be performed at regular intervals, often with the assistance of a gun fitter.[73]

Positioned in a fully developed gun pit and covered with camouflage netting, the gun was difficult to see from a distance until it fired. Four wooden poles supported a net that was threaded with strips of hessian scrim of various shades, usually laced to a pattern. For night engagement the crew removed the net for ease of working. It also helped minimise the characteristic fan-shaped blast marks in front of the gun—a tell-tale sign for photo interpreters. The gun was towed by a purpose-built tractor, the Marmon-Herrington Quad, which accommodated the driver and detachment of six with their kit, personal weapons and essential gun stores. The Quad's only real disadvantage was its vulnerability to destruction by fire if struck by bullets or shells.

An Ordnance officer ammunition specialist was available to each field regiment to advise the CO. The manufacturing source of fuses, shells and propellants was quite varied—British, Canadian, American and Australian—and their technical management was a corps concern. Each manufactured batch was labelled with a serial number/code and, in the case of propellants, the type (for example, Flashless Non-Hygroscopic: 'FNH' etc.) and other data. Where possible it was preferable to use the same batch within a battery/regiment to ensure consistency, particularly when calibrating guns. This also helped identify a batch that might produce more 'premature' detonations of a shell (than other batches), either in the breech or barrel of a gun, often with serious consequences for the crew. The major cause of 'prematures' was fuse malfunction, usually from a change in the chemical composition of the material within the detonator.

To generate gun data for firing, command post officer/gun position officer assistants (known as 'Acks') plotted the target location's grid reference on an artillery board, a gridded plane table. The location of the guns, set at one side of the board, was marked with a pivot, around which a graduated metal range arm moved. The centre of the zone the guns were to cover was called the zero line, and bearings (lines) in degrees and minutes read off a graduated metal arc either side of the zero line. The map range between the guns and the target was the basis for the calculation of the predicted range and line. Using range tables, the Acks performed five separate arithmetical sums either graphically and/or by hand to arrive at the 'correction of the moment' to be applied to the map range and one for the line, based on the current meteorological forecast, to produce predicted gun data for firing. When smoke shell requiring a fuse setting was required, two more calculations were needed. Troop gun data for firing battery and higher targets was cross-checked with their counterparts at the battery command post, either by phone or in person. Agreement had to be within 25 yards for range and five minutes of arc for line.

These procedures were followed for all higher targets such as barrages and Stonks. While barrages comprise several lines (lifts) of fire of certain length moving at a specified rate towards an objective, a Stonk was one line of fire from a regiment or, more often, a field and medium regiment. These were originally 1,200 yards long and 200 yards deep. Guns fired in parallel and there were conventions as to where each Troop of guns' fire was to be placed along the line. HQRAA allocated each regiment a length of the line and, at RHQ, the adjutant allocated batteries a length. Usually A and B Troops spread their fire across half the length right to left, and C and D, E and F Troops did likewise across a quarter of the length. Medium regiments with only four Troops followed a similar pattern, and their fire, with its heavier shell, was 'superimposed'. Thus the zone of the guns created a depth of fire. Based on the lethality of a 25-pounder shell, 550 yards would provide adequate density. The 1200-yard Stonks were too thin to be effective and subsequent Stonks were reduced to 600 yards and finally to 525 yards.[74]

At the OP, OPOs adjusted the fall of shot by referring to a line and bearing between them and the target, usually by ranging one gun until they obtained a 50 or 100-yard 'bracket' by ordering 'add' or 'drop', 'go right' or 'go left'. They would then order 'fire for effect'; that is, the number of guns and/or rounds of gunfire necessary to neutralise or destroy the target. OPOs also referred to cardinal points ('go north' etc.) or moved the fall of shot by using 'sweep' or 'search' by degrees left and right and adding or dropping hundreds of yards. In flat desert terrain, bold corrections were needed to establish a bracket. OPOs used 'ladder ranging' to save time, a technique that involved an OPO ordering a range and three or four 200 to 400-yard increments fired at 20-second intervals. This practice tended to disappear as the OPO became familiar with his zone.

The intensity of artillery fire in the front line sometimes posed a problem for the OPO. How did he differentiate the fall of shot from his own Troop, battery or regiment's guns from that of another OPO observing a proximate zone? In this situation, his OP 'Ack' was invaluable. After the passage of initial fire orders, the command post conducting the shoot would advise the time of flight (TOF) of gun to target as, for example, 'Able shot. TOF 25 seconds' and the signaller would repeat the words a moment later. The OPO would observe with or without binoculars and his OP 'Ack' would call 'Look in!' a few seconds before the shell bursts occurred. Sometimes 'salvo ranging' was used, in which a Troop or battery fired simultaneously to produce a characteristic fall of shot which distinguished it from others. On rare occasions there would be a shell burst from a gun (a 'swinger') that would fall in a different area to that of the rest of the fire unit. Investigation of these invariably showed data on the sights to be faulty, resulting in the gun firing 'off line'.[75]

A field regiment had a prodigious appetite for vehicles. For desert warfare, the regiment was issued with armoured/White scout cars, Bren Gun Carriers, 5 cwt and 15 cwt with four-wheel drive, motorcycles and Quads. The heavy lifters were 3-ton 4 x 4 lorries ('Blitz Buggy' Chevrolets), and each of the battery's three ammunition lorries carried 480 rounds, or 60 rounds per gun. The 3-ton lorries came in various adaptations including office, signals and bin fitted for stores, while some towed one-ton trailers (for water etc.). On the move, RHQ had 24 vehicles and 74 personnel, and its LAD another eight. The eight LAD men were dispersed amongst the batteries. A regiment on the move from a standing camp numbered around 154 vehicles and, with 100-yard intervals between the vehicles, occupied around 10 miles of road, although it was seldom a regiment's good fortune to have its full complement of vehicles.

The regiment's anti-aircraft defence comprised eight Bren light machine-guns. For self-preservation, the establishment allowed 47 pistols, 63 rifles, eight carbines (sub-machine guns), and seven Boyes anti-tank rifles. Like their infantry brethren, the gunners also accumulated a variety of extra weaponry for their personal use, most of which was German or Italian.[76] The attached Signal Corps personnel provided operational wireless links to higher formations, battery chargers and wireless set maintenance. A regiment would normally have three wireless networks for operations at Troop, battery and regimental level. When working with the infantry, artillery LOs used the WS 109:

Set WS Number	Frequency Range Kc/KHz/MHz	Operating Range (miles) Radio	Morse
WS 9	333 to 5 7	Mobile	20 Station
WS 11	5 to 15	2-7 Mobile	20 Station
WS 101	4.2 to 7.5	20 Mobile	50—500 Station
WS 109	1.5 to 12 MHz	20 Mobile	

The WS was a wet-cell powered super heterodyne sideband transceiver. Range depended on many factors, primarily power output, aerial used, ionosphere and local weather conditions. OPOs sometimes worked on regimental or divisional nets so they could, if authorised, shoot their own and/or other regiments' (field or medium) guns.[77]

The 6th and 9th divisions were supported by British medium artillery wherever Australian field guns were deployed. Due to the fact that a regiment's guns weighed between four and six tons and their shells weighed 60 to 90 pounds, a regiment consisted of only 16 guns/howitzers in two batteries each of two Troops. However, with attached personnel, it had a strength of 650 all ranks, the main increase in manpower comprising additional observation parties (to cover a more extensive zone than field OPOs), gun detachments (10 men), batmen and ammunition numbers in B Echelon.[78]

So how did a regiment operate effectively? Ultimately, this was a matter of following artillery doctrine, standing orders and training. No CO could afford to neglect these, nor could he neglect arrangements for quarters and administration. If these arrangements went awry, then operations and training would suffer. Taken in order, if a regiment was to undertake a tactical road move, the elements (composition of groups, speed, route, interval, halts, route marking, lights etc) would all be specified in the 'order of march' in Regimental Standing Orders. When the CO received his orders in manoeuvre, he would make a reconnaissance and ascertain where the gun areas would be located, usually in a specific zone. He would also determine where his observers would be located to cover their zone(s) to effectively support the infantry/armour. He would then call an orders group and issue detailed orders for each battery and for his second-in-command who attended to administrative matters.

One of the CO's more pressing decisions related to 'time and space' as his orders usually specified the latest time the guns should be ready for action. Line laying was the single most time-consuming event, and the CO had to visualise and estimate how long batteries and troops could spend on reconnaissance. Batteries and troops could provide fire support independently prior to line communications being established and before a common grid was 'surveyed in'. Survey matters were always important, as was liaison with infantry/armour at their headquarters. The second-in-command would worry about B Echelon, while the BCs and command post officers would want to know the technical details of gun areas, observation zones for deployment of OPs, target registration policy (active or silent), tactical plan for the operation being supported, cooperation with artillery reconnaissance (provided by the RAF) and many other factors.

For B Echelon, casualty evacuation and location of aid posts, ammunition points, water, ration and petrol points, vehicle evacuation policies, Military Police and traffic control were of prime importance. There was one great truism about preparation for desert operations—

the only method was focused, hard training, ensuring that drills became second nature and reactions to impromptu situations could be critiqued and lessons absorbed. This regime was not always popular with the troops, but training a regiment for war was not about popularity and Australian regiments were certainly well trained and well led.

The British/Australian Army regimental system embraced the rather paternalistic concept of the regiment as a 'parent' to its soldiers. Everything was predicated on preparing the soldier for his role in the regiment and providing him with food, clothing and shelter. The soldier was an asset to be preserved in a relationship based on give and take between the individual's values, motivations and interests and 'the system'. It was here that leadership played its part in directing a soldier's core values to those of his unit as a whole, be it either from his NCOs or officers, to a standard that would contribute to effective teamwork through synergies between different skills.

Another of the most important factors in regimental efficiency was paperwork. To make any part of the Army system work required a vast quantity of printed matter of many kinds. An enormous variety of forms covered every facet of mounting operations or exercises and the administration of Army personnel and equipment records. These forms were the object of scorn and derision for those confronted with them at the gunner level and, as a soldier progressed further up the pyramid, he encountered an increasing amount of paperwork. Basically, paperwork was the Army's way of recording all the information necessary for effective personnel and equipment management, and ensuring that higher headquarters had the data required for decision-making.

For vehicles and equipment there were log and similar books (each gun and vehicle had one), petrol, oil and lubricants forms, issues, indents and requisitions for all manner of stores and equipment and replacement parts. Gun ammunition was indented for on ordinary message forms by the Quartermaster. Small arms ammunition was similarly replenished as consumed.

Two important and frequently issued documents were critical to unit administration—Routine Orders Parts 1 and 2. The former recorded the more important systemic orders and instructions that applied to all or some units, and might include policy, a warning or directions for a particular activity that would find its way into a standard procedure. Routine Orders Part 2 covered personnel matters including those battery and regimental events such as promotions, punishments, acting ranks, schools and courses, discharges, etc. These might also include roll books, pay sheets, charge sheets, record of promotions and punishments, injuries and wounds, hospital records and so on.

Paperwork concerning operational aspects included loading lists when exercises and operations were mounted for tactical or other reasons. Each vehicle's commander was nominally in charge of and responsible for all the stores required. For example, the gun position officer's vehicle would hold most of the stores for a fully functioning command post, the remainder being carried by the troop leader. Replacements for lost and damaged equipment (if not repairable) were requisitioned from Ordnance Stores through B Echelon. Artillery and gunnery training manuals and technical publications, and their frequent amendment as a result of operations analysis, were vital for gun firing data. There were maps, orders and instructions generated for exercises and operations, training curricula, logs and signals books, command post target record forms, hostile battery and high frequency task lists and gun programs. The regimental war diary was written up each day by the adjutant, the quality of which depended on the incumbent's dedication and powers of English expression. Then there were 'returns' (status reports) sought by higher headquarters in their unceasing multitude, often daily. No account of the Army's activities would ever have existed without pencils, pens, paper, duplicator machines and ink, Army forms and carbon paper copies. They were the Army's birthright and an historian's food and drink.

Finally, a man's kit went with him—haversacks, bedroll, weapons and ammunition, bulk supplies of which were held 'on wheels'. Each man had two dog tags or identity discs which were worn on a thong around his neck and on which was stamped his Army number, name and initial(s) and religion. In the event of his death, he was buried with one and the other began a journey to Base Records along with his personal effects. Paperwork began with enlistment and ended with his discharge or burial. No-one could dodge it.

To feed the troops the Army way was to allocate a specific quantity (usually by weight) per soldier of a wide variety of foodstuffs and fluids to maintain good health under active service conditions. In the desert, the calorific intake was at least 4000 kilojoules of energy made up of balanced nutritional items. Trained cooks then had to prepare it in an appetising way—often a significant challenge given the conditions of the time. If this was impossible, as it often was, then hard rations were issued on a basis that suited the sub/unit size. The men then prepared their own meals. The rationing problems were numerous, with water the single most problematic issue. The siege of Tobruk could not have been sustained without adequate supplies of water. No army could fight without it and neither would vehicles operate for long. Water supplies made the most demands on a rationing system, more so if it had to be purified. The troops used mugs and dixies and had eating irons issued to them, as were razor blades and soap.[79]

Additional rations were officially available to the troops, such as Australian Comforts Funds parcels and food sent by families and friends. Each vehicle had a primus stove or equivalent (a petrol fire set in sand) for brewing a billy of tea, and most soldiers lived by the dictum that 'any fool can be hungry or uncomfortable'. These amenities were augmented by scrounging from whatever source was available, and the 9th Division scrounging champions were reputed to be the 2/2nd Machine Gun Battalion. This reputation was a source of much pride.

CHAPTER 6
PALESTINE TO BENGHAZI

Nothing succeeds like success.

Proverb

The newly arrived AIF units were introduced to offensive operations with Lieutenant General Richard O'Connor's Western Desert Force (soon to be redesignated XIII Corps) in late 1940. British armoured and infantry formations mounted continuous patrols close to the Egyptian frontier and launched constant small-scale attacks to unsettle the Italians. The Italian forces in Libya boasted considerable numerical strength but, militarily, were something of an unknown quantity. However, they responded quickly to the British raids with Marshal Rodolfo Graziani's five divisions attacking in strength on 13 September 1940, forcing the British to withdraw to the east of Sidi Barrani. By 9 December the Allies had pushed through with a counter-offensive at the Battle of Sidi Barrani and, a few days before Christmas, the Australians reached the front near Fort Capuzzo. Operations were concentrated around the Tummars and Nibeiwa, although it was not until an attack was mounted against Fort Capuzzo itself that the 7th Medium Regiment's historians noted 'the first coordinated shoot of the war in the Middle East'. Thus the Regiment secured the honour of firing the first rounds in offensive operations against the Italians. On that occasion all available artillery (eighteen 25-pounders and two 60-pounders) was laid on a large concentration of Italian soldiers which many observers described as a 'spaghetti parade'. The medium guns used some of their precious shrapnel while the others used HE.

The 6th Australian Infantry Division, which had assembled in Palestine, was commanded at a senior level predominantly by Militia officers with World War I experience. Many of the staff officers were members of the permanent forces who had seen active service, but there

2/1st Field Regiment gun of E Troop on Christmas Day 1940 at Bardia. The detachment shown is Sergeant Pearse, Bombardier Frankfort and Gunner Doug Hillcoat.
(AWM 005307)

were also large numbers of former Militia officers who had not. The Divisional Commander was Major General Iven Mackay. Mackay, Headmaster of Cranbrook School, Sydney, in civilian life, had trained the division since early 1940 and trained it well, despite crippling shortages of equipment. The division was still missing its full complement of field artillery when it began operations against the Italians in December. The 2/3rd Field Regiment was due to arrive in December and would not be ready for operations—even on a reduced scale—until later in January. To meet other exigencies, both field regiments (the 2/1st and 2/2nd) had been forced to adapt to each other's weapons and roles. When the time came for the reduced division to prepare for its baptism of fire at Bardia, only the 2/1st had 25-pounders (24) while the 2/2nd was equipped with 12 each of the older weapons (18-pounders and 4.5-inch howitzers). En route to Bardia, Herring's artillery, now primed for its first action, deployed to a position south-east of Fort Capuzzo which guarded the border.

Western Desert Force's commander, Lieutenant General Richard O'Connor, was an infantryman with a DSO and MC from service on the Western Front with the Cameronians in World War I. In the aftermath of the war, his career had followed a similar path to that of many senior officers in the theatre—regimental postings, staff college, India

Major General Iven Mackay, GOC 6th Australian Division, led his division with distinction across the Western Desert from Fort Capuzzo to Benghazi in 1941.
(AWM 005058)

and Palestine. O'Connor was short in stature and regarded by some of his colleagues as unconventional in thought with little experience in the use of armour. He was, however, possessed of two significant traits that were the hallmarks of a successful general. He was astute in assessing the strengths and weaknesses of subordinates and in persuading others—often in a roundabout way—to agree to his plans and orders by deflecting ownership of ideas. This approach, together with his characteristic modesty, made him popular with his men. His attitude to operational orders and instructions was based on the careful analysis of all contingencies and the structuring of his plans to take account of these, so that he was never surprised by the unexpected.

On his appointment to Western Desert Force, O'Connor made the acquaintance of Blamey, Mackay and Herring and their staffs, with whom he was very impressed. After the war, the Australian Official Historian, Gavin Long, wrote to Brigadier Latham in the British Cabinet Office observing that:

Brigadier and Commander, Royal Artillery (CRA) of 6th Division, Edmund 'Ned' Herring framed the artillery plans for the divisional battles of Bardia, Tobruk and Benghazi.
(AWM 020401)

[O'Connor] won over the Australians as few others have done; perhaps, in World War II, as no other UK or Allied commander did ... General O'Connor's vignettes of the Senior Australians are most interesting—remarkable that he was able to see them clearly in so short an acquaintance.[80]

The Australian gunners' experiences prior to Bardia provide some indication of the early desert campaign of the unblooded troops. With Lieutenant Colonel Lewis Barker at his 2/1st RHQ was Staff Corps Adjutant Captain J. 'Hans' Andersen. His battery commanders were Major C.N. Peters (1 Battery), a Staff Corps officer, and Major Frank Richardson (2 Battery), formerly of the 2/2nd and a Militiaman.[81] The Regiment occupied its first gun position at night on 23 December 1940. It was cold, windy and dusty with poor visibility and the gun positions occupied depressions in an otherwise featureless desert. No. 1 Battery was forward in the area designated by the Brigade Major, Major George O'Brien. Dawn brought the unpleasant realisation that the battery was in front of the infantry and could be clearly observed by the

Brigadier Lewis Barker, DSO, MC. Barker was the first Commanding Officer of 2/1st Field Regiment,
(AWM 062748)

enemy. While maintaining a low profile for the day before making a quiet tactical withdrawal at night might have been an option, Barker and his gunners, who had been waiting for just such a moment, reacted typically and traditionally: No. 1 gun of A Troop, 1 Battery, 2/1st Field Regiment, fired the opening rounds of Australia's field artillery war. Shortly afterwards, the Italians replied in kind, bracketing their position, firing airburst. There were no casualties despite the enemy's attentions and, during the night, 1 Battery quietly moved to the rear of 2 Battery. Nearby, the 2/2nd applied its training drills, mostly to the satisfaction of the CO, Lieutenant Colonel W.E. Cremor. Over the next few days, the enemy artillery continued to be active. Both British and Australian gunners observed that, while Italian artillery fire was accurate, casualties were light or non-existent, possibly because of faulty ammunition. The Italian shell-burst danger zone was a narrow cone to the rear of the burst while, in comparison, a 25-pounder produced a lethal 'daisy cutter' effect.[82]

Lieutenant Colonel E.A. Lee, RA, was attached to Herring's staff to direct the counter-bombardment arrangements for Bardia. The Australians were part of XIII Corps which comprised a substantial force of British field and medium artillery and which soon moved to take up its preliminary positions outside Bardia. In addition to the Australian guns, the artillery force for the projected attack comprised:

> 1st Regiment, Royal Horse Artillery (1 RHA): A/E and B/O Batteries
> 4th Regiment, Royal Horse Artillery (4 RHA): F and J Batteries
> 104th Regiment (Essex Yeomanry), Royal Horse Artillery (104 RHA): 339th and 414th Batteries
> 51st (Cumberland) Army Field Regiment, RA: 203rd and 307th Batteries
> 7th Medium Regiment, RA: 27th/28th Battery, A Troop 25th/26th Battery
> 68th Medium Regiment RA: 234th Battery

The 1st Regiment, RHA, was attached to the Support Group of the 7th Armoured Division which operated west of Bardia in a containing role. The 1st, 4th, 104th RHA and 51st Regiments were equipped with 25-pounders. The 7th Medium Regiment, RA, which was to share a number of military triumphs and tragedies with the Australians over the next two years, was armed with eight 60-pounders and eight 6-inch howitzers. Then commanded by Lieutenant Colonel J.H. Frowen, a bespectacled 'roly-poly' figure and a very popular CO, the 7th provided the bulk of medium artillery support. Already seasoned in operations under the command of Lieutenant Colonel G.H. Elliott, the 7th had transferred from India to the Delta in late 1939 and trained in Palestine and at El Alamein, where it joined General O'Connor's XIII Corps at Sollum in late 1940 with a battery (25th/26th) occupying sniping positions.[83] The 7th took part in the successful attacks at Nibeiwa, Tummar East and Tummar West. On several occasions it was on the receiving end of accurate counter-bombardment fire and sustained a number of casualties as a result of the attentions of the Italian Air Force. At Sidi Barrani on 11 December, a dust storm prevented a battery commander from engaging a large body of men advancing towards his OP. Fortunately for him—and them—the men were attempting to surrender at the time. Later in the campaign, an entire battery of 75-mm guns, their officers and other ranks, would be captured by a Troop commander.

The 234th Battery, 68th Medium Regiment, RA, also joined the corps at Bardia. It was armed with eight 4.5-inch guns, a conversion of the 60-pounder.[84] As part of the overall firepower available to the corps, there were two anti-tank regiments—the 3rd RHA and 106th RHA (Lancashire Yeomanry)—both armed with 2-pounder and 37-mm Swedish Bofors anti-tank guns. Each 6th Division brigade had contributed personnel to form three anti-tank companies, the 16th, 17th and 19th, armed (at that time) respectively with nine, two and two 2-pounders.[85]

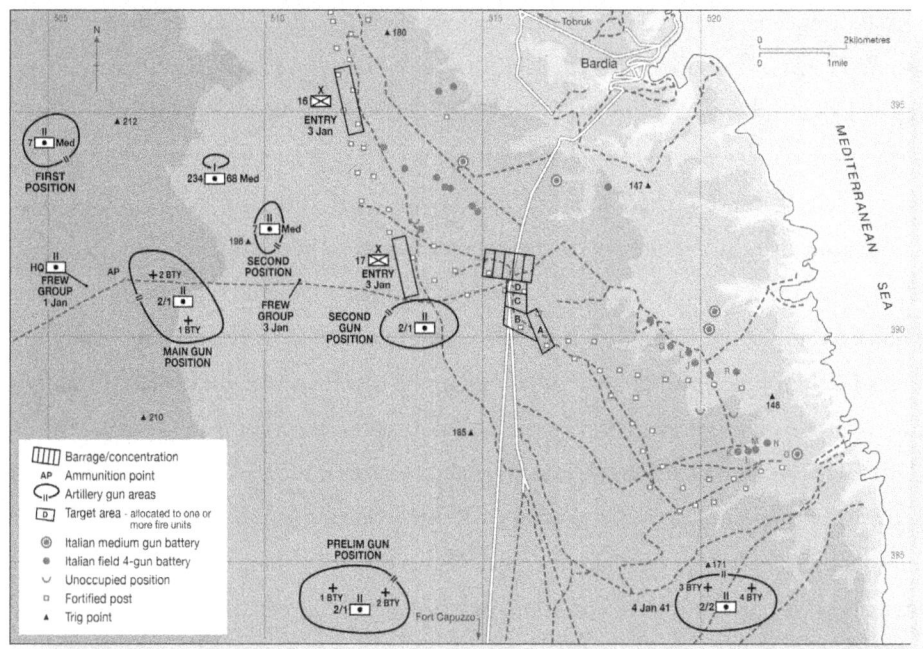

This map shows the dispositions of the attacking infantry and supporting artillery, and that of the Italian artillery, together with the barrage form to breach the defences for Phase I of the assault.

Source, Army History Unit

To breach the extensive Bardia defences manned by four Italian divisions would, in Mackay's estimation, require operations on a scale reminiscent of World War I. In form it would differ in that 'it had to be quick, with few casualties'. This meant his plan had to embrace speedy redeployment of his guns, particularly the second phase. Accordingly, O'Connor arranged for the provision of 500 rounds per gun.[86]

During these desert operations Herring had two British regulars to assist in his artillery planning. These were the CCRA, Brigadier Dibb, who had served with Herring in World War I in the Balkan campaigns, and Frowen of the 7th Medium, who served as the Medium Artillery Commander at Corps level (CCMA). Opposing O'Connor and Mackay were at least four divisions which boasted considerable artillery strength. Each division included a regiment of twelve 75-mm guns, with twelve (sometimes as many as 24) 100-mm howitzers. Each infantry regiment had eight 65-mm assault guns, twenty-four 47-mm anti-tank guns and thirty 81-mm mortars.[87] Bardia's defences were roughly the shape of half an ellipse oriented approximately north-south, its port close to the centre.

The artillery plan for the assault was the handiwork of Brigadier Herring, Divisional CRA. He divided his artillery four to one between the main assaulting brigade (the 16th) and the 17th

Brigade's diversionary attack. Herring divided his artillery into two groups, the first of which was Frowen's 'Frow Group' supporting the 16th Brigade (F Battery, 4th RHA; 104th RHA; 2/1st and 51st Field; and 7th Medium Regiments). The 2/2nd Field Regiment, J Battery, 4th RHA, and the 234th Battery, 68th Medium, supported the second group, Brigadier Stan Savige's 17th Brigade. Herring lost the 7th Medium's B Troop of the 25th/26th Battery's guns when the need for urgent mechanical repairs saw them return to Cairo. The 27th/28th Battery and 234th Battery were tasked for air cooperation shoots on hostile batteries including the biggest gun in the Italian armoury, the 150-mm 'Bardia Bill', as part of the counter-bombardment effort.

As Herring began the process of compiling his artillery plan, he noted that air photos showed only 96 enemy field guns. Herring also devised a deception plan involving the 2/2nd and the 104th RHA which sought to create the illusion that their guns were dug in and firing. This ruse used many hundreds of special devices that simulated muzzle flash. The gunners made dummy guns from captured Italian guns, old wheels, piping, anything that could resemble a gun from a distance. Cremor ordered camouflage nets left over all vacated pits except one, and an 18-pounder detachment from the 2/1st maintained spasmodic fire. As a measure of the success of this deception, the Italians heavily shelled the area soon after its preparation was completed.

In preparation for the attack on Bardia, gun positions were surveyed, positions prepared and camouflaged, and ammunition dumped and concealed. So expertly was the ammunition dump concealed by the 7th Medium that it could not be located. To supply ammunition to the gun positions required a high degree of organisation. Ammunition from an Ordnance Depot at Sidi Haneish was dumped at an ammunition point at Halfaya Pass, and regimental parties from each Troop (on this occasion) boarded a truck and lurched off into the night to perform one of the least popular gunner's tasks. Loading metal boxes of shells and cartridges onto a tabletop truck, off-loading and stacking it in dumps scattered to the rear of the gun position was back-breaking work. Its arduous nature was exacerbated by the fact that movement was restricted to a defined path (not the shortest distance between two points) to reduce its visibility to aerial reconnaissance. Once in the dumps, the ammunition was stacked according to lot and type (and fuse) and otherwise prepared for firing.

The Australian positions were closer to enemy observation than the mediums, so their preparations had to be more expert and they had to use more guile to conceal their intentions. Work on the gun positions was confined to the hours of darkness. As Gavin Long notes, 'On the eve of the battle they [Frew Group] were to be secretly concentrated in the shallow depression about 5,000 yards west of the point of entry. The positions were fixed [surveyed] by a troop of 6th Survey Regiment, RA, and the guns would open fire without preliminary registration. Thus, 120 guns in support of two brigades was on a scale only a little smaller than that produced in France in 1917 and 1918.'[88] Bardia was ringed by defensive posts, marked on Italian military maps and numbered. The 7th Medium was to record many attempts to unseat pole-squatting OPs that were part of the perimeter defences, lamenting the fact that 'no definite successes were recorded'.[89] This defensive system allowed a simplified allocation of fire units to conform to infantry objectives. Herring's fire plan aimed to neutralise enemy fire within 1500 yards of the infantry, the limit of the effective range of the Italian medium machine-guns, and mislead the enemy by firing a box barrage where the Fort Capuzzo road entered the perimeter. This location had been aggressively patrolled by the brigade to further mislead the defenders as to its intention.

Herring's (outline) plan for the 16th Brigade's attack, Phase 1, comprised:

Time	Targets	No. of Guns/Rate	Remarks
0530 - 0555	Posts in 45—47) 1st Line 43) 49)	52 25 pdrs 3* 12 25 pdrs 3* 4 4.5 in 2*	Incl. 8 6-inch Howitzers for 15 minutes
	Posts in 42) 2nd Line 44) 46)	12 25 pdr 3* 4 4.5 in 2* 4 60 pdrs 2*	At 0545 8 x 6 inch Howitzers lift on these posts
0555 - 0620	Posts 42, 44, 46 and posts to NE and S	36 x 25 prs and remainder of artillery, Rates* 3, 2 and 1	Field and medium artillery
0620 - 0655	Box around tank bridgehead	All artillery Rates* 2, 1 and 1.5	Field and medium artillery
0655 - 0730	Artillery concentrations move south on each post in turn to Posts 25, 24 and 26	12 guns on each post for 10 minutes	Artillery concentrations to move SE at a rate of 5 mph
0655 onwards	Enemy batteries	Selected medium batteries	

*Rate (of fire) was given in rounds per gun per minute—rpgpm—also noted as (for example) 'Rate 2' for 2 rpgpm. As part of the deception plan, the 2/2nd Regiment engaged Posts 4, 6, 7, 8, 9, 10, 11 and 13 from 0530 to 0545 hrs and, for the next 45 minutes, fired regimental bursts (5 rgf) on enemy batteries and Posts 22, 24, 25, 26 and 27.[90]

To augment the land artillery, three RN ships (HMS *Terror*, a monitor, *Ladybird* and *Aphis*, both gunboats) bombarded selected targets during daylight hours on 2 January 1941. The RAF was also involved, dropping up to 24 tons of bombs on the defenders from 31 December 1940 onwards. The RAF also performed artillery reconnaissance and tactical reconnaissance on D Day, providing three sorties at 7.00 am, 9.00 am and 12.00 noon, and being tasked during the battle with 'air shoots'.

A pivotal part of the fire plan was a creeping barrage of 20 lifts ahead of the infantry Start Line behind which the infantry advanced at 200 yards in three minutes, extending over 1500 yards. After that, further concentrations were fired by the 2/2nd at five other posts as Brigadier Allen's 16th Brigade infantry and their supporting tanks attacked and subdued their opponents. The enemy replied to the early bombardment by firing pre-registered targets and most of their shells fell behind the leading infantry. A high proportion of the enemy artillery was moved south as a result of the deception plan and this was where the Italian gunners directed most of their counter-bombardment fire. The enemy artillery was still in range of the medium guns, however, and was engaged heavily. The 7th Medium fired 180 rpg on the initial fire plan, turning the 60-pounders' barrels black. Nonetheless, some Italian gunners fought their guns well, although the same cannot be said for all of them.

The bizarre exploit of Captain W. Griffiths and Sergeant W. Morse of the 2/5th Battalion on 3 January in capturing two Italian batteries (one field, one light anti-aircraft), five tanks, a medical post and an artillery headquarters near Wadi Scemmas on D Day is worthy of mention. The artillery headquarters, sited in one of the numerous dugouts in the wadi, contained no fewer than 25 smartly dressed officers who had been well protected from the Allied artillery fire in what was described as 'considerable luxury'.[91] Griffiths' men were outnumbered 20 to 1 by the prisoners of war (POWs) they moved to the rear out of harm's way, not an uncommon occurrence as the battle progressed.

As the conflict escalated, the Italian forces were forced to fight with their backs to the sea. Some units, including gun batteries, did so with vigour, but there were others that were quickly converted from rallying points to assembly points for soldiers eager to surrender at the earliest opportunity. This outcome was consistent with the appreciation completed by O'Connor's staff in the lead-up to the attack which concluded that there was a high probability that the Italians would abandon Bardia after the debacle at Sidi Barrani.[92] Thus the scene was set for the attackers to show some audacity. Captain Norman Vickery, an OPO with the 2/1st Field Regiment attached to the 2/2nd Battalion, won the first MC for a gunner when, on the afternoon of 4 January, he was in the battalion area in his Bren

Gun Carrier and observed an Italian battery being shelled by his regiment, the OPO for which was Captain Dwyer. He approached the Italian battery from the rear, firing steadily and, in an act of brave if histrionic bluff, he encouraged around 1000 soldiers to surrender. He ushered them towards the 2/5th Battalion. Not far away, Lance Corporal Squires of the 2/3rd Battalion took his section to reconnoitre a wadi that housed a battery position. It, too, was crowded and 500 more Italians went 'in the bag'. The Italian guns were dug in and protected by wire entanglements, but this did not seem to have inspired the Italian gunners to fight more vigorously.[93] From the end of the timed program (7.30 am), OPOs engaged targets of opportunity, with the three assaulting battalions allocated F Battery, 4th RHA, 2/1st and 51st regiments for support, while the 2/2nd was to support the 17th Brigade. As there was insufficient artillery for a simultaneous attack by two brigades, the diversionary attack by the 17th Brigade (Brigadier Stan Savige) was not an unmitigated success. On this occasion, the Italian artillery was uncharacteristically effective.

The artillery plan began at 11.30 am and opened on Posts 24 and 25, then moved southeast on each post in turn, finishing on Posts 17, R 11 and an enemy battery GR 51923898 at 1.25 pm. Twelve guns fired for 10 minutes and 36 guns fired for the last 3 minutes on each post in turn. The rate of advance (for fire) was 100 yards in three minutes. For medium artillery on counter-bombardment tasks, Italian batteries east of Posts 19 to 24 were to be engaged for almost two hours, coordinated with the pace of fire from the field regiments. For Phase 2, the 2/1st Regiment supported the 2/5th Battalion and the 2/2nd supported the 2/7th Battalion.[94]

The 2/2nd was to incur the first gunner officer casualties of the battle when Major A.E. Arthur, BC 4 Battery, and Lieutenant J. Crawford (and Signaller Gunner P. Russell) were wounded, Arthur seriously. The men were observing fire from a courtyard when enemy infantry approached, throwing grenades. The Australians were taken prisoner but later freed, with Arthur saved in the nick of time from a leg amputation by a well-meaning Italian surgeon. Captain W.R.G. Hiscock assumed command of 4 Battery. On the same day, Lieutenant Bobbie Nethercote, an OPO with the 2/6th Battalion, was killed by shellfire and posthumously Mentioned in Despatches for his observation work.[95] Italian artillery continued firing despite the counter-bombardment program, and Frowen joined Cremor to organise more effective counter-bombardment tasks, as well as harassing fire tasks to the rear of the defenders.

Over the next three days there would be little rest for the infantry and artillery as Mackay's divisional advance continued relentlessly towards the town. In his post-battle analysis,

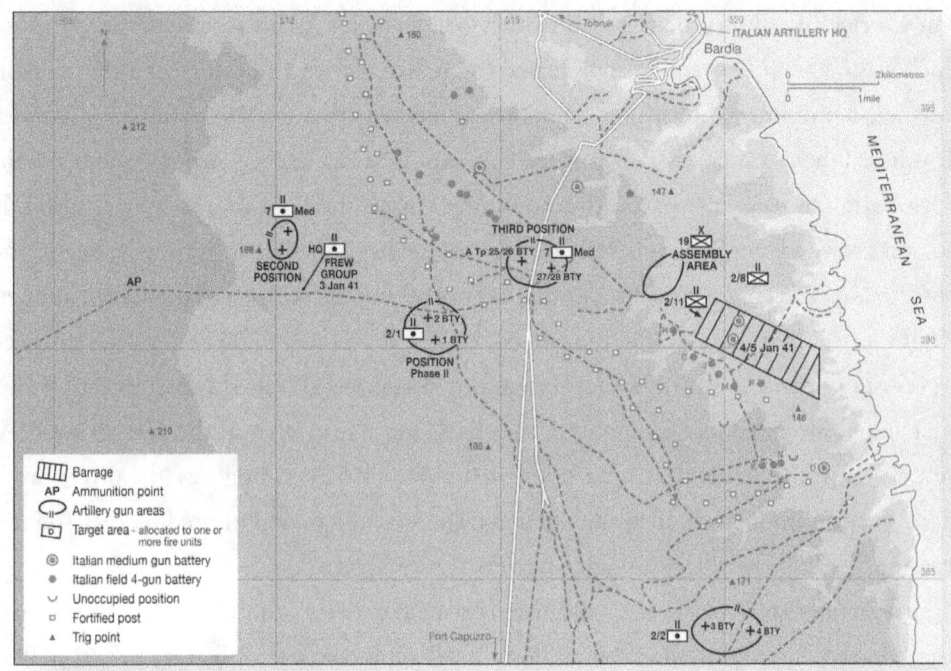

Having gained vital ground the 6th Division infantry are deployed for Phase II of the battle. The supporting artillery, including the medium guns, has moved to more favourable positions. Many Italian batteries were silenced.
Source, Army History Unit

Colonel Frank Berryman, the Division's senior staff officer and a World War I veteran, commented that the 17th Brigade's attack had been under-supported in artillery and that the infantry had suffered unduly as a result. On 5 January, in one of the last artillery actions, an Italian battery on the coastal escarpment engaged the advancing 2/2nd and 2/3rd battalions. It was silenced by F Battery, 4th RHA, firing over open sights. Following the Italian battery's surrender, the staff totted up the booty in captured weapons. Of interest to this account was the artillery tally which was more than double the number of pieces opposing it: seven medium guns, 216 field guns, 41 infantry guns, 26 heavy anti-aircraft guns, 26 coast guns and 146 anti-tank guns.[96] Thus ended the first significant battle and the first occasion for the forming of reputations in which many gunners, regardless of rank, performed spectacularly well. In the 2/1st Field Regiment, for example, 1 Battery BC, Major Peters, an officer not usually given to praising Militia troops, paid Captains G.Y.D. Scarlett and K.F. Dwyer a generous compliment. It was recorded in the War Diary that these officers 'carried out their duties as OP officers in a manner deserving of high praise showing coolness under shell fire and trying conditions ...

[their conduct] was admirable and a great credit to the regiment.'[97] The regiment had fired more than 6000 rounds HE and 133 smoke, and this tally was matched by the other units. In his report, Mackay noted that, at Bardia, 50 per cent of the ammunition was fired on counter-bombardment tasks and, in total, averaged 350 rpg or around 40,000 rounds for the battle.[98]

Of interest to artillerymen were the air shoots by the 7th Medium's A Troop on a particular Italian battery which was holding up the Australians' advance. The demolition of the battery was greeted with great delight by spectators. One last event that passed into artillery history was the capture of an Italian divisional commander and his staff by members of the 7th Medium's RHQ. Likely escape routes between Bardia and Tobruk were heavily patrolled and General Argentino and his staff of the *2nd Blackshirt Division* were captured as a result. The victors were delighted with the amount of vehicles, guns, wireless and telephone equipment and other useful stores left behind by the vanquished. Like good soldiers faced with imprisonment, the Italians had done much damage to such valuable devices as gun sights, ensuring that these would be beyond repair if not beyond improvisation. Nonetheless, some sights were repaired by the innovative Australian gunners and taken to Tobruk to serve the Allied cause.[99]

In summarising the outcome of a very successful attack against daunting numerical odds, General Mackay praised Herring, noting that his artillery fire kept the pressure on the enemy by virtue of its accuracy. Gavin Long's summary of the battle was similarly laudatory: 'It was a victory for bold reconnaissance, for audacious yet careful planning, for an artillery scheme which subdued the enemy's fire at a vital time, and a rapid and continuing infantry assault which broke a gap in the enemy's line.'[100] All commanders from general to colonel sent congratulatory messages down the line to the division, brigades and regiments, Mackay issuing a Special Order of the Day. Barker, for his part, was very satisfied, but stressed that 'we review difficulties of the past ... so that we can overcome them before they arise.'[101]

From an artillery perspective, the use of survey, predicted fire, effective communications between guns and observers, the accuracy of delivery and its timing, particularly observed shooting, all stand out. So too does the fact that infantry battalions and companies had their artillery advisers 'in their pockets', a vital ingredient in desert battles. Cooperation between the RAF, Army and artillery also contributed to the destruction of hostile batteries at critical times, as did naval gunfire on one occasion. This suggests that some of the doctrinal ghosts of the BEF's Flanders campaigns had been exorcised by the participating British units.

Last but not least was the human factor—the 'New AIF' as Long called it. The volunteer army of the 6th Division had been well trained despite enormous equipment shortfalls that ranged from guns to telephones. These problems had been largely neutralised by good leadership, training and innovation, plus the augmenting of establishment levels with captured material. Their next battle, Tobruk, and its subsequent siege four months later, were to conclusively demonstrate the abilities of the 'New AIF', particularly in the artillery battle.[102] Brigadier Herring noted that the Australian and British gunners had been unknown to one another a week before the battle, yet they cooperated and established mutually beneficial relationships as their campaign progressed. They showed 'coolness and steadiness that would have done justice to veterans of many campaigns.' Importantly, they also felt that they had gained the confidence of their infantry.[103]

General O'Connor ensured that his formations maintained their momentum and, as the 7th Armoured Brigade was situated close to Tobruk, he sent it west in an attempt to 'persuade' the Italians to abandon the port city or, if that didn't work, to bar their westward retreat. The brigade reached almost as far as El Adem and succeeded in cutting the coast road and providing a blocking force should the Tobruk garrison decide to evacuate. Another armoured blocking force was in place to the east, and the 6th Division was preparing to assemble for its attack towards the south-east sector. Intelligence estimates suggested that there were 25,000 enemy soldiers and 200 guns in the defended area. Tobruk was the Italian Army's second major fortress, designed and installed (using much concrete) with a perimeter of 128 posts with interlocking fields of fire 27 miles long in an elliptical shape with Tobruk harbour at the 'centre'. This same area in Sydney would extend in a shallow arc from Palm Beach through Strathfield to Cronulla.

On 7 January 1941, Mackay ordered the 19th Brigade to take up positions on the eastern side of the fortress with two field regiments (the 2/2nd and 104th RHA) in support and a cavalry squadron and the machine-gunners of the 1st Battalion, Royal Northumberland Fusiliers. Soon afterwards, these troops were on the receiving end of high frequency fire from the Italian artillery. Time was Mackay's main enemy as he had stipulated that 500 rpg must be in hand before he launched his attack. Two days later, the 16th and 17th brigades and their artillery arrived, and then began a period of planning, reconnaissance and preparation for D Day. During the divisional artillery move to Tobruk, three of the 2/1st's tractors had overturned; fortunately this proved to be a minor mishap rather than a calamity for Barker.

All 6th Division guns moved into preliminary positions east of the town astride the Bardia road. They were limited to 5 rpg per day for counter-bombardment tasks until more ammunition could be brought forward. The Gladiators of No. 3 Squadron, RAAF, were stationed at nearby Gambut airfield, ready to protect the RAF Lysanders of Nos. 6

and 208 Squadrons (Army Cooperation) of 202 Group, its headquarters co-located with General O'Connor's headquarters. Brigadier Dibb was appointed Corps Artillery Adviser to O'Connor prior to taking up the position of CRA 6th British Division. They selected the main gun area south of Mackay's proposed point of entry into the defences of Tobruk, leaving some presence near the Bardia road ostensibly to mislead the defenders. Major General Leslie Morshead arrived at this time (17 January 1941) and spent the next five days 'much in Herring's company' until the town was captured.

Herring had compiled a detailed plot of all the Tobruk defences based on maps captured at Sidi Barrani and Bardia and verified by aerial photographs and observations by reconnaissance aircraft. Needless to say, the counter-bombardment work was vital, the hostile battery data obtained by pilots risking their lives as heavy anti-aircraft guns took pot shots at them. Gun regiments set up flash spotting OPs and identified enemy batteries and their arcs of fire. This complemented the aerial reconnaissance and provided vital input into Herring's plan. But the proof of the pudding was in the eating—all enemy batteries within range would, in the event, be neutralised by the mediums and guns/howitzers under Herring's command.[104]

The evidence suggests that, for both the Bardia and Tobruk assaults, artillery planners had remembered a truism of World War I—the importance of survey in improving the accuracy of predicted fire. It was fortunate, therefore, that a survey Troop was available to assist in planning. One of the two survey Troops in a survey battery of the 4th (Durham) Survey Regiment, RA (TA), was commanded by Captain J. Bird, who had trained it prior to its deployment to the Middle East where it became 1 Troop, 6 Survey Regiment. The Troop acted as an independent sub-unit, performing with great competence. Its reputation was cemented at the Battle of Sidi Barrani and the Troop then moved to Bardia, arriving ahead of the artillery reconnaissance groups at Tobruk and joining the action all the way to Agedabia.[105] Bird attached his troop to the 7th Armoured Support Group and established a divisional grid at Tobruk in short order. This removed one 'unknown' for HQRAA planners who issued him with their survey priorities for 6-inch howitzers, 25-pounders and other medium guns.

Prior to the attack, the 7th Armoured's Support Group and the 1st RHA's 25-pounders were to fire high frequency tasks for four days. The plan involved the infantry (with tank support) punching a hole half a mile wide in the southern ring of posts, then fanning out east and west from the general area of the three-way road junction. This was four miles from the Start Line and five from the town and port. Their opponents, the *61st (Sirte) Division* and the artillery of the *17th (Parma) Division* could muster a massive total of 208 heavy, medium and field guns, 48 heavy anti-aircraft, 12 coastal guns and 24 anti-tank guns.

Mackay's plan put H Hour at 5.40 am on 21 January 1941 and he allotted his artillery in a similar pattern to Bardia, except that he now had more guns facing him. A creeping barrage would support the 16th Brigade after which all guns would turn on the 24 known Italian batteries. This cannonade would be preceded by a naval and air bombardment of three hours' duration (3.00—6.00 am), partly to generate noise to conceal the movement of tanks, troops and guns. Mackay and Herring had 146 guns and 20 howitzers. This included a battery (the 5th, under BC Major R. Bale) and RHQ (under Lieutenant Colonel Horace Strutt) of the newly arrived 2/3rd Field Regiment with twelve 25-pounders, latecomers to the action through no fault of their own.

Like Mackay, Herring divided his assets in much the same way as he had at Bardia. Under his command he had J Battery, 4th RHA, 104th RHA, the 2/1st and 51st Field Regiments. Frowen commanded F Battery, 4th RHA, 7th Medium, 211th Battery 64th Medium and 234th Battery, 68th Medium Regiments. The zero lines for these units ranged from 320 to 360 degrees and their positions were around 3000 to 7000 yards from the perimeter. Under Cremor's command, the 2/2nd and one battery (No 5) of the 2/3rd Field Regiment were sited on the eastern side, and 6000 rounds of ammunition had been dumped for their use. On 20 January at 6.15 pm all the guns except the Cremor group moved to their battle positions.[106] Herring's fire plan opened at 5.40 am with 88 guns from south of the perimeter firing on a line of posts from 52 on the left to Post 59 on the right, a distance of 2500 yards, and around 600 yards from the Start Line. Wide dispersal of the guns ensured excellent coverage of the targets. The creeping barrage advanced at 200 yards every two minutes for 12 lifts. At 6.05 am the guns engaged Post 53 (outer line) and Posts 54, 56 and 58 (inner line) for 15 minutes. At 6.20 am all guns fired hostile battery tasks for one hour.

Some of the artillery highlights of the attack illustrate the types of support provided and the role of supporting infantry and armour. The 16th Brigade's 2/1st Infantry Battalion, supported by armour and artillery, swept inside the eastern line of posts. The battalion covered more than four miles in two and a half hours, and was then switched to the western sector. The 2/2nd Regiment's 3rd (Major R.F. Jaboor) and 4th batteries (Captain W.R.G. Hiscock) supported their battalions in the final clean-up of the area before the second phase began. The 4th Battery position enclosed an Italian gun position by-passed by the infantry, so the gunners rounded up another 100 POWs. By 11.00 am the 2-pounder anti-tank guns of the 3rd RHA were 3500 yards inside the perimeter—these Italian gunners were far less

King George VI inspects 2/3rd Field Regiment in Britain in the autumn of 1940 where the regiment was part of German anti-invasion force defending the south east of England. His Majesty is escorted by Lieutenant Colonel Horace Strutt, Commanding Officer. The unit sailed for the Middle East in November and later joined 6th Division at Tobruk.
(AWM 04569).

resolute than their colleagues at Bardia. G Troop of the 104th RHA, taking up an advanced position, was engaged as it supported the 2/3rd Battalion which had turned left inside the breach. A subaltern took a two-gun section to a flank and, with 20 rounds over open sights, neutralised the battery (it surrendered). There was a spirited counter-attack from the south towards the 2/3rd Battalion, but this was held by the 2-pounders of the 3rd RHA and the Australians continued their advance.

At the scheduled time of 8.40 am, Brigadier Robertson's 19th Brigade began the second phase under a creeping barrage fired by 60 guns, the opening line of which was around 1000 yards long and followed a track near Bir el Mentescha. The barrage also comprised 12 lifts of 200 yards, the final line of which dwelt on the fortifications of Sidi Mahud, the junction of the Tobruk, El Adem and Bardia roads. Concurrently, the remaining guns engaged hostile batteries for an hour. The 19th Brigade had less resistance to overcome as it advanced behind its creeping barrage, during which half the guns were allotted to it and the other half to counter-bombardment fire. One battalion (the 2/11th) was lifted onto its objective by accurate fire, reaching the escarpment beyond the Bardia road without a casualty. Robertson had to wait until early afternoon before the 84 guns allotted to him for the advance on the town could be 'surveyed in' and were ready to fire. F Troop, 104th RHA, blew up the magazine of an 'especially troublesome LAA [light anti-aircraft] battery with a lucky shot, which took their gun crews with it.'[107]

The 2/2nd Field Regiment's role was to support the 17th Brigade (again) with a feint box barrage on the Bardia road, the main entrance to Tobruk. In all, 44 guns were tasked with this role. Thanks to the feint, the Italians complied by shelling the road for two hours by which time the infantry and tanks had passed through the real gap. Lieutenants Tatchell and Cox were FOOs with the 2/7th Battalion and Mair and Crute with the 2/6th. At 10.30 am, RHQ moved to just inside the perimeter near Post 75. During this time Cremor directed his regiment in observed shooting, clearing some wadis of skulking enemy. On 21/22 January, a tricky night occupation of a position was completed which reflected highly on the gunners' previous training in Palestine. The regiment went into bivouac on 23 January.[108] By dusk on 21 January, four battalions (the 2/8th, 2/4th, 2/11th and 2/5th) were sited overlooking Tobruk harbour and the area to the west from Forts Palestrino and Ariente, awaiting Mackay's orders for the next day. The following day the enemy garrison, both naval and military, surrendered to Brigadier Robertson.

Herring's counter-bombardment arrangements were meticulous and contributed significantly to the successful outcome of the battle. All the medium guns, together with the 4.5-inch howitzers of the 2/2nd Regiment, were allotted to counter-bombardment programs. He had established a good working relationship with Lieutenant Colonel M. Yates, the Corps Counter-Bombardment Officer, who had with him Captain M.C.W. Dumaresq and Sapper Hancock, Royal Engineers (RE), a surveyor. As in the Bardia attack, the allocation of hostile battery target lists was predicated on whether these guns could cover the forming-up positions (FUPs) and start lines for the assault. The second priority was hostile batteries covering the 'break in' phase. Italian gunners used little camouflage and guns were placed in sangars and emplacements that were easily detected on air photos, while those dug in were often connected by tunnels. Where digging was well executed, the gun platforms showed little parapet and were harder to detect than those in sangars and emplacements.

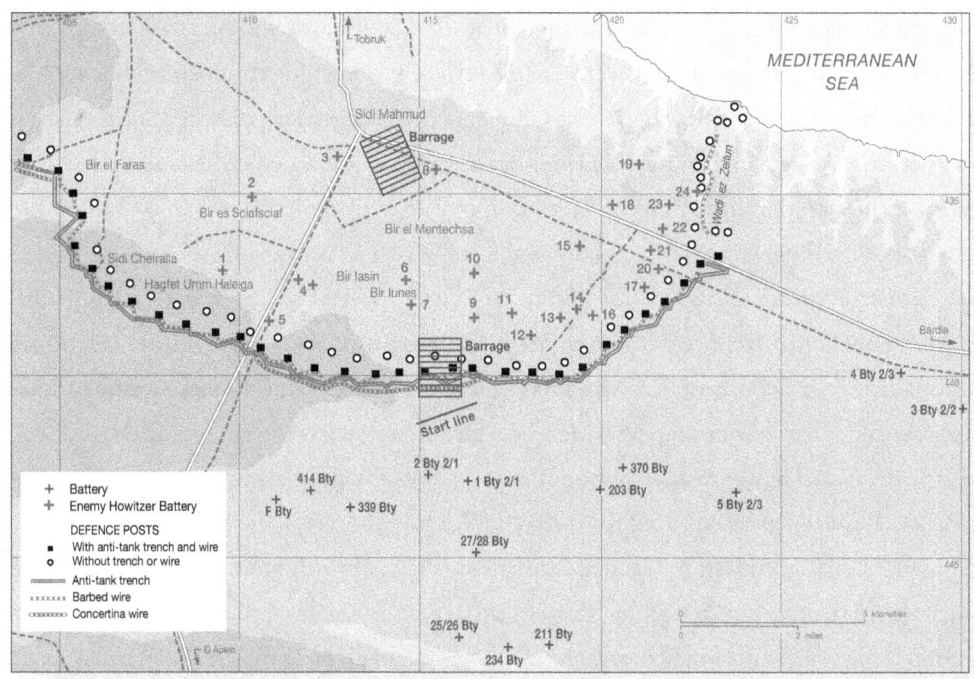

General O'Connor deployed most of his artillery south of the town, and as a diversion, deployed other units at the eastern edge of the battlefield. 6th Division artillery provided the barrage while other RA field and medium guns engaged Italian batteries.

Source, Army History Unit

The success of the operation was also a result of the excellent cooperation between the artillery reconnaissance RAF Lysander pilots of 208 Squadron who regularly ran the gauntlet of hostile anti-aircraft fire to direct shoots, reconnoitre and fix battery coordinates. The significance of these arrangements was quantified by Yates following the capture of Tobruk. For the two-phase attack, Yates had compiled a list of 24 hostile batteries. His follow-up inspection to ascertain the accuracy of British and Australian gunnery revealed stunning results. Eleven of the batteries had sustained direct or close hits that had immobilised the gun and/or crew. Rounds had fallen 'plus' of the target in four cases (average 40 yards) and six 'minus' of the target (variance of 50 to 200 yards). Most of the latter came from the 2/2nd's worn 4.5-inch howitzers. One 'over' round hit an ammunition dump which caused an enormous explosion.[109]

Overall, 25-pounder ammunition usage averaged 223 rpg. The medium gunners were also heavily involved with comparable ammunition expenditures. Their most anxious moment occurred during the occupation of their first position. The 25th/26th Battery was engaging Italian hostile batteries at a range of around 6000 yards and could clearly see their shell bursts. The gun position officer spotted a truck approaching the position but took no notice until it was identified as Italian. Gunners (between rounds) reached for their rifles when it stopped, a mere 200 yards in front of the guns. Out jumped eight Italians with their hands in the air—the battery's second lot of captives in a fortnight.[110]

Tobruk fell, yielding yet again a sizeable tally of POWs (25,000), 208 guns and an assortment of stores. It took until March to empty the POW cages. All that remained of Mussolini's army were five of the original 12 divisions that had formed the garrison of Cyrenaica. Once again, both RAA and RA units were singled out for special praise for their accurate and effective fire support which worked, as one observer noted, 'to make the tanks more effective'. These remarks were not merely self-serving platitudes uttered by senior officers. A divisional casualty count of 49 KIA/DOW, 306 WIA and two missing after one augmented infantry division assaulted four enemy divisions with more substantial resources and defence structures, was nothing short of extraordinary. Such a toll of casualties had been the fate of World War I infantry assaults in less than half an hour on the Western Front. As Bidwell notes of both Bardia and Tobruk, much of the success was attributed to the relatively 'old-fashioned' method of attacking with artillery: 'there was a lesson here, harassing fire, creeping barrages, concentrations and a heavy counter-battery program by the 166 guns [at Tobruk] now available which was to pass unnoticed. It is not fashion that matters in war, but the number of living infantrymen the guns have assisted onto the objective.'[111] Tobruk was to be the last of its kind for 18 months.

O'Connor's subsequent advance to Derna was not to be as painless as his staff had anticipated and, in fact, Mackay thought that to take Derna on 26 January would be 'a gift for Australia on Australia Day'. Initially, an advance guard of all arms was to push forward, but this force was soon augmented so that the 19th Brigade could 'attack and occupy the town'. However, once the airfield was captured, it became clear that infantry would be required and a coordinated attack was mounted. The ensuing fight was more a series of skirmishes than a battle although, at times, Italian artillery fired at its opponents as if determined to use all its ammunition, while Mackay's division (and the 7th Armoured) were both beginning to feel the restrictions imposed by logistics. Australian artillery was rationed to 10 rpg per day and O'Connor could not mount his optimal pursuit for lack of transport, fuel, replacement tanks and other stores. On 26 January, Barker continued to move his troops and batteries forward (as did Cremor) and, when a sizeable smokescreen rose from the town, both men read this as the sign of another withdrawal. It was not to be. Enemy guns fired from Fort Piemonte and more enemy guns, hitherto undetected, opened up, engaging A Troop and scoring a direct hit on Sergeant A. Pearse's No. 2 gun after a 50-yard bracket gave the Italians the range. Three members of the detachment were killed and three wounded. It was the AIF's first lost gun.

A two-battalion attack was now launched and the 2/1st was prepared to support the assault as an aerial report was received which described a column of enemy vehicles moving forward. The attack was suspended while the CO scrambled to a high point and engaged the column using an improvised form of fire order signals directed to those of his guns that remained in view.[112] The column dispersed before more accurate fire was possible. One of the regiment's observation parties was captured and the brigade had to rethink the assault on Derna. There was little joy for the gunners under these circumstances, although the actions of the 2/1st Field Regiment supporting various battalions to clear a fortified escarpment on 31 January were noteworthy as an example of fire and movement in supporting an infantry assault. No 2 Battery was supporting the 2/4th on the flank just as the 2/11th struck unaccustomed resistance. The battery quickly redeployed to assist the 2/11th and its supporting cavalry squadron. Derna finally succumbed on 31 January, and the regiment's casualties were two KIA, one DOW and seven POW. The 2/2nd's experiences in support of the 17th Brigade were similar, although the skills of their drivers were tested crossing the Wadi Derna after the road was blown.

The artillery's advance through the tangled hills from Derna to Giovanni Berta in support of battalion groups (cavalry, infantry and engineers) based on several battalions of both brigades appears to mark the start of a discontinuity of command and cooperation. The advance was characterised by a loss of the unity ensured by the support of artillery batteries for their 'organic' unit (for example, the 2/1st Artillery normally supported the 2/1st Infantry). This was not wholly a bad development and demonstrated a laudable adaptability. On the other hand, infantry officers were sometimes selective in displaying confidence until it was earned under combat conditions. As this was not a factor of any moment until the division was withdrawn, there was no harm done. The division's progress along the coast road was routine in artillery terms, although the gunners occupied numerous positions with OPs and remained prepared to assist the infantry. The 17th Brigade marched 75 miles in three days in battle order without firing a shot, testimony both to a severe shortage of transport, the battle fitness of Savige's brigade and the absence of retreating Italians. The Italians remained one step ahead until they were bottled up at Beda Fomm and utterly destroyed. The Australians were cheered somewhat by the change of terrain from barren desert to hills, trees and towns, one of which (Barce) surrendered to Captain Vickery of the 2/1st.

Captain Norman Vickery, 2/2nd Field Regiment, was the first RAA officer to be decorated for bravery in World War 2 for his actions at Bardia.
(Private Collection)

The Australians reached Benghazi on 7 February.[113] In the course of this long, successful advance they had made mistakes and learned much. They had coped with an alien terrain and trying conditions of bitter cold and freezing temperatures. They had negotiated a featureless, road-free 400 miles mostly without calamity. More than this, however, they had affirmed the doctrine of their employment as they had been trained and had performed well. Their casualties were minimal compared to the infantry with whom they shared the hardships of the campaign. To 29 January, the 2/2nd lost one officer killed, two wounded and two missing at Bardia, while another five were accidentally wounded. At Tobruk and Derna two were accidentally wounded in each town. Influenza was by far the greatest scourge, with no fewer than three officers and 28 other ranks hospitalised for illness.[114]

The real excitement of the chase westward and the artillery action fell to the 7th Armoured Division which, since its advance in November, had received several new units, but remained desperately short of tanks in serviceable condition. O'Connor sent them through Mechili,

Tecnis, Msus and Antelat in a desperate gamble to catch and neutralise the retreating Italian *10th Army*. Their success was the result of a brilliant display of cooperation between armour, artillery and infantry, all with martial virtues well to the fore. According to Bidwell:

> The dazed Italians withdrew and decided to evacuate Cyrenaica, but they found waiting for them the thinnest and smallest finger of the 7th Armoured's left-reaching arm at a speck on the map labelled Beda Fomm, far away to the west. It was made up of the 11th Hussars, the 2nd Battalion, the Rifle Brigade, C Battery 1st RHA (25-pounders) and nine little 37-mm portee anti-tank guns of the 106th RHA, all under command of Brigadier J.A.L. Caunter, a veteran of the Armoured Experimental Force and a member of the Royal Tank Regiment. This small force established a blocking position on the road which led the Italians to Tripoli and safety, with the Yeomanry guns cunningly defiladed behind the hillocks and dunes on both sides of it. For a day and a night a huge Italian column—it was estimated at 11.5 miles long and included 30 tanks—attempted to smash through, but by the afternoon of 5 February they were all 'in the bag', although not without hard fighting on the part of the blocking force. The Yeomanry claimed 27 tanks, one sergeant cunningly taking six from behind after they had passed his position. The 106th RHA were awarded two Military Crosses, one Distinguished Conduct Medal and three Military Medals, and the CO of the Rifle Brigade battalion had the battery paraded before him after the battle was over and thanked them.[115]

This success had an unfortunate sequel in terms of doctrine for the use of mixed arms forces. Those tacticians in favour of deep thrusts of armour to the enemy's rear to paralyse and create panic considered this 'classic armoured warfare'. However, both infantry and tanks still had to depend on artillery (in whatever form) to strike a mortal blow based on concentration of fire. The gunners, watchful observers of the battle, were cognisant of the problems of synchronising the efforts of all three arms. The British General Staff agreed—artillery would not detract from the utility of armour by limiting its freedom of action. This axiom was enshrined in training manuals and proven in the impact of the 7th Armoured tanks on the outcomes of the Tummars and Sidi Barrani. It was proof also that, against a technologically weaker, irresolute army, British doctrine was effective. In Carver's opinion, a less fortunate outcome was that, at a senior level, O'Connor's easy victory created a mindset that discouraged cooperation between armour, infantry and artillery. A contributing factor was that the military virtue of dispersion in the desert was regarded as evidence of unit and formation 'desert worthiness'.[116] In the not too distant future, however, it would be shown to be flawed in practice, albeit not theory, since the Germans had thought through practical problems with greater efficiency than the British.[117]

And so, their arrival in the town of Benghazi ended the Australians' glorious baptism of fire in the Western Desert. They had played a key part in General O'Connor's success with the cooperative links established between their infantry and artillery as well as the British artillery. O'Connor's force was well balanced considering the condition, morale and equipment of his adversaries, although this was to change abruptly. Its tactical, perhaps strategic Achilles' heel was logistics. Many times during those three months he could have sent lightning thrusts into the rear of his enemy had he sufficient resources and trained manpower. Nonetheless, his achievement was masterly: 20,000 prisoners, 112 medium tanks, 216 guns and 1500 wheeled vehicles was booty indeed. General Tellera had been killed and General 'Electric Beard' Bergonzoli captured.[118] In a piece of appalling luck, O'Connor himself was to be captured and spend over two years in an Italian prisoner-of-war camp. During this period he was replaced by less gifted generals who had the misfortune to be pitted against one of the best in the German officer corps: *Generalleutnant* Erwin Rommel of the *Deutsches Afrika Korps* (DAK).

Thus we leave Benghazi and the 6th Australian Division, Major General Mackay and Brigadier Herring, the two major figures in this brief period of Australian history and their infantry and artillery. O'Connor later acknowledged that the Australian success was attributable to the fact that 'Mackay [and his division] was the right man at the right time.' The determined, introspective mathematics teacher and headmaster, pedantic and fussy, had shown many military virtues and had driven his troops hard.[119] He was to be subject to a far greater test in the weeks to come in Greece and Crete.

Lieutenant Colonel Barker issued a Special Order of the Day to his 2/1st Field Regiment. Barker told his men, 'Harder struggles may be in front of us but the end is certain. ... We shall not fail.' For many of the regiment, the end was certain, albeit not the end their commander envisaged. Cremor's homily to his regiment focused on lessons learned, an indirect reference to his 'bastardry' training exercise of the previous November. The two regiments' replacements were not the artillery of the newly formed 9th Australian Division which was still in Palestine. From the highest direction of the war came orders for most of the units of Western Desert Force/Cyrenaica Command to proceed to Palestine, the Delta, Abyssinia or to Greece (then Crete). Allied artillery was never granted another such opportunity to demonstrate its prowess, and many gunners eventually went into captivity, their expertise and vitality lost for four years.

On 9 February, Brigadier J.J. Murray (20th Brigade, 9th Division) arrived to take over from Brigadier Allen and, over the next few weeks, units of the 6th Division returned to the Delta, destined for Greece. We turn now to Major General Leslie Morshead's 9th Infantry Division and their desert war.

Chapter 7
The 9th Division in the Western Desert

Diligence is the mother of good luck.

17th century proverb

To use a human analogy, the 9th Division had a difficult and somewhat complicated birth compared to the other three AIF divisions formed in the period 1939–41. As a result of decisions taken by the Australian Government—with occasional interference from the British War Cabinet and Chiefs of Staff—the 9th Division's subordinate units began 1941 spread from Britain to Darwin. Originally these units had been allotted to the 6th, 7th and 8th divisions but, as a result of German and Italian successes in Europe and Africa, they were soon transferred to the 9th.

The well-trained Australian 18th and 25th brigades, commanded by Major General H. D. Wynter, were shipped by convoy to the Western Desert from Britain via the Cape in November 1940. Wynter had been appointed divisional commander in October and, with Brigadier Morshead (18th Brigade), flew from Cape Town to Durban and on to Palestine to ensure that administrative arrangements were in place for the arrival of the two brigades. They soon encountered their first significant hurdle—no field, anti-tank or anti-aircraft artillery, engineers, signals or service corps units had been included in the divisional structure.

Lieutenant General Blamey objected to the mixing of well-trained infantry with newly enlisted personnel who required considerable training. He insisted that the newest division in his command be comprised of units already formed. Thus the 24th Brigade, 2/3rd Anti-Tank Regiment and the Engineer Field Companies from the 8th Division, along with two field artillery regiments already in theatre as Corps Troops, were allotted to the newest AIF division. The 24th Brigade arrived at Julis, Palestine, in January, the 25th in mid-March, less one battalion which remained in Darwin as the garrison force.

In January and February there was more upheaval. Wynter fell ill in January and was succeeded by Morshead whose brigade went to Lieutenant Colonel George Wootten. Morshead assumed command on 29 January 1941. Three weeks later, the 18th and 25th brigades were transferred to the 7th Division and the 20th and 26th brigades from the 7th joined the 9th. General Wavell then ordered the 18th Brigade to Tobruk, where it was joined later by the 20th, 24th and 26th. At the time, Headquarters British Troops Egypt referred to this shuffling as a 'temporary expedient'. The outcome (affirmed by time) was that, by April/May, the 9th Division's arms comprised:

Artillery:	2/12th Field Regiment
	2/3rd Anti-Tank Regiment
	8th Battery 2/3rd LAA Regiment
Engineers:	9th, 2/3rd, 2/7th, 2/13th Field Companies
	2/4th Field Park Company
Infantry:	2/1st Pioneer Battalion
18th Brigade:	2/9th, 2/10th, 2/12th Battalions
20th Brigade:	2/13th, 2/15th, 2/17th Battalions
24th Brigade:	2/28th, 2/32nd, 2/43rd Battalions
26th Brigade:	2/23rd, 2/24th, 2/48th Battalions
Medical:	2/3rd, 2/8th, 2/11th Field Ambulance
	2/4th Field Hygiene Section

The 2/2nd Machine Gun Battalion and the 18th Indian Cavalry Regiment were eventually added as divisional troops. The division's two other field regiments (the 2/7th and 2/8th) were attached to the 22nd Guards Brigade until September 1941.[120]

This 'temporary shuffling' occurred in a climate of uncertainty which deepened in late 1941 when two momentous events severely tested both Morshead's patience and his leadership. The first was the withdrawal of the 6th Division from Cyrenaica to form 'Lustreforce' commanded by General 'Jumbo' Wilson for the defence of Greece and Crete. The second was Japan's victorious southern march through Indochina and the Dutch East Indies with its consequent impact on Allied global strategy in general and on the Australian Government's strategy in particular as it now faced a difficult choice between deploying troops to the Middle East and diverting them to the Far East.

Lieutenant General Leslie Morshead, GOC 9th Division, discusses issues with Colonel Henry Wells, his principal staff officer (GSO 1) during the Tobruk siege.
(AWM 040583)

In terms of experience in command and battle, Leslie Morshead ranked with Blamey, Gordon Bennett and Iven Mackay—a quadrumvirate of citizen soldiers who had survived World War I and soldiered on in Australia's Militia to equip themselves for higher command. By profession, Morshead was a schoolmaster and cadet corps enthusiast who taught at The Armidale School from 1911 to 1913. But it was to the warrior mould that Leslie Morshead was truly born. A Gallipoli veteran who saw arduous service for six months and was wounded in Lieutenant Colonel Braund's 2nd Battalion, Morshead was repatriated to Australia but remained in the Army. He had quickly come to the attention of Major General John Monash who was raising the 3rd Division for service in France and who gave him the task of raising and training the 33rd Infantry Battalion which he led with distinction. As a schoolmaster, Morshead was a cultivated man, a musician and choirmaster; on the battlefield he was a combative, aggressive and determined leader. His credo on martial virtues was well known to the officers of the 33rd and is germane to the battles that follow in this narrative. C.E.W. Bean remarked that Morshead's philosophy was that of a martinet although, to his credit, he distinguished between the necessary and the worthless. Discipline was his touchstone around which all else revolved. He also regarded order, cleanliness, personal appearance, hygiene and the welfare of other ranks as paramount. He had no patience with 'time-servers'—as far as he was concerned, officers were on duty 24 hours a day. Morshead valued good communication, particularly the precise assessment of situations, tactical or administrative, and their correct transmission to subordinates. Having been on the receiving end of many inaccurate verbal reports, he was reluctant to accept a message at face value.

Morshead summarised his policy as a battalion commander (and his battalion was regarded as one of the best in the AIF) as 'the best man, no matter who he is, is going on top. I have adopted an impersonal attitude—it is the only way—any other policy would be childish or fatal.' He believed that 'character was higher than intellect' and that hard, fair commanders produced the best soldiers. Within his battalion he never forgot a name, nor did he stint on praise when it was merited. Morshead was the only battalion commander who lasted the distance in his division, from his battalion's raising in Armidale in January 1916 to March 1919. He was wounded twice, gassed once and came home with the CMG, DSO, Companion d'Honneur and was Mentioned in Despatches six times. Yet greater achievements beckoned.[121]

Morshead's interpersonal style earned him the sobriquet 'Ming the Merciless', an appropriation of the principal character in the *Flash Gordon* comic strip set in outer space, stereotyped as an evil, mustachioed villain. The latter two words were soon dropped as he put his stamp on his divisional command. Leadership, discipline, knowledge or confidence begotten of experience was the central thrust in the training of his brigades. Morshead did not waste words, quickly saw through poorly reasoned plans and explanations and his gaze, demeanor and clipped speech implied resolution, authority and the exercise of command in its broadest sense.[122]

Following World War I, Morshead left teaching and went into business, although he continued his military service with the Militia. By 1939 he was a senior brigadier commanding the 5th Infantry Brigade before being offered command of the 6th Division's 18th Brigade. By the outbreak of World War II, Morshead had accumulated nine years in command of battalions and six in command of brigades. He had kept abreast of changes in military thinking, attending British Army manoeuvres in East Anglia in 1937. These involved regular formations, including armour, and he noted the difference between his trench warfare experiences and a tactical situation in which brigades were many miles apart. In an address the following year, Morshead told his audience that '... artillery was [now] far more mobile and fire power greatly developed. This modern, speedy movement, this mobility, demands correspondingly quicker thinking, quicker decisions and certainly quicker actions.' As a student of military history he was interested to learn whether any trends had emerged from the Spanish Civil War. In 1937 he heard the British Garrison Commander at Gibraltar voice his opinion that the light tank had proved a disappointment. Morshead's own experience with tanks on the Western Front had been at Accroche Wood where he observed that they were 'offensively efficient'. He noted that first generation purpose-built anti-tank guns would change their 'bogey' status. Of the effect of field artillery, his original viewpoint that 'there was little evidence to suggest its value as an anti-tank weapon in an indirect fire role' he revised soon after his visit to Britain.[123]

On 3 September 1939, when Prime Minister R.G. Menzies announced that Australia was at war, the army's 'mobilisation wheels' were already turning. Blamey was appointed commander of the 6th Division and three renowned veteran leaders were his choice for his three brigades: Morshead, Allen and Robertson. Morshead moulded the 18th Brigade with his training regimes and, by May 1940, it was ready to embark for the Middle East. With the stunning successes of the German forces in Poland, Norway, the Low Countries and France, a decision was taken to divert the 18th Brigade's convoy to Britain to bolster British defences in the event of a German invasion. When this threat subsided, Australian troops were convoyed to their original destination, the Middle East.

In 1918 as a young battalion commander, Leslie Morshead had seen firsthand the problems of unit and formation 'loyalty' that arose when the 36th Battalion merged with his 33rd. Now, as the fledgling 9th Division took shape, similar issues confronted him. The men who would comprise his division varied in outlook and circumstance. The low-numbered NX, VX, QX, TX, SX and WX men who had enlisted in the early days of the war were characterised by eagerness for active service and a strong motivation to oppose the tyranny of Hitler's Nazis. The remainder of the 9th Division comprised men with a generally higher average age who were regarded quizzically by the 6th and 7th Division soldiers as 'deep thinkers'. Morshead also recognised the need to brush aside the 'temporary expedient' tag bestowed by Headquarters Middle East (HQME) staff if he was to weld this disparate body of men into a combative, efficient force of arms. Morshead was just the man to do this.

Despite the rapid changes of plans, deployments and redesignations of Australian units at home, in the Middle East and elsewhere, the important matter of key staff postings in the division was not neglected. Lieutenant Colonel C.E.M. Lloyd, a Staff Corps officer who had been commissioned in 1918 and had completed a law degree at the University of Sydney between the wars, was Morshead's Chief Staff Officer (GSO1). Lloyd was described as 'a strange mix of bluntness and friendliness ... with a dislike of humbug and a desire to come straight to the point. He is no respecter of persons and is essentially a realist who sees a job to be done and goes about it in the most direct way.'[124] He was a good organiser and protected his commander from non-essentials. His GSO2 was Major Thomas White, a Queenslander and Duntroon graduate, whose métier was staff work.

Morshead's most important artillery advisor was Brigadier Alan Ramsay, a Victorian secondary school teacher who also lectured at Teachers' College. Ramsay had served in World War I for a short period as a signaller, then as a gunner and NCO with the 22nd and 4th field brigades of the 2nd Division Artillery. He had served a long and hard apprenticeship from 1915 to 1918 at, amongst other places, Noreuil, Third Ypres, First Bullecourt and Passchendaele and finally the Somme. He was commissioned in 1918 and awarded the Meritorious Service Medal (MSM). Ramsay served in the Militia in the aftermath of the war, rising to field brigade command (the 10th) in Melbourne and later appointed CRA 4th Division Artillery. His success in raising the 2/2nd Field Regiment at the start of the war had not gone unnoticed—he had dropped rank to do so—and it was recorded that his strength was 'to pick officers, train troops to work as a team and weld them into a perfectly coordinated fighting team.'[125] His colleagues considered Ramsay 'broad minded, unselfish, just and far-seeing, determined and imperturbable'—precisely the qualities required for the immense task he was later to oversee.[126]

Ramsay was soon appointed CCMA and later CCRA of 1 Australian Corps, a mixture of field and medium regiments. When the 9th Division Artillery was formed from the 2/7th, 2/8th and 2/12th field regiments, he was appointed CRA to Morshead. As will be related, the division's artillery was split up and HQRAA 9th Division functioned at a reduced level. Major Jim Irwin (ex 2/7th) was appointed BMRA in June 1941, attended Haifa Staff College and subsequently handed over in May 1942 to Major Hylton Williams. Irwin later went to Divisional Headquarters as GSO2 in place of White. When the 2/7th finished its duty tour at the School of Artillery, Al Maza, in December 1941 and found the remainder of the division in Syria, all Ramsay's units were finally under command for the first time. The formation of the 9th Division's field regiments and their movement to Tobruk and Mersa Matruh prior to their withdrawal to Syria in October are worth recounting given their role in the Western Desert campaign.

The 2/7th Field Regiment, RAA, was originally designated an Army (Corps) field regiment nominally associated with the 7th Infantry Division, AIF. The 2/7th was raised in April 1940 and commanded by Lieutenant Colonel Thomas Eastick, former CO of the

13th AFA, an Adelaide-based Militia unit. The regiment was to be formed from the 13th AFA and from the 3rd AFA from Perth, Western Australia, another Militia unit—until Army Headquarters dropped a bombshell. The 2/7th was to become an anti-tank regiment. Eastick was outraged that many highly technically proficient and award-winning field artillery personnel should be squandered in an anti-tank role. So impassioned was the young commander that he took the next train to plead his case at Army Headquarters in Melbourne.

By the time he arrived in Ballarat, however, he discovered that the decision had been revoked.[127] His second-in-command was Major John O'Brien, a Melbourne engineer who had attended the Long Gunnery Staff Course at Larkhill prior to the war at his own expense, and who later commanded the 2/5th Field Regiment before becoming Director of Artillery in MGO Branch at Army Headquarters. Major Jim Irwin, the 13th Battery Commander and later second-in-command in Syria, was the first Militia officer in the 2nd AIF to be appointed BMRA. The first Adjutant was Captain A.W.R. Geddes (Staff Corps) and the Regimental Sergeant Major, Warrant Officer Class 1 (WO1) Weakly of the AIC.

The 2/7th RHQ and 13 Battery comprised mainly South Australians, while 14 Battery consisted almost entirely of Western Australians. The welding of highly individualistic soldiery from the west with the more serious, less 'matey' types from South Australia was a good test of Eastick's leadership. Each battery had three Troops, two of 18-pounders and one of 4.5-inch howitzers. The regiment's training for war was split, with RHQ and 13 Battery (A, B and C Troops) based at Woodside and 14 Battery (D, E and F) at Northam. When the regiment embarked on HMT *Stratheden* in November to sail for the Middle East, their embarkation also followed state lines. RHQ and 13 Battery embarked at Port Adelaide while 14 Battery joined in Fremantle. Their attached 63rd LAD (EME) also embarked.[128] Having arrived in theatre, the 2/7th trained assiduously despite a shortage of 18-pounders and 4.5-inch howitzers, attended 'schools of all descriptions, did route marches, rifle and guard drill.' In May 1941 it moved 180 miles to Mersa Matruh where the regiment dug in as an element of fortress troops comprising the 1st South African Division, two infantry brigades and other artillery (field and anti-tank). In July the 2/7th finally began receiving its 25-pounder gun/howitzers and was eventually fully equipped with 24 guns.[129]

The 2/8th Field Regiment, RAA, was raised from Victorian and Tasmanian personnel, many of whom had served in the Militia, and was originally allotted to the 8th Division. Its first CO was Lieutenant Colonel Alan Crisp, DSO, CdeG, VD, a World War I veteran who had served with the 6th Field Brigade. He was commissioned on 18 August 1914 and served with brigade and divisional artillery headquarters until 1919. He was Mentioned in Despatches three times. His second-in-command at the time of the raising of the 2/8th was Major Bruce Klein (Staff Corps) who was seconded from HQRAA I Australian Corps. The 15th and 16th batteries (each of three Troops) comprised Crisp's command. The former (and RHQ) trained at Puckapunyal, Victoria, and the latter at Brighton, Tasmania, under Klein's stern gaze. The unit followed the same regimes of basic and corps training, and came together for the first time at Puckapunyal before sailing to the Middle East in November 1940. Once in the desert, the 2/8th reorganised into three batteries (the 58th comprising the third) each of two Troops in much the same fashion as the 2/7th. The regiment was supported by F Troop, RA Signals, and the 64th LAD (EME). After Crisp was posted to another role, the regiment was commanded by Lieutenant Colonel Walter 'Tinner' Tinsley, a Staff Corps officer who had served in the 1st AIF and subsequently graduated from Duntroon in 1923. He was tall and distinguished, gregarious

Lieutenant Colonel Bruce Klein was the first BMRA of 6th Division. He served as Counter Bombardment Officer in Greece and Crete before going to as CBO Tobruk Garrison for a period during the siege.
(Private collection).

and courteous, but a stickler for correctness. In the 1930s his career had taken him into survey, locating and artillery reconnaissance. Major R. Johnston was BC 15 Battery and Major A. A. Salter BC 16 Battery; Captain Ken Mackay (Staff Corps) was Adjutant and Captain D. Holt, Quartermaster.[130] The regiment trained in a similar fashion to the 2/7th and also first saw action at Mersa Matruh. Later, the 2/8th was to become part of the 22nd Guards Brigade operating in the Halfaya-Buq Buq-Sidi Barrani area.

The 2/12th Field Regiment, RAA, was originally formed in Victoria in April 1940 as the 2/2nd Medium Regiment of I Corps Troops and commanded by Lieutenant Colonel Shirley Thomas William Goodwin. Due to a shortage of medium calibre ordnance, the 2/2nd Medium Regiment was redesignated the 2/12th Field Regiment on 21 October 1940. Eleven 18-pounders were borrowed from the 2/8th Field Regiment for a week's crash training course. Many of the original members of the regiment were Militiamen

Lieutenant Colonel Goodwin, Commanding Officer of 2/12th Field Regiment and commander of the Australian component of the Tobruk garrison artillery during the siege.
(Private Collection)

from the 2nd and 6th medium brigades and PMF soldiers from the Port Phillip Defences. Raised on 60-pounders, they regarded their 'new' weapons as 'peashooters' and never used them in action. Ironically, they 'inherited' 60-pounders for their service in Tobruk, where they were the mainstay of counter-bombardment artillery during the siege. Goodwin was a veteran of World War I, having had the distinction and misfortune to be taken prisoner at Gallipoli when, as an artillery spotter, his aircraft was shot down and he was captured by the Turks. Prior to his command he had been in charge of the Port Phillip fortifications. He was known to his family as Bill, but to the troops he was 'Buddha'. This sobriquet arose from the fact that he had an imperturbable countenance that never showed the slightest emotion, regardless of the situation. His second-in-command was Major Geoffrey Houston, a 2nd Medium Brigade Battery Commander who would, as it eventuated, command the regiment at El Alamein. His nicknames were 'Gentlemen Geoff' and 'Boy'. The latter was bestowed for his frequent use of the phrase 'well done, boy'. The position of adjutant changed several times (Captain R. Jones was the first), but the RSM, WO1 Fred Smith (AIC), soldiered with the regiment for most of the war.

When the unit sailed on the *Stratheden* it numbered 731 all ranks. The 23rd Battery Commander was Major R.A. Milledge and, commanding the 24th Battery, was Major A.A.C. Carter. The regiment was camped at Qastina and Ikingi Maryut before being deployed to Tobruk Fortress when General Morshead was persuaded by his aide-de-camp (ADC) not to send the 2/7th to Tobruk in a role more suited to a medium regiment as the 2/12th, which had begun the war as a medium regiment, could do the job. Later, the 2/12th was also reorganised into the 23rd, 24th and 62nd batteries. The regiment was supported by G Troop, RA Signals, and the 61st LAD (EME).[131]

As noted above, the 9th Division artillery was called to the fore when HQME was sent to mount the campaign in Greece (and later Crete). It meant splitting the offensive capacity of the division, and was the first of many decisions to dismember structurally sound formations and disperse them willy nilly around the theatre. For its part, the 2/12th was to feature in the siege of Tobruk, while the other regiments were destined for Sidi Barrani.

Chapter 8
Cyrenaica and Tobruk

Artillery was largely useful for either siege warfare or the protection of camps.

Machiavelli
The Art of War

On 27 February 1941 the 9th Division began its westward trek from its bases and camps in Palestine to Cyrenaica by road and train, with the 20th Brigade on the vanguard. As they travelled west, the men took their first look at the famed 'Western Desert' and also encountered the scars of war for the first time. By 4 March they had reached Tobruk, having passed the second fortress to fall to an advancing British and Commonwealth army led by Lieutenant General Richard O'Connor. In June the previous year, as GOC Western Desert Force, he had confronted most of the 500,000 troops the Italian High Command had stationed in Libya and Ethiopia. His force was 50,000 strong and also included the 6th Australian Infantry Division. O'Connor's reserved, low-key persona concealed boldness, a contempt for orthodoxy and a keenness to flirt with mobility now that he had units of the Royal Tank Regiment (RTR) under his command.

The desert battleground was the very antithesis of O'Connor's own country and the land where he had fought 25 years earlier. Distances were immense. There were few villages, fewer towns and even fewer ports. Except for the rurally occupied green 'hump' of Cyrenaica, the land was uninhabited, removing the problems of civilian control and casualties (apart from the Bedouin tribesmen). There were oases, of which Siwa was the largest, and Giarabub the most strategically important. The terrain, at first glance flat and desolate, possessed a strange beauty to a discerning eye, not unlike that of the Australian desert. There were subtle shades in rock and sand, camel thorn bush and minor undulations that became tactically important in defence and attack. Elsewhere there were sand hills, swamps near the coast, and escarpments bordering the Mediterranean with tactically important gaps or passes at Fuka, Halfaya and Sidi Rezegh. Three ridges would pass into Australian military folklore—Miteiriya, Ruweisat and Alam Halfa. A railway line and sealed road hugged the coast as far as Sollum, but otherwise ancient camel tracks used as trade routes became important landmarks for navigation. At night the stars shone with a brilliance never seen in northern climes, while at any time the *khamsin* (sandstorm) could bring observation and movement to a standstill. In military terms it was like a polo or rugby field, with no place or need for trench lines. The 'touch lines' were the Mediterranean to the north and the Qattara Depression in the south. The try lines and goal posts were the Suez Canal to the east and beyond Benghazi to the west.

This would become a contest of logistics. Success brought a lengthening supply line and the Mediterranean became the most important factor in the equation for both sides. There were no crops or water and all supplies had to be husbanded carefully, particularly in summer. Of the combatants, the Italians had more experience in desert campaigning than the British although, as the common axiom held, an Englishman could live anywhere. The Italians had mapped large areas of Libya and Cyrenaica and the British had excellent maps of Egypt. As desert warfare involved greater distances than Europe, large-scale maps were vital and map-reading skills essential for most ranks, particularly drivers.

O'Connor began his campaign modestly, responding to Italian General Rodolfo Graziani's advance to the Egyptian frontier to threaten the Suez Canal. A British force—which became XIII Corps on 1 January 1941—had already taken the Sofafi Camps, Nibeiwa and the Tummars and, in early December, Sidi Barrani fell, followed by Bardia in the New Year. The Bardia assault involved the massed firepower of field and medium artillery in a two-phase attack. At Bardia alone the Italians lost over 40,000 men (killed, wounded or captured), 400 guns, 13 medium and 115 light tanks and 706 trucks. O'Connor repeated his triumph three weeks later at Tobruk when the 6th Division, supported by its own and RA artillery regiments, achieved a degree of fame for their élan and efficiency in overpowering a garrison of 25,000. More booty fell into Allied hands. O'Connor then sent his tanks west via Mechili, Msus, Antelat and Beda Fomm, while the Australians advanced along the coast road through Barce, Benghazi and Beda Fomm. It was a copybook campaign and a great feat of arms and generalship by any measure. In 10 weeks British and Commonwealth forces had advanced 500 miles, destroyed 10 Italian divisions, taken 130,000 prisoners and captured 400 tanks, 1,290 guns and two fortresses. O'Connor's losses were 476 killed, 1,225 wounded and 43 missing.

An unfortunate outcome of O'Connor's successes was that Hitler ordered the despatch of an army corps of armour, the *Deutsches Afrika Korps* (DAK), and an air division to Africa to bolster Mussolini both militarily and politically. He also ordered troops and aircraft south through Yugoslavia to Greece to threaten the Suez from the north, hoping that Turkey (and even Spain) would join the Axis. This accentuated the strategic importance of Syria with its Vichy French administration, and the Allies quickly determined that it had to be neutralised to create a buffer between the Suez Canal and Turkey. General Sir Archibald Wavell (GOCME) now found himself in a very awkward situation.

On 24 March, the commander of the DAK, General Erwin Rommel, having conducted extensive reconnaissance in front of and behind the British front since his arrival in February, launched his first attack against the British at El Agheila. O'Connor had become unwell and returned to Egypt as GOC British Troops, Egypt. His command passed to Lieutenant General Phillip Neame VC who had earned his VC in World War I and cemented his World War II reputation in Abyssinia fighting the Italians. A tall, good-looking, manly figure, he had the unenviable task of confronting fresh German troops who had good equipment, sound tactical doctrine and commanders who could think on their feet, lead from the front and were energetic. The British consolidated their forces as did the Germans, landing their units at Tripoli and bringing them forward. When assembled and oriented, Rommel considered his forces sufficiently strong to launch an attack, despite the fact that his division was incomplete. Defying Hitler's orders he advanced, prompting Wavell to order O'Connor to return to the Western Desert as an 'adviser' to Neame.[132]

At this point the 9th Division was deployed along the coast east of Agedabia. Divisional Headquarters was sited at Bir el Tombia, Headquarters 20th Brigade at Bir el Ginn and the 2/15th and 2/17th battalions forward at Marsa Brega with the 2/13th at Beda Fomm behind the British defensive front line. This account will not follow Cyrenaica Command and the 9th Division's fortunes in detail, but rather highlight some of the more critical command arrangements, operations and artillery themes that arose during the following three weeks until Lieutenant General John Lavarack was ordered by Wavell to take over Cyrenaica Command at Tobruk.

The first contact between the DAK's *5th Light Division* and British armoured cars came on 24 February. The British were forced to withdraw from their weak defensive line at El Agheila and, in a desultory series of delaying actions, the 2nd Armoured Division conducted an orderly withdrawal to Antelat in the face of superior forces. This was in accordance with both Wavell's intentions and Neame's plan. The most serious problem in western Cyrenaica, where Morshead's brigades were positioned, was the monumental task of acquiring sufficient equipment—particularly communications equipment—to fight efficiently.[133] Morshead reported serious equipment shortfalls, particularly in tanks, transport and essential stores, much of which was designed for use in Europe and not suitable for the harsh conditions of the desert. Their reduced serviceability levels would later place the division in a perilous position. However HQME eventually provided the extra men and material to the British formation, bolstering its defence against the German armour. Despite this the campaign was to be dogged by shortages.

This was not to be a happy period—not just for the 9th Division, but for Neame's entire command. The 20th Brigade arrived at Benghazi on 8 March and Neame later told Morshead that it would come under command of the 2nd Armoured Division, at which time Headquarters 9th Division would move back to Gazala. Given that the other two brigades were still en route from Palestine and there was no organic artillery, Morshead was unsurprisingly appalled. His 26th Brigade eventually arrived in mid-March and paused at Tocra. His screen comprised the armoured cars of the 1st King's Dragoon Guards, most of them machines captured from the Italians (as were the tanks of the 2nd Armoured Division), the light tanks of the 3rd Hussars and the medium tanks of the 6th RTR. His artillery support was provided by the 1918-vintage 18-pounders and 4.5-inch howitzers of the 51st Field Regiment, RA, tried and true weapons, but no match for the modern German armaments now facing them. In addition, Morshead had only 15 anti-tank guns and 19 LAA guns for his allotted area. At Marsa Brega, Morshead foresaw disaster if his brigade could not be withdrawn to the Barce area as deployment of the division in the face of the panzers would be catastrophic. To defend both routes eastward to the rear in Cyrenaica required more transport assets than were available, notwithstanding the quantity captured during the advance in January.

Major General Michael Gambier-Parry arrived from Crete on 20 March to command the armoured division and was forced to conduct a rearguard action against well-trained troops while contending with poor communications, flawed intelligence and deficient planning. Benghazi fell to Rommel on 4 April. Morshead relied on RAF reports and his own visits to the front for intelligence on Rommel's plans and preparations rather than the reports sent to him by HQME.

The artillery of the frontier defence was gradually augmented by armoured and RA units as well as the 26th Brigade (2/24th and 2/48th battalions). The 24th Brigade, now commanded by Brigadier A.H.L. Godfrey, arrived in Tobruk on 26 March. On 4 April, the

2/13th Battalion at Er Regima, despite all its equipment deficiencies, fought a courageous delaying action on an extended front until nightfall. Having inflicted heavy casualties on the DAK's *3rd Reconnaissance Unit* of the *5th Light Division*, the 2/13th withdrew 10 miles to Barce losing 82 men, captured during the action.

Events now moved swiftly, with Rommel seeking to envelop the right (coastal) flank and, at the same time, destroy the British armour on the inland route. Wavell flew from Cairo to Barce where the 9th Division had taken up positions, to conduct a personal assessment of the situation. He agreed to provide transport for the withdrawal of the 20th Brigade, retrieving it from a perilous position. Morshead had issued orders for the defence of Barce, although he had some doubts whether Neame would issue a 'stand and fight' order. The 9th Division's first coordinated deployment was aimed at holding the escarpment at Tocra, from the sea to the Wadi Gattara. The 26th Brigade took the right flank, with the 20th sited on the left, and they were supported by the 1st and 104th RHA with their 25-pounders, two RA LAA batteries (the 1st and 37th) and a signals section. Morshead observed that the two supporting RHA regiments were 'very excellent' and noted that his own cavalry support with its worn, unreliable equipment, grossly inferior to that of the Germans, compared very unfavourably with the equipment of the RHA. He reserved his spleen for the Divisional Support Group Brigade Major whom he termed 'quite the most stupid man I've known in or out of the Army.'[134]

Morshead's disappointment over the cavalry's lacklustre performance—primarily the result of incorrect doctrine adopted for operations with infantry—was to dog him until November 1942. It was with some chagrin that he noted the contrast with the German assault on Cemetery Hill at Marsa Brega on 31 March which was characterised by the well-coordinated use of machine and field guns, tanks and aircraft against the Support Group of the 2nd Armoured. On this occasion, the British were supported only by the 339th Battery's 25-pounders which were to have 'a busy day.'[135] That said, the British forces had achieved the mission set in Neame's plan. Wavell had instructed that Rommel's northward thrust be held with a small force while the armour pivoted on Antelat. However, a reappraisal of the situation prompted a general withdrawal which was ordered on 3 April.

The practice of using artillery support in Troop and battery size for 'targets of opportunity' was generally instituted on 30 March and references to concentrated artillery fire from any regiment are relatively sparse until the occupation of Tobruk later in the month. This was the experience of the 7th Medium Regiment, RA, with the 234th Battery under command. A few days later they withdrew eastwards from Giovanni Berta and continued to withdraw from one position to another all the way to Al Maza in Egypt! At one point they were almost called to join the battle outside Derna, where they dropped trails, waiting for fire orders that never came. Earlier, at Mechili, the attached battery had arrived after the encounter battle was concluded.[136]

Rommel's tactics prevailed and, by 3 April, Neame and O'Connor's dual command, which was to last only two days, held the fate of the entire 9th Division in its hands. Morshead demanded the immediate withdrawal of his division to Gazala and O'Connor reluctantly agreed. Battalions from the division protected their withdrawal.

Wavell's choice of Neame had been a poor one. In short order he had alienated the GOC 2nd Armoured Division (Gambier-Parry) and Morshead, the latter by not consulting Blamey in Cairo, selecting an impossible defensive position for the division at Beda Fomm and writing a lengthy letter to Morshead criticising discipline within his division. This Morshead took as a serious reflection on himself.[137]

By 4 April General O'Connor had realised that the command situation was unsatisfactory and asked Neame to resume command while he (O'Connor) returned to his advisory role on Neame's headquarters. This proved unfortunate as both British generals were captured between Msus and Mechili while withdrawing east on the riskier inland route when their staff car ran into a German detachment behind British lines. Based on information received that two generals were prisoners in a wadi near Derna, Morshead asked the CO of the 1st King's Dragoon Guards to mount a rescue operation. The CO demurred, arguing that it was not a role for armoured cars. In the event his second-in-command took a section of cars out into the night but he returned empty handed.

Gambier-Parry was at Mechili on the inland withdrawal route with the remnants of his command and ordered a breakout at dawn on 8 April. The 2/3rd Anti-Tank Regiment's 10th and 11th batteries (Lieutenant Colonel E. Munro) were heavily engaged in this action and many of the gunners were killed, wounded or taken prisoner.[138] If ever there was a tactical situation crying out for massed artillery support from field and medium guns it was at Mechili on 8 April. At this point, Morshead was advised by the CO of the 104th RHA, Lieutenant Colonel E.J. Todhunter, that he (Todhunter) had been appointed CRA of Morshead's forces.[139] Morshead was unaware that the remnants of the 2nd Armoured Brigade and its artillery were 100 miles away and he was unable to establish reliable wireless connections with his units. He had no staff or communications and eventually found his way to the headquarters of the armoured division where, in difficult conditions, he tried to piece together the situation.

As dusk fell on 8 April, Morshead found himself in the curious position of being the only general in Cyrenaica. He took command and, as history records, issued orders for the orderly withdrawal to Tobruk along the Giovanni Berta—Martuba Road and his division lived to fight another day. In Tobruk, General Wavell had positioned the division's 18th and 24th brigades along the perimeter defences and installed Lieutenant General John Lavarack as GOC Cyrenaica Command.

Through the tempestuous days of the withdrawal to Tobruk, the British artillery did all it could to support the infantry. The 51st Field Regiment's howitzers and 18-pounders, while outgunned by the German artillery, were also hobbled by a shortage of ammunition. In fact, the regiment's aged guns and howitzers were so worn that their zone of fire rendered them unsuitable for use in supporting static fire tasks as they were a danger to their own troops. The barrels of the RHA's 25-pounders were starting to bulge and, although not a serious impediment at this stage, caused increasing concern until they were eventually replaced in the safety of the Tobruk fortress. On 4 April at Er Regima, an 18-pounder Troop from the 51st Field Regiment had dropped smartly into action and engaged and mostly destroyed a German armoured car reconnaissance group in front of the Australians. There, supporting the 2/13th Battalion, they had expended their small ammunition allotment in what proved to be merely a temporary deterrent to thrusting enemy armour. The 1st RHA was also tasked to support Australian battalions although, at the time, it was more common for a battery to support a brigade, as with LAA artillery.[140]

The air support situation was not much better. By the time the RAAF and RAF squadrons were withdrawn from harm's way they could muster only a small number of aircraft between them. No. 3 Squadron, RAAF, and No. 6 Squadron, RAF, had followed and supported O'Connor's forces all the way to Agedabia. During the withdrawal they engaged *Luftwaffe* Ju87s and Ju88s harassing the 9th Division and the 2nd Armoured Division at Barce and Mechili. The German *X Air Division* had been particularly active since its arrival to support Rommel's army. While Rommel reconnoitred, his pilots assessed the capacity and tactics of the Allied air force, albeit not without loss. Ultimately, the sheer numbers of German aircraft prevailed and the absence of air force presence over Tobruk was to be a constant topic of conversation amongst its defenders until squadrons were rested, reinforced and re-equipped.[141]

The withdrawal also saw many parties fall foul of German units and ambushes on their way to the safety of Tobruk. On the night Neame and O'Connor were captured, the 9th Division Headquarters group was travelling towards Morshead's tactical headquarters with

Lieutenant L.K. Shave in command. The party consisting of the intelligence and cipher sections was halted by other vehicles. A German soldier poked his weapon into the back of the truck and ordered the Australians out. Shave slid out of the front seat and went to the back of the truck, shot the German, returned to his seat and drove off without further ado. In a confused withdrawal, mostly undertaken without maps, the Australian and British engineers found themselves busy denying assets of value to the enemy with their carefully considered demolitions.

Vehicles descending the Derna escarpment became a tempting target for the *Luftwaffe* on the exposed face and in the defiles about the town where the German aircraft wreaked havoc. Around the Derna airfield there was a spirited action between a German reconnaissance battalion and a scratch force from the 2nd Armoured Division's Support Group involving four tanks that were the sole survivors from that formation. During the withdrawal the anti-tank gunners and LAA guns pounded the Germans. The principal 'enemy' of the withdrawing troops, however, was fatigue, particularly among drivers. The 20th and 26th brigades and their supporting artillery, the 1st RHA and 51st Field Regiment, arrived at Tobruk late on 8 April and the infantry immediately established defensive positions.[142] The 103rd and 104th RHA reached Tobruk in relatively good order, driving off several tanks that approached the perimeter.

On the whole, the field and medium artillery available to Cyrenaica Command was ineffectually used. The reasons for this are consistent with the general tenor of the four-week campaign. Neame, as far as can be determined, did not have a CRA to coordinate his artillery, probably because he had insufficient staff and signals support. Likewise, Morshead did not have a CRA, although he used his field regiment commanders as advisers. The 2nd Armoured Division (also without a CRA) and the 9th Division were, for most of the time, sited some distance apart, which explains why concentration was difficult, if not impossible. The episode involving Todhunter and the fog of war surrounding the 2nd Armoured Division as Rommel probed its front is a salutary example of how little artillery battlefield intelligence reached the general staff at every level of command. Even had there been artillery advisers at these senior levels, they may have had little influence over plans and decisions. Much of the decision-making was probably effected at meetings to which advisers were not invited or which they were unable to attend. That the 7th Medium Regiment did not fire a shot for a month reinforces the notion that—with the exception of Morshead, his brigadiers and Brigadier Latham, a gunner and support group commander—the general officers and their staffs did not know how to use artillery effectively. The tank commanders of the armoured corps had yet to be trained effectively in the use of artillery and doctrinal flaws in their training and tactics were among the underlying reasons for their reverses. The lack of a suitable command structure with efficient communications added significantly to the gunners' burden. Assessing these and other factors in the disastrous campaign, the staff in Cairo suggested that a change to divisional structure was required. It would come none too soon.

Morshead finished this episode of his military education with an enhanced reputation. This brief campaign had exposed the enormous differences in command, control and leadership between British generals and battalion commanders and their German counterparts. The latter, whose officer corps was often characterised by 'Teutonic thoroughness', were trained very differently. Junior officers went into battle knowing what a battalion commander should know—that he had to resolve the tactical situation in which he found himself, innovating and taking risks if necessary. The British methods were the very antithesis. Training for higher rank was a graduated affair and caution a more frequent determinant of action and problem-solving. Staff work also focused on covering several contingencies or restricting freedom of action by referral to higher authority. Thus, junior British officers surrendered the initiative to a foe whose internal wireless communications and battle intelligence were superior. If the German communications failed they resorted to drills until the lines were re-established. While there were exceptions to these generalisations on both sides, in Rommel the Germans and Italians had a commander who was an energetic risk-taker and skilled tactician. His tactical plans were based on surprise (a potent military principle) which alarmed even his own high command, and he expected his subordinates to behave in a similar manner. These were men who demonstrated the Morshead dictum of 'confidence born of experience' while Neame and his successors merely exhibited an unsettling confidence.[143]

Morshead reported to Headquarters Cyrenaica Command on 8 April to find that General Wavell had arrived and brought Lieutenant General John Lavarack with him. Wavell instructed Lavarack and later Morshead that Tobruk was to be held for two months with his four brigades, three field regiments, one anti-tank regiment and four anti-tank companies. He had three anti-aircraft regiments, light and heavy, with searchlights, coastal artillery and an assortment of Italian guns. These last, variously known as 'Mr Clarke's Guns' and the 'Bush Artillery', were manned by Australian troops from various units. Wavell's instructions also related to the conduct of the defence, reinforcement, logistics etc. On 14 April Lavarack handed over to Morshead as Fortress Commander and flew back to Cairo with his headquarters.

The 2/12th Field Regiment gunners used many captured Italian guns, Howitzers and ammunition during the seige. This weapon was part of the 'Bush Artillery' of the Salient.
(Parsons Collection)

Morshead's armour was a mixed collection of 45 light and cruiser tanks and he 'spent several days in inspecting the defences of the fortress.'[144] He had witnessed the 6th Division's assault in January and gained a good grasp of the essential features of the topography and Italian defensive schemes. His brigadiers, Godfrey, Wootten, Murray and Tovell, all possessed a sound understanding of the construction of the fortress, having visited the Italian defences earlier. Wavell had decided to merge Cyrenaica Command with Western Desert Force under a new commander, the extraordinarily named Lieutenant General Noel Monson de la Poer Beresford-Peirse, an outspoken, extroverted, cheroot-smoking gunner who had considerable experience in India and more recently as GOC 4th Indian Division. However, he had little experience of handling large formations as events would soon reveal.[145]

Morshead was fortunate that an Australian Staff Corps Officer, Lieutenant Colonel T.P. Cook, had been appointed to command the Tobruk sub-base, succeeding Godfrey in this role. Cook performed two important tasks during his tenure. First, he made an appreciation of the situation and organised the defences accordingly, dividing the defenders into three components: a mobile striking force, a mobile reserve, and the remainder of the garrison which was tasked to place central posts on all roads through the inner perimeter and reconstruct the Italian roadblocks. His second 'tour de force' was to adopt General Neame's instructions regarding the use of captured Italian field guns and run a training program for infanteers (usually personnel in headquarters and B Echelon who were supernumerary) to learn how to load, lay and fire the weapons with least chance of injury to themselves.[146] The 'Bush Artillery' went into action almost immediately, repelling a German attack at the western end of the perimeter near the Derna road. Guns sited in the 2/28th Battalion area took the honours.

Morshead put the defenders to work strengthening the defences, as many of the posts had not been finished, wired or mined. He issued an unequivocal order to his garrison: there would be no surrender and no retreat. This was a simple exposition of Morshead the confident warrior who had shown superior tactical insights to his betters and peers during the withdrawal, and it set the scene for an *offensive* defence of the port.

The tactical employment of artillery at Tobruk posed some problems for its commander and his CRA. The principal dilemma was domination by fire of space between infantry units and sub-units predicated on the defensive post system installed by the Italians and later improved by the Australians. 'Conventional' warfare involving brigades/divisions adhered to the 'rule of thumb' that battalion flanks should be no more than 800 yards apart. Anything more was considered foolhardy, inviting defeat in detail. At Tobruk, the distance between battalion flanks averaged 2000 yards. Morshead believed that the disadvantage of distance could be neutralised if there was sufficient artillery and ammunition, minefields, or by clever use of the ground. To designate Tobruk a 'fortress' was to describe a somewhat useless piece of real estate with an emotive noun. The Italians had not made a fortress of it. There were no constructions and facilities of the type associated with the famous Maginot and Siegfried Lines—shelters, buried line communications, subterranean magazines and shelters or concrete and earthworks to discomfit an attacker. Morshead countered these deficiencies with a mobile defence, mobile reserves and a policy of aggressive patrolling.

Thus Morshead employed a number of measures to maintain the initiative, notwithstanding his enemy's aggressive tactics. With air supremacy the enemy could monitor daily changes in the fortress's dispositions. This capability was a double-edged sword in some respects, as concealment or disclosure of assets could either accurately reveal the defensive measures employed by the garrison or form the central pillar in a strategy of disinformation. A case in point, as at Bardia, was the construction of dummy guns and gun pits. To deny observation to enemy ground forces, the defenders deployed a 1500-yard hessian screen around the perimeter to obscure a Troop of guns. It was all part of the intelligence game.

Hill 209, the high ground of the Ras el Madauuar, was vital ground which afforded views of most of the salient south and west of Palestrino. This part of the perimeter was designated the RED line. At the end of April a second line of defence, the BLUE line, was created some two miles to the rear. It was protected by barbed-wire entanglements, anti-tank and machine-guns covering likely approaches and both anti-tank and anti-personnel minefields. The engineers constructed some minefields with a wheel-spoke pattern instead of a linear one to further confuse an attacking force. Platoon posts were placed around 500 yards apart. The field artillery was positioned behind the BLUE line and sited to deny enemy access to the Fort Palestrino area. Troop gun positions were arrayed in a crescent pattern instead of the stepped pattern from the pivot No.1 gun. This gave the guns better anti-tank fields of fire.[147] A third defence line, the GREEN line, was closer to the town, around three miles from the harbour, and passed through Fort Solaro.

On 11 April Rommel launched his forces against the perimeter in three columns. Ignorant of the layout of defences and topography until they acquired some Italian maps, they confidently mounted a series of attacks on the 24th Brigade front on the south-eastern side of the RED line. A combination of artillery, machine-gun fire, tank and anti-tank weapons rebuffed the assault. The guns of the 1st RHA fired 500 rounds of HE in 90 minutes, a rate that would become unsustainable in the future except on rare occasions. As Maughan notes, 'The German soldiers, who had been taught by their experiences in Europe to believe that boldness and a disregard of risks alone would suffice to carry them to their objectives, were soon to shed their illusions before Tobruk.'[148]

Morshead also received the welcome news that another field regiment was on its way to reinforce the garrison. The 107th RHA (South Nottinghamshire Hussars) commanded by Lieutenant Colonel W.E. Seely and equipped with 25-pounders, also comprised the 425th and 426th batteries. However, of greatest concern to Morshead and his new CRA, Brigadier L.F. Thomson, RA, was the lack of medium and heavy guns given that the 7th Medium Regiment had moved to Egypt. They were somewhat mollified by the availability of RN warships and captured Italian coastal defence artillery under the command of Lieutenant Colonel E.O. Kellett.[149] The Germans soon realised that the garrison was using these weapons and began deploying their forces and guns out of range of the three general gun areas. At an average range of engagement from their OPs of 2000 to 3000 yards, this allowed a panzer group in an attacking formation to take some comfort from the fact that the British artillery could not reach it at the vital time when lorried infantry were debussing. However, this comfort was to be short-lived: the 2/12th Field Regiment, RAA, had been despatched

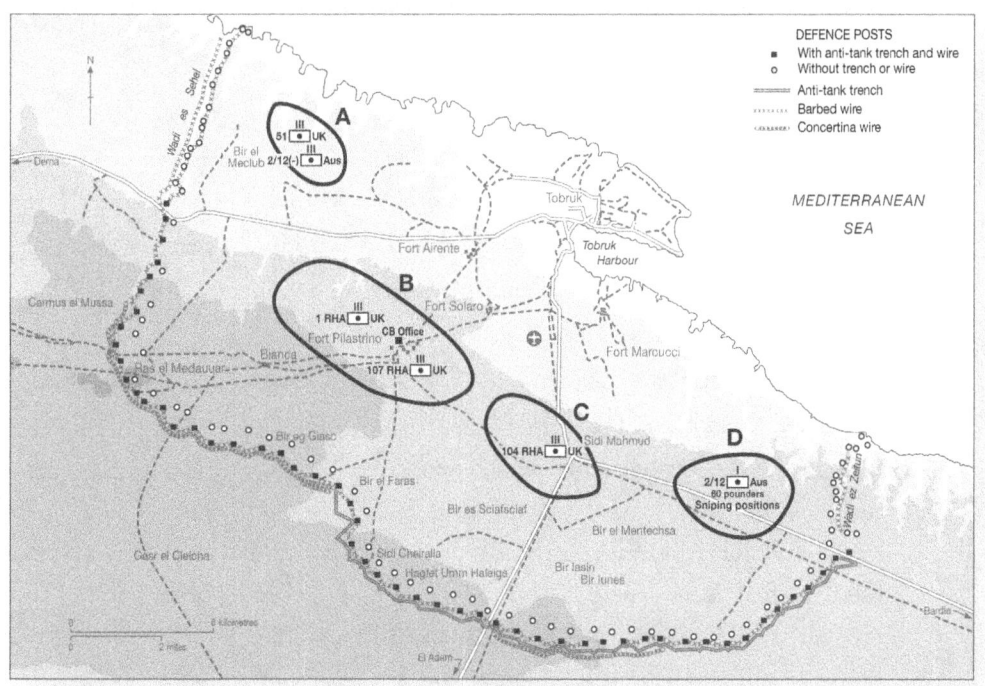

The Tobruk salient was more than 40 miles long and the British and Australian artillery frequently changed position to avoid German and Italian counter battery fire. Gun positions were restricted to general areas that suited an attack - defence posture against Rommel's potent panzers.
Source, Army History Unit

from Palestine, although it would provide only modest assistance.

The 2/12th Field Regiment was mobilised in Melbourne from heavy and garrison brigades as the 2/2nd Medium Regiment before being converted to a field regiment and allotted to the 9th Division. On 6 May the 2/12th arrived at Tobruk by sea in dribs and drabs, with the advance party of five officers and 82 other ranks led by the regiment's second-in-command, Major Geoff Houston, the first to arrive. B, C and D Troops followed on 7 and 8 May. The CO, Lieutenant Colonel Shirley Goodwin, landed a week later with his RHQ group, by which time D Troop with its 60-pounders had been in action for at least eight days, as had B and C Troops with their antiquated but still effective 4.5-inch howitzers (passed to them by the 51st Field Regiment, now equipped with 25-pounders). On 20 May A Troop arrived, while E and F Troops landed on 3 June. Such were the fortress's logistic demands that it took almost a month to assemble the regiment. In the meantime, gun detachments were seconded to the 51st Field Regiment to gain experience and to assure them of a meaningful role in the defence. Several gunners became casualties while filling this role.[150]

With the arrival of the 2/12th, Morshead and Thompson's artillery comprised:
BMRA: Major J.C. Smith, RA

Field Artillery

1 RHA, Lieutenant Colonel S. Williams: A/E and B/O batteries

104 RHA,* Lieutenant Colonel A.G. Matthews: 339 and 414 batteries

107 RHA,* Lieutenant Colonel W.E. Seely: 425 and 426 batteries

51 (Army) Field Regiment, RA, Lieutenant Colonel J.S. Douglas: 203 and 307 batteries

2/12 Field Regiment, RAA, Lieutenant Colonel S. T. W Goodwin, 23, 24 and 62 batteries (A, B, C, D, E and F Troops)

'Bush Artillery'—various units/detachments.

Anti-tank Artillery:

3 Anti-Tank RHA: J and M batteries

149 Anti-Tank Regiment, RA*

2/3 Anti-Tank Regiment, RAA: 9, 10, 11 and 12 batteries

16, 20, 24 and 26 anti-tank companies

Anti-aircraft Artillery

8 Battery, 2/3 LAA Regiment

4 AA Brigade, RA

51 HAA Regiment, RA

14 LAA Regiment, RA

Survey: No. 1 Composite Battery, 4 (Durham) Survey Regiment, RA,* Lieutenant Colonel J. Whetton

*Territorial Army units of the RA.

Morshead's CRA, Brigadier Thompson, had previously commanded the 4th Field Regiment, RA, in Palestine with the 5th Indian Division. He was promoted brigadier and posted as BRA Cyrenaica Command, thence CRA 9th Australian Division, for the duration of the siege. He designated three gun areas to cover the 27-mile frontage of the perimeter. The western sector, astride the Derna road, was occupied by the 51st Field Regiment, augmented later by the 2/12th Field Regiment on its arrival. The southern sector astride the El Adem road was home to the 1st and 107th regiments, and the 104th Regiment in the eastern sector sat astride the Bardia road. Their opponents were well endowed with artillery and it is worth examining the German formations facing this augmented infantry division.

On his arrival in Africa, Rommel and his DAK were initially placed under command of Italian General Italo Gariboldi, reflecting the sensitivities of the Italian High Command.

This was to end in October when Rommel became Commander *Panzerarmee Afrika* (PAA), Lieutenant General Crüwell became GOC of the DAK and the Italian generals reverted to command of their respective formations. The panzer division of 1941 was the result of careful deliberation in the aftermath of World War I, and was a balanced formation consisting of:[151]

- a recce battalion of 48 armoured cars, 2 x 75-mm howitzers, 3 x 37-mm anti-tank guns, 3 x 81-mm mortars
- two panzer regiments each of two battalions, each with 44 Mark III and 22 Mark IV tanks
- an artillery regiment of three battalions equipped with, in total, 24 x 105-mm gun/howitzers and 8 x 150-mm howitzers
- an anti-tank battalion of 12 x 37-mm and 18 x PAK 50-mm guns
- two panzer grenadier brigades, each of two battalions of lorried infantry and 6 x 150-mm gun/howitzers, either self-propelled or towed
- an engineer squadron
- various signals units

The German 105-mm gun fired a heavier shell than the 25-pounder (33 pounds) and had a split trail while the 25-pounder had a platform enabling a 360-degree traverse. This latter feature upgraded its utility when it was thrust into an anti-tank role, which happened increasingly as the desert war progressed. In medium artillery there was little to choose between the German 150-mm sFH 18 and the 5.5-inch gun/howitzer. However the 60-pounder that the 2/12th (and the 7th Medium Regiment initially) took to the desert had a limited traverse and a maximum elevation of just over 21 degrees compared to its successor's 45 degrees. Its 60-pound shell was 35 pounds lighter than that of the sFH18, although its maximum range was comparable at 15,000 yards. Unfortunately, few of the 4.5 or 5.5-inch weapons were available and thus they were introduced incrementally into service in the desert.

On 14 April the Germans introduced their dual-purpose anti-aircraft/anti-tank 88-mm Flak 18 gun into the land battle and, in one graphic account, the crew of a 'long barrelled gun' was cut down by rifle and machine-gun fire.[152] As already noted, the British 2-pounder gun was inferior to the PAK 38 50-mm, but superior to the 37-mm, where it had a 20 per cent advantage in penetration power.[153]

Several historians have noted that the Australian soldier showed more élan in attack than the Briton.[154] At Tobruk there was ample scope for both views. On 12 April the BC of B/O Battery, 1 RHA, Major R.G. Loder-Symonds, was in his OP at Post R32 in the same area as

Captain J.W. Balfe's company of the 2/17th Battalion. All that day the enemy threw tanks and infantry at the defenders. The men of 1 RHA engaged them vigorously for 1½ hours in the morning, RAF aircraft strafed them further out and, when a hot, dusty wind blew up in the afternoon and German infantry approached to within 500 yards of the wire, Loder-Symonds hit them again. At 10.00 pm, when the sun set and the moon rose, yet another attack was directed at Lieutenant Colonel Crawford's men of the 2/2nd Field Regiment. Balfe himself acted as OP for the RHA guns that night—their own BC, exhausted after a long, hot day and 400 rounds of 25-pounder, dispersed the German lorried infantry.[155]

The Easter Battle of 14 April, when Rommel launched the *5th Armoured Regiment* (38 tanks) followed by the *8th Machine Gun Battalion* with assault, anti-tank and 88-mm (General Purpose—GP) guns at Captain Balfe's D Company at dawn, is also worthy of mention. No better example can be found in the annals of the siege of the use of field and anti-tank artillery to defeat a superior armoured force. All the British field regiments fired HE concentrations on German tanks, troops and machine-gunners in the early stages of penetration at Post 38 so that the infantry of the 2/17th could kill or capture the enemy infantry when they became separated from their tanks by fire. When the Germans, whose plan envisaged a sweep north of the post, attempted to penetrate the outer line of posts, their tanks came in range of direct and indirect fire of the field artillery and anti-tank guns, fixed and mobile. Their tank numbers were reduced dramatically and the confusion caused their infantry and machine-gunners to retreat.

The Easter attacks had given the defenders experience, confidence—and a warning which they were to heed to the letter. Ultimately, however, the Australian infantry and gunners had responded in a way that Morshead believed to be the insurance necessary for the defence of the fortress. Rommel appeared to have miscalculated and underestimated the quality of his opponents and this shook the confidence of his command. This would not have been possible without a competent artillery arm, particularly the superb performance of the 25-pounder in a secondary role firing HE at Mark III panzers over open sights. The episode was also important for the fact that the '88' had made its first appearance in a field role. Historians agree that, in an encounter battle with many 'highlights', two stand out. The first concerns the shooting of the Chestnut Troop of A/E Battery, 1 RHA, which halted the tanks. The second comprises the resolute infantry of C and D companies of the 2/17th Battalion. While, over the previous three months, the RA field regiments (except 107 RHA) had come to know the 9th Division during their supporting role in the move from Benghazi (and their advance to Tobruk with the 6th Division), this Easter battle was the climax and it made a lasting impression. As historian Shelford Bidwell writes:

The notion of Australian soldiers being brave, undisciplined amateurs is quite mistaken. Australians place a Victorian emphasis on manliness and courage. They are blunt and outspoken to the point of incivility, and have no use for the answer that turneth away wrath: wrath is something they take no pains to avoid. They have no respect for artificial rank or status unbacked by performance, and invented the coarse word 'bullshit' to describe unprofitable and flat-catching activities, but they are when necessary well-drilled and smart soldiers. They are strongly aggressive and competitive. 'Sportsmanship' in the English sense they have no use for. An Australian first of all reads the book of rules carefully and then plans to win: a game lost is not a game enjoyed.

These attributes, all desirable in soldiers, were strengthened by a recent but strong martial tradition. Their discipline was strict, as in all good armies. True, it is doubtful if an Australian private ever asked leave to speak, and in their unbuttoned moments of leisure they tended to give the provost-marshal and his men trouble, but no more than half a dozen British regiments that one could name. Only strict discipline and hard, skilled training could have produced so early in 1941 infantry with the qualities the Australians displayed in Tobruk. The Australians, however much they value equality and independence despise half-hearted work or amateurishness. The Royal Artillery can, therefore, feel sober satisfaction in the fact that they earned the respect of this army of pugnacious democrats. The Australian historian speaks of the 'mutual confidences and esteem that in only two days had sprung up between these Australian infantry and the British gunner supporting them.'[156]

The Australian gunners and doubtless many infanteers had an abiding admiration for the discipline of the Tommies and Jocks who were stoic in adversity, matter of fact in their approach to orders, adaptable and humorous, and cherished a deep pride in their regimental system, however incomprehensible it may have seemed to other armies. A *laissez-faire* approach to command and ambivalence to order and authority matched the 'Jack's as good as his master' style of the Australian. Commissioning officers from the ranks in the field was a practice that astonished some of them. That said, there was an underlying motivation in that, while they respected the German soldier as a well-trained and committed troop who dearly liked to fight, the British Commonwealth forces never lost sight of the fact that they were fighting against an inherently evil regime. Thus, only one outcome was worth contemplating.

Only one RAA field regiment took part in the siege of Tobruk—the 2/12th. The importance of medium artillery in counter-bombardment was amply demonstrated at Tobruk, although its conjunction with sound ranging and flash spotting was somewhat adventitious. Initially, Morshead proposed to allot the counter-bombardment role to the 2/7th Field Regiment. His ADC, Captain J. McKeddie, reminded him that another of the division's regiments, the 2/12th, had been raised in Victoria in the AIF I Corps as the 2/2nd Medium Regiment, equipped with 4.5-inch howitzers and 60-pounders. Much to the relief of the 2/7th, the 2/12th was given the counter-bombardment role.

At the same time as these major decisions were being effected, 23 Battery (Major R.A. Millege) and 24 Battery (Captain A.C.C. Carter) were busily integrating their troops into the defensive gun areas. This was no simple task, as it was the first time their organic field artillery had supported the division. However the gunners' first impression of life in Tobruk concerned not the defensive scheme, but the quality of the tea—an atrocious-tasting brew 'made from equal parts of tea, sulphur and sand mixed with hot swamp water'. Following hot on the heels of this alarming discovery was the realisation that they were to be engaged in the seriously strenuous business of digging gun pits, command posts and 'doovers' (scrapes in the ground). Eons of time had removed the soft sand and deposited it elsewhere, leaving a base rock that defied pick and shovel. An alternative to the exhausting task of digging was to build a sangar of rocks or—even better—persuade the engineers to do the hard work with their machinery. Finding a reverse slope not too far away from the gun post helped somewhat as the shrapnel ricocheting on the stony ground was guaranteed to set the adrenaline pumping. The frequent dust storms during the warmer months that were universally loathed were welcomed by the contrary gunners. These storms shut down operations until they cleared, as observation was usually very limited.

One of the most serious constraints on the fortress artillery was a restriction on ammunition usage. Like all necessities it came by RN, Royal Australian Navy (RAN) and merchant ships, together with other stores and food, compared to which it arrived in modest proportions. In sum, around 40 tons of supplies a day were required to keep the garrison operating. Of that total, 25-pounder ammunition comprised 11 tons per day, 4.5-inch amounted to 1.75 tons and 60-pounder 1.5 tons per day.[157] Thus there was never any possibility of engaging targets with the generous rate of fire which had been the norm during the advance through Cyrenaica.

The enemy did not labour under these difficulties and directed hundreds of rounds per day at specific targets, particularly the guns, while the nights frequently saw the infantry peppered with mortar fire. The 2/12th was never out of range of the enemy guns and, to their credit, most of the gunners became quite used to the constant shellfire. Men learned from experience to discriminate between a gun report aimed at them and one firing right or left:

'Groups of men sitting around chatting and apparently ignoring the sounds of gunfire would suddenly stop talking, leaving a word suspended in mid-sentence while they carefully listened. Everyone dived for cover if the unique "coming your way" hostile bark was recognised. The man who ducked unnecessarily lost prestige; but he who dived too late risked much more.'[158] At one point C Troop, stationed north of the Derna road, was on the receiving end of a lengthy shoot—their reply consumed their entire ammunition allocation.

The OP was manned for two days, with OPOs rotated back through the gun position for their two days' respite. The OPO and OP Ack prepared a 'panorama' to record target and topographical data. In this sketch the vertical distance (angle) of the ground under observation was exaggerated, and the zone they covered marked left to right in degrees. Enemy positions were marked, as were tracks, wire, vehicle hulks and the line, range and angle of sight of previously engaged targets. Thus an OPO could call for fire from his Troop or battery within seconds of identifying a target. At the OP it was far more hazardous:

> In the OP the arrival of a mortar bomb was accompanied by a sibilant hiss as the German observer established a short 'bracket' of bombs. His target was the OPO, his signaller and his OP Assistant, who were all crouched in a connecting trench covered by sandbags supported by timber. It took a direct hit and Signaller Keith Ford said of it, 'I remember a blinding flash of flame, an enormous explosion, dust and debris everywhere. I knew my feet were gone. I couldn't feel them and I couldn't see them. At first we could hear nothing, our ears deafened by concussion; slowly sounds began to penetrate and each checked that the other was alive. The mortaring stopped so we continued observing from an open weapon pit nearby.' Ford still had his feet.[159]

Those manning the OP had to endure daily 'hates' from guns or more often, mortars, and the resilience of these men drew deserving praise from Morshead.

Gunners have recognised for years that it is better to give than receive and, as soon as practicable, captured enemy artillery and its plentiful stocks of ammunition were employed. Once E and F Troops had joined the siege, Goodwin was able to take stock of his resources (guns, ammunition, men and technical data) and implement plans for their utilisation. He had a number of practical problems to solve concerning the operation and reliability of his Italian ordnance. They included firing mechanisms and recuperators with mechanical problems and a shortage of dial sights. Goodwin decided to equip E Troop (Lieutenant Bromley) with 75-mm (Cannone da 75/27) and F Troop (Captain V.L. Young) with 105-mm guns. E Troop gunners saw more action than their brethren. Their 75-mm ammunition stocks were plentiful and the gun itself proved a useful supplement to the fortress artillery with a 14-pound shell and a maximum range of 10,240 metres. F Troop's 105-mm guns proved problematic and the Troop was switched to 100-mm howitzers. All the Italian ordnance technical data was in metres for range and 'mils' for vertical or horizontal angular measurement. Goodwin expended much technical effort in producing range tables (ballistic handbooks) for use in command posts. Thus shooting data could be sent to the OP Office for each target engaged.

All the troops were involved in engaging enemy targets of opportunity, subject to ammunition stocks, whether from their 'home' positions or sniping or roving positions 'surveyed in' so that targets could be engaged quickly and the pit then vacated. There are numerous references in the history of the 2/12th to the scale of return counter-battery fire. A Troop (Captain D.H. McDermott), manning its 149-mm howitzers, fired 13 rounds for an important calibration shoot. The enemy replied with 300 rounds, none of which caused any casualties, although one round struck the EME truck and destroyed its equipment.

Throughout June and July the scale of enemy counter-battery fire escalated, and no gun position in the salient was immune. The gunners fired at tactical targets and the Germans fired at the gunners. When that seemed to produce no diminution of fire, the Germans resorted to dive-bomber assaults on the gun line and these attacks were particularly heavy in the first three days of May. Surprisingly, there were very few casualties relative to the number of incoming rounds and bombs and the gunners noted a high incidence of faulty shells. C Troop, then located ahead of E and F Troop, reported a proportion of shells that exploded prematurely over their gun position when E and F Troops were firing. Nor was this restricted to Italian munitions. Many German impact fused shells despatched at the defenders did not explode. However 'blinds', as they were known, could be lethal when they arrived on rocky terrain. The troops called them 'mad dogs' as they skittered and bounced over the landscape in a random fashion, so much so that the men considered them just as terrifying as those that exploded. By mid-June the regiment was manning ten 4.5-inch howitzers (B and C Troops, Captains D.L. Holmes and H.P. Hamilton), four 75-mm, three 100-mm and one 149-mm from the other troops, all well forward in the salient. The howitzer, designed to demolish entrenchments with its plunging fire, was useful against heavy machine-gun positions. OPs were their principal targets. Unfortunately for the OP parties and infantry in forward positions, as one was 'neutralised' another one invariably surfaced.

In the forward posts, changeover times—'Spandau-policed' by the enemy guns—generally occurred after sunset (7.00 pm) as troops could move about freely above ground for relief, reinforcements and maintenance. In post R8, for example, relief occurred between

8.00 and 10.00 pm on alternate nights. There was a convention that each side respected the other side's 'right' to an uninterrupted changeover period. Men moved quickly through known fixed lines of fire. A burst of Spandau fire signified the end of the period. The Spandau machine-gun was mounted in a pit and was very accurate when firing on fixed lines.

Counter bombardment or counter battery has a special place in the artillery world and assumes greater importance if the enemy's guns and howitzers are of longer range, use a heavier shell, are more plentiful and are more mobile. If the enemy has equivalent flash spotting, survey and sound ranging capabilities, the counter-battery battle for artillery supremacy becomes one of guile, scientific method and tactics. The key elements of organisation comprise a survey regiment, a Counter Bombardment Staff and Office, Computing Centre and an Air Photograph Interpretation Section (APIS). These are normally sited near a corps headquarters, and commanded by a Counter Bombardment Officer (CBO). At Tobruk a suitable organisation was created at Fort Palestrino, the first incumbent a lieutenant seconded from Headquarters 4th Indian Division. He was followed by Lieutenant Colonel Bruce Klein who arrived three days after the landing of a composite sound ranging and flash spotting battery of the 4th Survey Regiment, RA (TA).

The 4th Survey Regiment was a Territorial Army unit raised in 1936 at Gateshead, Tyne and Wear, County Durham. It provided battlefield survey data (fixation and orientation), flash spotting and sound ranging Troops to locate enemy artillery. The 4th was to have a long association with Australian gunners in the Middle East. Towards the end of May 1941, HQRAA, I Australian Corps, obtained agreement from GHQ to second a detachment of sound ranging personnel with a composite battery of the 4th Survey Regiment, RA, to the Tobruk fortress for experience. The six selected were Sergeants I.L. Braid and E.F. Spreadborough, Lance Sergeant J.B. Donaldson, Bombardier P.S. Ramsford and Lance Bombardiers V.W. Hercus and D.P. Lampe. They boarded HMAS *Vendetta* for the night run to Tobruk, sharing the deck with ammunition, rations and desperately needed stores of all description. The detachment arrived on 6 June in the middle of an air raid and was collected and hospitably received by its host battery (soon to be known as No. 2 Composite Battery).[160]

A number of techniques were employed by the unit to locate enemy artillery including the use of skilled surveyors. Survey applied to topographical maps enables points on the ground such as trigonometrical stations (trig points) to be fixed by measurement of horizontal and vertical angles and distance with great accuracy. A survey troop first arrived in Tobruk in early 1941 to establish a divisional grid. This grid enabled guns in positions marked on this grid (and higher grids) to provide coordinated fire. The accuracy of the 'fix' determined the accuracy of gunfire. Survey troops established bearing pickets (BPs) in the area which provided the coordinates and height of the point indicated by the grid bearings (in degrees and minutes) to visible, identifiable features.

Flash spotting involved three or more OPs across a wide front whose positions were fixed to a matter of fractions of a yard (or metre). OPs were manned day and night and equipped with a binocular range-finding theodolite. Gun flashes (bright at night, less so by day) from enemy batteries became the source of an angular measurement from the OPs. The angles were reported to a Troop headquarters where they were drawn from each OP on a gridded plane table. The point at which the lines intersected provided the location of the gun (Troop or battery). An error of one yard in fixing each OP could mean a 30-yard error in fixation.

Sound ranging used the physics of sound waves and their frequency to locate (fix) the source of the sound. Guns and howitzers create sound waves of different frequencies when fired. Sound ranging troops laid out five or six sensitive microphones which formed a base. Each microphone was accurately surveyed and was connected by wire to an advanced post (AP). An observer in the AP 'operated' (activated) the microphone when he saw the gun fire. The microphones were connected to a pen recorder which recorded the distinctive sound wave from the gun or guns the moment it passed over the microphones. The outline of a wave form was recorded (traced) on special paper. As the sound wave passed over each microphone at a different time interval, the recorder produced a wave form. Guns, howitzers and other battlefield noises possess tell-tale wave forms that can be traced on the paper. An experienced sound ranger can read much into these traces. The data collected by flash spotters and sound rangers was collated by the CBO and his small staff. Together with other data from OPs, aerial photographs, patrol reports, prisoner interrogations, captured documents etc., the CBO can produce an accurate, timely assessment of the number, calibre and type of artillery confronting friendly troops.

The survey regiment's detachment included an attached RAF meteorologist to provide the gun regiments the latest information for predicted fire targets. All data were used to compile counter-bombardment task tables and hostile battery programs. Artillery units provided the bulk of battlefield intelligence as they had the most comprehensive communications network. The CBO was located with the Troops of the Composite Battery of the Survey Regiment at Fort Palestrino.

General Morshead had an augmented infantry division and all branches of artillery, engineers and other resources that appeared generous at first sight. However, his assessment of 'relative strengths' was based on a critical estimate of the strength and efficiency of the

PAA's artillery. In numerical terms the fortress was heavily outgunned (three to two) in field and medium—Morshead, in fact, had only eight medium guns. The RA/RHA field units behind the perimeter were armed with 25-pounders, except for the 51st Field Regiment which was initially equipped with a battery of 4.5-inch howitzers and another of 18-pounders, MK V. The 4.5-inch howitzers were designated for static fire tasks only and the 51st was not permitted to take on a support role until equipped with 25-pounders.[161]

Each sector had its own counter-bombardment artillery group commander. Within these large areas, gun regiments had fixed, roving and sniping positions surveyed in. The 25-pounder regiments also filled an anti-tank role in the defence plan. While their primary role was supporting the infantry brigades, with Brigadier Thompson's arrival, counter-bombardment considerations were soon added to the gunners' agenda. Thompson formulated an 'active' counter-bombardment policy which involved the collection and collation of hostile batteries and mortars; the coordination of resources; the submission of a counter-bombardment intelligence report; and the compilation of morning and afternoon shelling summaries.

Initially, Thompson's CBO was Major Fryett, RA, followed by Lieutenant Colonel Bruce Klein, RAA (who arrived with the Australian surveyors) and later, as the siege continued, Lieutenant Colonel S.T.W. Goodwin of the 2/12th. In order to improve the effectiveness of the fortress artillery, which amounted to that of almost two divisions, A Flash Spotting Troop and R Sound Ranging Troop from the 4th Survey Regiment, RA, under the command of Major L. Kellett, were ordered to Tobruk and embarked on HMAS *Waterhen* at Alexandria on 2 June. They were accompanied by their RAF Meteorologist, Flight Lieutenant Bradshaw, and his section.[162] They became Klein's eyes and ears and his means of energetically implementing a comprehensive counter-bombardment reporting system in which high quality collation and analysis could be achieved. By 9 June his system was fully operational and capable of providing the 1st RHA with a hostile battery list showing locations and characteristics of his adversaries' artillery.[163]

Prior to this Kellett, who was slightly wounded during an inspection of the perimeter, had decided that the sound ranging base would be deployed facing west and A Troop's flash spotting base oriented south/south-east. Enemy fire came mainly from these sectors and was influenced by the topography. South-west of the fortress were deep hollows which rendered flash spotting nigh on impossible, while the terrain to the south/east favoured it. Kellett is credited with arranging for three tall towers to be built for the flash spotters by the Royal Engineers who constructed them from tubular steel scaffolding 50 to 100 feet high. Heavy timbers formed a platform at the top and this was surrounded with sandbags. Observers climbed to the platforms by rope ladder, sometimes screened with hessian so that enemy

OPs could not detect people climbing or descending during daylight hours.[164] From the outset, the Counter-Bombardment Office and the 2/12th cooperated closely and continued to do so until October 1941 when the division was relieved.[165]

Kellett's specialists quickly produced excellent location results. Communications between A and R Troops and the Counter-Bombardment Office and the two dedicated counter-bombardment fire units, R Troop, 1st RHA, and the 2/12th Field Regiment's 60-pounders of Captain M. Feitel's D Troop were given top priority. Captain Oliver of the 2/12th was Liaison Officer, an arrangement which enabled very close coordination to be achieved in neutralising enemy guns, particularly 'Bardia Bill', which shelled the harbour.

Nonetheless, Tobruk was evolving into an artillery battle and counter bombardment became a preoccupation at HQ RA. The 60-pounders' range of 16,000 yards made them the most effective counter-bombardment weapon in the fortress in terms of combatting the harassing fire of 'Bardia Bill'. This large-calibre enemy gun's priority target was Tobruk harbour and, when it fired, the 60-pounders retaliated immediately. The enemy gun proved an elusive target and Feitel's guns adopted roving positions in a 'cat and mouse' battle of counter-bombardment chess. Within seconds of the flash of a gun it was possible to retaliate and quieten it, allowing the unloading of vital stores in the harbour to continue uninterrupted. Thompson frequently cursed Bardia Bill as 'a considerable nuisance'. The unit's historian records that 'the big guns moved from square to square, the 60-pounders seeking but never achieving dominance. With a range of 14,600 m, it competed with German 150 and 155-mm guns with ranges to 18,500 m. By sneaking forward to the wire and operating as sniping guns, D Troop gunners compensated for the range differential.'[166]

But it was not long before German counter-measures began to dint the Australia gunners' successes. The first German innovation was the change to a flashless propellant. This initially caused some concern but, through sheer diligence, the flash spotters continued to locate the guns. The sound rangers also had their problems. The enemy introduced the practice of firing with split-second delays between detonations. Combined with adverse meteorological conditions, this made interpretation of the recorders' film extremely difficult. However, Major Fewkes, a physicist who had arrived in August with his CO (Lieutenant Colonel J. Whetton), arranged for the compilation of a log of every gun report which a team of his experts then analysed and collated to locate the offending weapons. The Geordie sound rangers had developed a 'sound ranging comparator', a device which allowed own guns to be ranged by their bursting shells on hostile batteries. Details of its construction were readily provided and the Australians returned to their unit very much more *au fait* with their craft.[167]

Bardia Bill, was the name given to an Italian 150 mm Heavy gun that was first used against the 6th division at Tobruk during its attack, and again during 9th Division's siege. Its shells were much respected by the troops.
(AWM 040453)

The Regimental Survey Officer of the 2/12th, Lieutenant J.O.F. Wittus, organised three flash spotting posts to augment those of the 4th Survey Regiment. Such was the intensity of the enemy artillery fire that the survey section became well practised at determining bearings to the guns and also 'flash to bang' calculations to give their approximate range. It was only possible for the flash spotters to operate at night because their 'rather large' binocular theodolites were easily visible to the enemy.

The extent of haze and dust over the fortress determined shooting activity by both sides. Best visibility and peak times for engagement were from 7.00 to 9.30 am and from 5.30 to 8.15 pm.[168] These were the optimum times for observed shooting, although there were a few 'bombards' (specific calls for fire on enemy guns) by observed fire. Gun regiments generally used predicted fire based on meteor data from the 4th Survey. As enemy activity usually comprised harassing fire and was almost continuous during the night, this placed great strain on all concerned. There was no specific RAF allocation due to lack of suitable aircraft for counter bombardment, but occasional artillery reconnaissance and air photos were provided. However, in September, arrangements were made with RAF Middle East Headquarters to use Hurricane fighter-reconnaissance aircraft in an artillery reconnaissance role. In due course, a lone fighter arrived at Tobruk, piloted by Flight Lieutenant Geoff Morely-Mower, RAF, of 451 Squadron, RAAF. Provision of the fighter had been made possible by the construction of camouflaged underground hangars to conceal the presence of aerial reconnaissance from the *Luftwaffe* whose nearest airstrip was at Sidi Rezegh, a short distance away. Morshead's air force adviser was an Australian serving in the RAF, Wing Commander Eric 'Digger' Black. Morely-Mower met Morshead at his headquarters and was briefed thus: 'You're here because I can't see far enough beyond the perimeter, particularly where there is a depression ... where a considerable force of tanks could build up. I'd like you to do some counter battery stuff ... and give them some good old-fashioned revenge.' Morely-Mower was warmly welcomed by the gunners, delighted to have this added capability, and who had not seen a friendly aircraft literally for months. He and another pilot conducted air shoots against German and Italian guns until their withdrawal, noting that the '88s' were easy to spot because of their 'castle-like sandbag emplacements.'[169]

The strain of the siege on the troops was acute. Flash spotters and sound rangers changed roles from time to time and the calibration of worn guns was a constant task. The sound ranging base worked best with a five-microphone 1500 metre sub-base (which ran north-west to south-east) and prompted the Germans to move their guns some 2000 metres back from their original positions. The sound ranging troop successfully ranged naval artillery on troublesome targets in the Derna road area from the two RN gunboats HMS *Gnat* and *Ladybird* which fired from Tobruk harbour from time to time. The locaters' problems generally involved adverse meteorological conditions (high winds, dust storms etc.), ground waves, an echo effect, sand in the microphones and a shortage of grids and film. On their best day, they found 15 locations. Lieutenant Colonel Klein reported that 100 hostile batteries, 40 pole OPs and 20 mortars had been located in the six-week period from 1 July to 15 August. He concluded that the best counter-bombardment policy was an active one.

As for the flash spotters atop their tall, spidery towers, the unit historians recorded that 'morale was always high'. Major Kellett, with his A Flash Spotting Troop and R Sound Ranging Troop, was the catalyst for a subsequent change in regimental establishment which occurred in September. Two composite batteries now comprised each regiment, designated Nos. 1 and 2, with the original Survey Section of RHQ retained in its original position.[170]

The 2/1st Survey Regiment detachment occupied positions in a disused concrete water tank without a roof on the western side of the fortress and shared the same tough existence as

60 to 80 feet high Flash Spotting Towers were a feature of 8th Army salient at Tobruk and El Alamein. Observers were protected by sandbag breastworks on platforms atop their lofty towers. In good visibility observers could see 20 kilometres. They were of great help collecting timely battlefield intelligence.
(IWM E 15621)

the remainder of Morshead's garrison. Six weeks later they boarded HMS *Decoy* and returned to their unit having gained much valuable experience. During their time with the garrison they had manned the forward posts, operated and maintained the equipment and worked in the plotting centre. They shared all the hazards of constant shelling and air raids with their RA colleagues and returned unscathed and with much locating savvy.[171]

Before the sound rangers and flash spotters swung into action, the General Staff and Brigadier Thompson's BMRA sought a novel solution to counter-bombardment fire from an Italian battery on Carrier Hill, south of the Derna road. Given the shortage of ammunition the decision was made to take offensive action and send a tank/infantry/anti-tank force to assault the battery position with field gun assistance from the 51st Regiment. The sally from Post R5 would be led by Captain W. Forbes, 2/48th Battalion, while two other raids on either flank by the 2/23rd and 2/17th battalions would be mounted simultaneously. As Forbes' force approached Carrier Hill with three tanks from the 7th RTR grinding away, RAF Lysander army cooperation aircraft flew over the area to provide observation and background noise. Notwithstanding being observed and fired upon, within an hour the force had taken Carrier Hill. Not every detail of the operation went according to plan, but Forbes directed his carriers and infantry with consummate skill and surprised the battery, which had directed its attention at the tanks. Despite moderate opposition, the Italians were overcome and 368 prisoners brought back to Tobruk. The raid yielded four anti-tank guns, motor transport, gun sights (much in demand for use by the Bush Artillery), maps etc. It was probably the most audacious episode in counter-bombardment history to date. The other raids by the 2/23rd and 2/17th were less successful for a variety of reasons.[172]

In August, the intensity of German artillery fire increased, with Shelreps (reports of shelling/mortaring provided in a standard format from an observer) reporting 200 rounds falling on a troop position as 'somewhat normal'. On 15 August, 982 rounds fell on the fortress. Aerial photos showed a build-up of ammunition dumps and gun positions which heralded the pursuit of a more active policy by the enemy. By mid-July, the 24th Battery diarist noted that his battery had fired over 33,838 rounds. Captain Feitel's troop had fired some 1970 shells, and 5680 rounds had been despatched from the much-maligned Italian 100-mm guns. Klein noted that the enemy counter-bombardment artillery was haphazard and primarily restricted to area targets rather than precision shoots. It changed positions frequently, used dummy flash munitions and sometimes elaborate concealment, although it was usually readily spotted. Enemy guns were seldom engaged by observed bombards. Subterfuge is an essential part of the counter-bombardment art and it is worth recording some of the unique methods of the Tobruk siege.

To blind enemy observers and screen gun positions (of whatever kind) the gunners arranged for trucks to be driven in front of the guns to raise a dust cloud through which the guns fired. They also replied in kind by firing different calibres simultaneously. They destroyed landmarks (buildings at El Adem road junction, a fort at Palestrino) and telegraph poles, and sited dummy tanks near the Derna road. The Counter-Bombardment Office discovered that enemy reconnaissance aircraft used four frequencies, two each for speech as well as Morse Code. The Australian signallers enjoyed jamming enemy transmissions in both modes. Against Morse they tapped out a continuous stream of 'V for Victory' or, alternately, some well-chosen phrases in plain language.[173]

The German artillery and *Luftwaffe* cooperated in shelling and bombing the harbour area simultaneously, apparently one of the few policy areas in which they agreed. The defenders believed an active counter-bombardment policy the best form of defence and the gunners had the RN and RAN ships and organisation to thank for having the ammunition to pursue it. The infantry and artillery behind the perimeter were due for more of the same, thanks to Adolf Hitler. The Australians had to endure another dreadful month of shelling before their relief in September.

According to some historians the British occupation of Tobruk and its signal success against the first full-scale attack sent Hitler into a mental frenzy only replicated when Stalingrad defied its capture later in the war. He ordered a siege train comprising five artillery units to be formed to crush the defenders under command of *Generalmajor* Karl Boettcher (*Artillerie Kommandeur 104* or Arko 104) consisting of:

- 9 x 210-mm howitzers (German)
- 36 x 105-mm howitzers (Italian)
- 12 x 150-mm guns (French)
- 84 x 149-mm guns (Italian)
- 12 x 120-mm naval guns (Italian)

Arko 104 was grouped under command of the Italian *XXI Corps* (*Bologna, Trento, Parma, Brescia* and *Savoia Divisions*) and fully operational by August.[174]

Outside the wire, in the 5000 metres separating Posts R7 to S7, the Germans sited more than 200 machine-guns and 72 anti-tank guns—a ratio of 3:1 for every 150 metres. While this may be interpreted as a compliment to Morshead's warrior spirit, it subjected the garrison to even heavier fire. Whereas in July the daily average of rounds arriving was 650, of which 450 fell in the 2/12th sector (Goodwin's command), it peaked at 1500 on one nightmarish day in August. This prompted a demand for more and heavier Bush Artillery to be pressed into service. Goodwin formed a detachment from RHQ to man two 149-mm howitzers which they fired from a safe distance using a long lanyard. He juggled his gun 'mix' to maintain pressure on the enemy and became involved in the set-piece attack on Post R7 in August which every regiment was tasked to support. Goodwin was in charge of counter-bombardment fire for this important assault and used nine 4.5-inch howitzers, eight 75-mm, three 100-mm, four 105-mm, two 149-mm and two 60-pounders.[175] The unreliability of

Italian ordnance proved a sore trial, as did the diminishing amount of ammunition to serve the guns. Some of the 75-mm guns were rationed severely as only 4000 rounds remained in the fortress. As September approached, enemy counter-bombardment fire continued unabated with daily tallies of more than 500 to a high of 982 rounds. The 100-mm guns of the 2/12th Battalion could only manage 340 rounds at best. Ammunition expenditures by the regiment by calibre were not recorded, but some idea of usage can be gained from the fact that, by mid-July, 24 Battery had fired 33,838 rounds comprising:

    E Troop 75-mm: 16,615
    F Troop 105-mm: 9573
    D Troop 60-pounder: 1970
    D Troop 100-mm: 5680

The final (approximate) figure was 56,000 of which the 75-mm guns accounted for 27,500. The total estimate was as high as 180,000 rounds, to which must be added the Highland Division artillery (X, Y and Z Troops) total of 40,000. A Troop alone despatched 28,156 rounds.[176]

Artillery support for the infantry was always an element of planned operations such as the 2/28th Battalion attack on Ras el Madauuar in early August. The attack was unsuccessful and a disappointment for the garrison. The root cause was the late arrival of the infantry at the Start Line and the loss of concentrated fire. Mention has also been made of the inability of the fortress artillery to bombard enemy artillery positions because they lacked numerical superiority. The planning 'rule of thumb' was a minimum ratio of 5:1, although a more even ratio was preferable. About half the artillery ranged against the fortress was Italian, which Thompson decided to target rather than the Germans, so as to induce a state of nervousness in the fragile Italian troops.

Life in the fortress was spartan and a severe test of character, adaptability and motivation. Nor was it for the faint-hearted, for survival depended on two main factors—the RN maintaining a lifeline to Alexandria and the garrison's supply organisation, and discipline. Under Colonel Goodwin's leadership and his common sense approach, the regiment became a cohesive force, notwithstanding its dispersion in each of the artillery gun areas to meet the tactical exigencies of the time. The lack of organic artillery transport and the existence of other 'establishment' necessities which meant that many NCOs and gunners experienced a change of occupation also challenged the regiment's cohesiveness.

Living conditions in the garrison were extremely difficult. The old axiom that 'an army marches on its stomach' is, in fact, a military truism. Water was severely rationed, initially to one quarter of a gallon (1.25 litres) every two days. Later, in the autumn, the ration was doubled, much to the delight of the troops. From this allocation each man shaved every second day and washed himself in his shaving water employing what was commonly referred to as an APC (armpit and crutch). The remaining water he drank. Battery and troop cooks had four gallons of water for every 30 men. The garrison's main staple was Tobruk bread, weevils and all; it was doughy and became tougher by the day. The more inventive developed a method of expelling the weevils using a primus stove. English wafer biscuits offered a bread substitute, but were also full of weevils. Meat came as tinned meat and vegetables (M and V), as did tinned herrings, called 'goldfish' after the colour of the sauce and skin. Rice three times a day substituted for potatoes. Food (and the dead) brought rats that scavenged boldly. The distribution of an orange at one mess parade evoked loud cheers.

Personal hygiene was difficult in the summer months when small abrasions and cuts festered and became ulcerous, often lasting for months despite daily treatment with Gentian Violet antiseptic. Some men were reduced to wearing a 'lap lap' made from mosquito netting to seek relief from lesions in their crotch. At irregular times troops were transported to the beach to bathe, a much-anticipated event. While beach baths sometimes coincided with *Luftwaffe* sorties, bathing was still described as 'pure heaven'.

Sleep, or lack of it, was something each individual adjusted to as the sounds of war were unrelenting and never far away, except during a *khamsin*. Men slept in their 'doovers', sangars or underground. Either way, fleas, scorpions, rats and snakes were constant companions. However, the worst pestilence was the flies, which increased to epidemic proportions in the summer months. They swarmed anywhere and everywhere and, for most men, eating food without ingesting a random quota of flies was nothing short of miraculous. Given the primitive toileting arrangements, their continued multiplication was assured. All these blights, singly or severally, drained the robust health of the men who arrived from the Delta in good condition. Falling shrapnel from constant anti-aircraft shells was dangerous and wearing a steel helmet at all times was mandatory. Driving a truck raised dust which swiftly drew enemy artillery fire. Italian aircraft dropped an array of 'attractive' devices, designed to look like money boxes, thermos flasks etc, as booby traps. Hazards were omnipresent at the gun position.

The spiritual wellbeing of the 2/12th was entrusted to Padre Wilfred 'PK' Collins who conducted services seven days a week and was a popular figure in the unit. The regiment's casualties for the siege totalled 25 killed and 31 wounded. On 19 September the 2/12th was relieved by the 144th Field Regiment (Surrey and Sussex Yeomanry), RA. The 144th's commander, Lieutenant Colonel H.T.M. Clements, described the equipment handed to his men as 'the most extraordinary collection of junk with which any British regiment ever went into battle'. He had every right to be indignant; he had left a well-equipped 25-pounder regiment in the Delta. Other 2/12th guns went to the Polish 1st Carpathian Regiment.[177] When the 2/12th reached El Amiriya in Egypt, the gunners progressed from rags to riches, taking over almost-new 25-pounders, 36 tractors and 114 assorted vehicles.

Morshead's strength in field artillery was adequate, although the same could not be said for his medium artillery; nonetheless he was well and truly outgunned by Rommel. While the Germans could concentrate a superior force at any time or place and use their medium and heavy artillery brigades for maximum effect, Morshead and Thompson had half a dozen pieces which, in any reasonably based defensive fire plan, would have been barely sufficient to engage one hostile battery, let alone as many as Rommel had available. Rommel's disadvantage was having to divide his force between engaging the fortress and pressing on towards the Delta and, in so doing, lengthening his logistical tail, at least until he could capture Tobruk, the only decent port. Bidwell notes that the 'chief interest from an artillery point of view in the siege lies in the efficiency of all the arrangements, the confidence and respect that developed between the Australians and the Royal Artillery, and also the first signs of the ascendancy over the battlefield of the 25-pounder gun.'[178]

The gunners laboured under enormous difficulties. There were insufficient wireless sets (the artillery was issued 17 instead of 60) and these were unreliable and thus seldom used. Line laid above ground was frequently cut by bomb and shell splinters and spent projectiles of various sizes. Ammunition was short, with the exception of that captured and returned in kind. This affected all field regiments, although offset by Brigadier Thompson's dispositions. From any of the three gun areas he could concentrate all his artillery on any threatened break-in point. As it was, British OPs were frequently deployed outside the wire in offensive mode with Australian infantry protection. Enemy parties reconnoitring the defences were engaged at the first opportunity.

Another disadvantage Morshead suffered, and one he could do little about, was air support and reconnaissance. The absence of air support puzzled the fortress staff who noted that the *Luftwaffe*, which raided Tobruk 800 times between April and November, made only 100 attacks on Egyptian airfields in the same period. Morshead did not have the same quality intelligence as Rommel received from his special wireless intelligence company (*U621*) and from aerial observation or photography. This was somewhat alleviated by the alert OPOs scanning their front continually and by deep patrolling by the infantry. These patrols drew praise and admiration from the British gunners, as they usually involved night sorties and sometimes lasted 24 hours. During daylight the men would 'lie up' and gather data before returning. It required a special kind of courage compared to that faced by an infanteer in attack or defence. Lieutenant Colonel Seely, CO 107th RHA, wrote to Brigadier Thompson:

> May I ask you to place on record the enormous extent of deep patrolling by the infantry ... which has proved to be so invaluable. Without ArtyR and air photos we were set an apparently hopeless prospect of correctly deducing the enemy dispositions ... beyond our zone of observation. It was only through the fearless and meticulous investigation of terrain out of view that we have a clear knowledge of his defences and organisation. It is through this that we gunners have been able to strike deeply and accurately in depth.[179]

The Germans, on the other hand, had Henschel observation and other aircraft that flew over regularly, just outside LAA gun range, either directing fire or taking photographs. Not all returned to their base. However much the enemy believed they dominated the tactical skies above the fortress, it is evident that the aggressive counter-bombardment policy based on the locating assets exerted an influence well beyond its numbers. This was predicated on the expertise nurtured since World War I and also confirmed the judgement of CIGS General Alan Brooke not to disband the survey regiments. They proved invaluable on this static front, despite the fact that no-one on any General Staff could have predicted that static fronts would recur. This was the high point in the war for Thompson and his staff, including Klein and the 2/1st Survey Regiment's detachment.

How well did Morshead's artillery adhere to the well-known tenets taught at Staff Colleges and enshrined in Artillery Training Pamphlets (Vol. III) and *Field Service Regulations*, that 'primer' for officers of all corps and services? Concentration is one of the more obvious principles, for this uses the zone (dispersion of fall of shot) of the gun to cover an area. It also connotes the simultaneous application of fire which produces a 'shock' effect. Within the fortress, communications were such that this could be almost guaranteed. Economy of force was allied to concentration and mainly served 'means to end' criteria in addressing a desired result—neutralisation or destruction. Surprise is yet another aspect of offensive action that enables a force's main artillery strength to be marshalled for decisive blows. By using predicted fire, concealment and deception, gunners achieve results which very often defy the relative numerical strengths of guns. Bardia and Tobruk were good examples of this principle.

Cooperation was the keystone at Tobruk. Artillery can achieve little by itself, and even less without the confidence of the arms it is supporting. As events unfolded after Tobruk had been lost to the Germans in June 1942, artillery command and control in the desert war drifted from its principles and the outcomes then became a foregone conclusion.[180] During the siege, one of the major lessons learned was confidence in weapons. The 25-pounders had contributed their share, albeit a small one, to the destruction of 39 German tanks with anti-tank artillery and minefields, compared with seven by the fortress's small tank force. However, their ability to stop armoured thrusts was never precisely determined during the siege and commanders at all levels were reminded that misplaced confidence in them would be hazardous.

It is appropriate to include, albeit briefly, the influence of anti-tank gunnery on the 9th Division's occupancy of Tobruk. The 2/3rd Anti-Tank Regiment remained at Mechili with some elements caught by Rommel's speedy advance. Their role was to support the infantry with direct fire aimed at approaching tanks. Had this regiment and anti-tank companies been armed with a superior weapon, the whole of the Middle East war may have assumed an entirely different complexion. They did as well as they could during the withdrawal to Tobruk and, once inside the wire, acquitted themselves in the best traditions of the regiment. Thompson rated the Australian anti-tank gunners' staunchness the equivalent of the British regulars of J and H batteries of the 3rd RHA. Anti-tank artillery was ineffectual in night operations, although it accounted for approximately 15 to 20 armoured vehicles and numerous personnel for the loss of seven killed, 32 wounded and 20 captured. The 24th Anti-Tank Company suffered 43 casualties and the 20th nine.[181]

An account of Operation Battleaxe (15—17 June 1941) has quite deliberately been excluded from this chapter. Morshead was never enthusiastic about the operation and, while much detailed staff work was devoted to plans and contingencies, the battle concluded in two days. The inevitable tank versus tank engagements left the British armour shattered, while the Germans were able to recover their tank casualties quickly. Once again, the flaws in British doctrine for armour-infantry cooperation were patently obvious. No Australian artillery was engaged.

If Morshead was disappointed in any aspect of the siege, it was that Rommel had established a salient with good observation within his perimeter at Ras el Madauuar from early May. In the early stages of the break-in, Colonel Lloyd had great difficulty defining the Germans' gains from his infantry companies, battalions and brigades due to the severing of line communications. On many occasions, however, artillery observers' messages provided situation reports which helped commanders make tactical dispositions with greater confidence.[182]

Whenever he could spare the time, Morshead visited his infantry and other units, aware that his appearance and presence were important in maintaining morale. But it was his direction of the battle that was important, given his brief in April from Wavell and Lavarack. By aggressive action, such as deep patrolling, the 2/12th's 60-pounders taking on 'Bardia Bill' or set-piece attacks, he kept Rommel on edge and bought time for Wavell to bolster the defence of Egypt. He also effected the historic first defeat of German armoured forces in the war. He welded the Australian, British and Indian troops under his command into a superb force in which cooperation was built into every action, planned or impromptu. His achievement should be considered in terms of the disadvantages he faced, particularly in lacking his establishment of weapons, adequate supplies of all kinds and fully trained troops when ordered forward to Benghazi and later in Tobruk.

Morshead made great demands on his troops who maintained their aggressiveness in contact with the enemy on a reduced ration scale in a hostile climate for five months (April–September) without respite while suffering air raids of a frequency that would test the stoutest heart. Admittedly, he was lucky that his infantry had not been surrounded during the hasty withdrawal from Benghazi. By any measure Morshead was the best commanding general in the Western Desert and he was to achieve even greater successes in a year's time. He was awarded a KBE in April along with Blamey and Lavarack, while Captain Maurice Feitel of D Troop, 2/12th Regiment, received an MBE for 'his outstanding service, energy and determination and high morale under fire'. For his part, Lieutenant Colonel Goodwin received plaudits from Brigadier Thompson and Lieutenant Colonel A.G. Matthew of the 104th RHA who wrote:

> My dear Goodwin, we watched you amidst a maze of difficulties and with mixed equipment hurrying to our assistance as a very much needed reinforcement. I shall always feel that your arrival was a turning point in the defence of this place.[183]

As is well known, reports of the heroism of Australian infantry were seldom out of the Australian media, thanks to the detailed reporting of accredited war correspondents. At Tobruk, Chester Wilmot made his reputation and wrote his account of siege operations. It was not his fault that it came to be regarded as a wholly Australian show. However, Morshead's infantry was sufficiently astute to realise that credit was also due to the British artillery. The South Australian battalion, the 2/48th, produced a news-sheet in Tobruk which noted, 'We must not forget that the greater part of the praise for what has been done here ... is due to the British artillery regiments. They have done a magnificent job. We would like to express our most sincere and heartfelt thanks for what the British gunners have done for the Garrison.'[184] Perhaps the best epitaph for the 2/12th Field Regiment was penned by war correspondent John Hetherington, who wrote in 1942:

> I wonder if any man who served in Tobruk, would, if given his choice, turn the clock back and will himself out the campaign? I don't think so. Take 2/12th, one unit among many, one part of a hard, bright weapon that broke every effort by Rommel to lay hands on Tobruk ... No artillery regiment could have had a more versatile role ... British soldiers, including Australians, always need all their ingenuity in the early phases of war. They have to make shift with odds and ends of equipment that would cause less robust spirits to give up the ghost. That was the way of 2/12th![185]

The other two divisional artillery regiments had very different campaigns until they joined the 2/12th en route to Syria in October. Their vastly different experiences form the subject of the next chapter.

With the relief of the Australian 9th Division in September and October 1941, Tobruk was held by South African and British troops under Fortress Commander Major General Klopper, a South African. The port fell to Rommel's PAA on 21 June 1942 and the reasons for its capitulation have been the subject of much debate. One reason offered was 'that Klopper was no Morshead'. Certainly the fall of the garrison was, in many ways, a direct function of Klopper's deployment of artillery and its control, in that the guns were sited too far forward and could not concentrate their firepower; the CRA lacked a unified command and communications system; there was excessive ammunition at the gun positions; and a lack of coordination of field artillery and anti-tank fields of fire.[186] This is some measure of the influence of artillery on the outcome of a battle.

Chapter 9
The 9th Division artillery (2/7 and 2/8 Field Regiments) in Egypt–1941

Go where glory waits thee,
But while fame elates thee,
Oh! Still, remember me.

Thomas Moore

The role of the 2/7th and 2/8th field regiments, the remaining two elements of the 9th Division's artillery, was entirely dependent on equipment availability, in particular 25-pounders. While the 2/12th had been sent to Tobruk on the strength of its familiarity with medium guns, both the 2/7th (Lieutenant Colonel Tom Eastick) and 2/8th (Lieutenant Colonel Walter Tinsley) had to wait for their moment of glory. The tactical situation confronting GHQ was, simply put, that Australia's 9th Division in Tobruk was making it difficult for the Germans to ignore the garrison. Rommel's options were more plentiful than Morshead's but, in a move typical of him, he decided to push east where he met British forces and ran the gauntlet of Operation Battleaxe, launched from 15 to 17 June. In the aftermath of the operation, the British established a line that stretched 25 miles from North Point (Sofafi) to Sidi Omar, west of the escarpment, and then ran in a crescent shape from Sofafi to Sollum. The 22nd Guards Brigade and the 4th Indian Division occupied the escarpment along with South African armoured cars and the 7th Armoured Division's Support Group. The rear of the British defence was based on Sidi Barrani, 30 miles further east from Halfaya Pass.

Operation Battleaxe was one of the first operations designed to confront the German and Italian forces that had bypassed Tobruk and were now heading east. Wavell had been kept occupied with planning issues since April and, by June, which marked the start of his Syrian campaign, he was ready for Battleaxe. The operation itself was planned by Beresford-Pierse, now commanding XIII Corps. Between the 7th Armoured Division and the 4th Indian Division, there were some 180 cruiser and I tanks and 25,000 soldiers, the latter belonging to Beresford-Pierse's former command. The divisions were commanded by Major Generals Creagh and Messervy respectively. The Germans had constructed strong defences in the Capuzzo-Halfaya area with the coast as their eastern flank, although British intelligence was unable to estimate the exact enemy strength. Beresford-Pierse planned a frontal attack on the Halfaya position with armour making a left hook towards Capuzzo. It was an orthodox plan of attack by an orthodox person who, up to this time, had never commanded armour. The wild card was the possibility of German reinforcements from Tobruk, some 80 miles away. The British had few, if any reserves on which to call and risked critically weakening the Ninth Army's Syrian garrison with any diversion of its troops.

One extraordinary aspect of Beresford-Pierse's plan was the complete absence of artillery preparatory or supporting fire. With the benefit of hindsight, this seems incomprehensible for a gunner and, unsurprisingly, his frontal attack proved a dismal failure. This was due principally to the effectiveness of newly arrived German PAK 50-mm anti-tank guns and several 88-mm GP guns employed in the same role which severely mauled the supporting armour. Beresford-Pierse's flanking formations met with more success until they were stopped near Sollum having taken Capuzzo. He then made a serious error of judgement in his use of armour and, as a result, the 4th Armoured Brigade lost 98 of its 104 tanks. Armoured tactics in support of the attacking infantry also proved flawed. Rommel timed his counter-attack perfectly and employed superior armour and artillery in his assault. The forward field regiments attempted to halt the enemy advance using concentrations of fire. This merely deflected the direction of the *schwerpunkt* ('concentration of force') and failed to halt the advance.[187] It was expensive in ammunition and comprised a belated attempt to use firepower to influence the battle.

By this time the Germans had refined the type of artillery support that could be provided by their 88-mm guns. Their victory was yet another salutary lesson on the use of guns against tanks and infantry that went unheeded by the British. During Operation Battleaxe, the *21st Panzer Division* had deployed 13 tanks in three different positions: five at Halfaya and four each at Hafid Ridge alongside the *8th Panzer Regiment*. The 88-mm guns were to be used for the immediate support of the motorised infantry. When employed close to the panzers, the guns would provide harassing fire in conjunction with the motorised support group.[188]

On 17 June both British divisions returned to their former positions to lick their wounds. Historian Shelford Bidwell likens Battleaxe to the 17 April attack by the *5th Panzer Division* on the 9th Division at Tobruk, albeit in reverse. The British could have drawn some important tactical lessons from their successes at Sidi Barrani and Tobruk and applied these to later campaigns but, for various reasons, they failed to do so. CCRA XIII Corps, Brigadier Siggs, was one of two experienced gunners at the helm and yet the artillery was dispersed. In addition, by placing all his armour on his left, Beresford-Pierse implied that his tanks could and would better the German panzers.

A post-battle critique postulated that artillery concentrations did not of themselves stop or destroy tanks but that, if they were in a defile, they had a better chance of doing so. If, on the other hand, the leading panzer group was being engaged by anti-tank guns, then HE concentrations were the answer. In launching tanks on his left flank as a key part of his plan, Beresford-Pierse and his armoured advisers were applying, disastrously as it eventuated, the lessons of I tanks at Arras in 1940 in mastering the German 37-mm anti-tank guns and securing a convincing tactical win without being crowded by field artillery.

For his part, Rommel both learned and applied some tactical lessons. He established strong positions of well-stocked battalion groups south-west from Halfaya to Sidi Omar to protect his right flank and reorganised his forces into two panzer divisions. He also learned the artillery lesson of using his guns to support tanks in attack and defence, and he directed his commanders to be as mobile as their tanks. In future, British wireless communications were to be monitored so that the Germans could 'read the battle' and react appropriately.[189]

Battleaxe was also noted for the emergence of the Armoured Support Group (lorried infantry with field, anti-tank and LAA assets) as an independent mobile force. In effect, both infantry and tanks were denied concentrated artillery fire to beat the anti-tank gun. During the German counter-attack the RA lost its first battery (C Battery, 1st RHA) to the panzers. The battery had no armour-piercing shot, only HE. The panzers' method was to concentrate their fire on one gun in turn, using machine-guns and self-propelled 150-mm howitzers and field guns. In this way they serially demolished each Troop (or battery). Another episode, concerning the 31st Field Regiment, RA, evinced a somewhat contradictory lesson. The 31st was at Capuzzo heading west when a 'tank alert' signal compelled one battery of the regiment, equipped with towed 25-pounders, to drop trails and prepare to engage the Panzer Mk III and IV tanks. This was the type of action that would became commonplace in the desert as formations reacted to Rommel's thrusts, were dispersed or surrounded and destroyed. One of the detachment commanders later wrote of his experience:

> Out came the ammunition boxes from the limber (HE—we had no armour-piercing) and we quickly filled ammo boxes with sand and built a wall under the gun shield. The quads went off to the rear. Soon we could see something moving in the heat haze, but our drill was very strict; all the detachment laid down and kept still. The layer put on 600 [yards] and zero deflection, leveled the piece and lay down too. I alone kept my head above to watch. There they were. Five tanks 1,000 yards in front ... then I counted 18. They moved slowly ... using machine-guns ... their fire coming into the gun line and becoming quite severe. The men wanted to shoot back. I had to speak to them to make them lie still. 800, 700. They crept closer. I picked my tank and moved the gun to cover it. 600! 'Engage!' shouts the Troop Commander, 'Take Post!' Up jumps the detachment. 'Tank, front, 600,' I order. '600 set,' calls the layer, and 'On!' 'Fire!' All six guns of our troop fired at once. We could see four tanks burning. One tank charged No.1 gun but three of us engaged him and he didn't get far. They slugged it out for about 10 minutes, then backed away, still shooting. Eleven tanks wrecked or burning, but we hadn't time to talk about it. Orders came to limber up and we were off again, listening to small arms ammunition exploding in the tanks as we went past.[190]

This No.1 Troop of the 31st Field Regiment was fortunate on this occasion but, as part of the 4th Indian Division, lost three of its four Troops in December. What is incongruous at first sight is that a Troop commander was in front of his guns, detached from his armour and out of sight of his lorried infantry when this action took place.

The fallout from Operation Battleaxe was enormous and focused on deficiencies in training, tactics, leadership and fundamental doctrinal weakness in all-arms cooperation. Beresford-Pierse was sacked, Wavell was made Commander-in-Chief of ABDACOM (American-British-Dutch-Australian Command) in the Far East and Lieutenant General Sir Claude Auchinlech promoted and appointed GOCME.[191] As if a profound neglect of tactical doctrine was not sufficient, the next manifestation of muddled tactical thinking emerged in the concept of the 'Jock column'.

As a lieutenant colonel in the 2nd Armoured Division in April 1941, 'Jock' Campbell had been detached with a collection of armoured vehicles and supporting arms to harass German columns advancing east near Mechili. This he did successfully. In rapid time he was promoted brigadier, OC Support Group 7th Armoured, and quickly persuaded his superiors to use his columns to attack and wreak havoc on supply points and enemy columns.

The Indian Army had a long and worthy record of sending columns of all arms (cavalry, artillery and infantry) to deal with restless natives. This doctrine, if it can be dignified with that term, was simple and had the added benefit of using mobility, surprise and the gun. There was an element of hit and run gangsterism about it and, had it remained at a minor action level, it would have died a natural death. It was unfortunate that Campbell, a VC winner at Sidi Rezegh, influenced the wrong people—the Staff Officers. Thus the 'Jock column' became a further step along the road to breaking an organic division into its component parts. The 9th Division was a natural target by virtue of being *in situ* and as a Commonwealth force. The 'Jock column' was used on four occasions: the 2/13th Battalion was attached to the Tobruk garrison; the 2/12th Field Regiment was also sent to Tobruk, while E Troop of the 2/7th was despatched to Siwa Oasis. In July three columns, Fait, Hope and Char were created within the 22nd Guards Brigade using the 2/7th and 2/8th field regiments.

Jock columns initially comprised a squadron of tanks or armoured cars, a 25-pounder troop (four or six guns), an anti-tank troop (four 2-pounders) a company of infantry (lorried) and sometimes a troop of LAA guns for protection. Its *modus operandi* centered on field guns, while the column's flanks were protected by the anti-tank guns. The OP was also 'protected' by anti-tank guns. Yet, as Brigadier A.L. Pemberton later wrote, 'Provided the enemy did not advance directly towards the guns, but moved round them with boldness and in open formation, a troop of 25-pounders could not hope to stop him.'[192] Even Campbell postulated that Jock columns could neither capture ground nor hold it. Boldness and speed of manoeuvre were of the essence and, in concept, this suited the temperament of the Australian soldier. However, in the annals of the regiment's history, their employment was minimal and the results achieved minuscule. On the other hand, most gunners agreed that the employment of Jock columns was a better option than remaining in a static position.[193]

Over the next six months Auchinlech was to preside over more operations, none of which achieved the promise of his tactical plan. He had inherited a concept which was regarded as a mild panacea for desert operations and which involved two Australian field regiments in a questionable example of the application of artillery doctrine at the time. In August, three Jock columns barred the old coast road, Victory Road (a bitumen highway to the south of Halfaya), and the marked tracks that ascended the escarpment to Sofafi respectively. Major R. Johnson, BC 15 Battery (2/8th Field Regiment), was appointed commander of Char column and the battery remained in this column until relieved by 16 Battery (Major M. Ralph), although Johnson remained in command. On 2 September the 2/7th moved into the area and 13 Battery was placed under command of Faith column commander and 14 Battery under Hope's commander. The 16th Battery remained with Char. These columns saw some action and fought several engagements, but the effect on the enemy was quite minimal, although the operation did have the advantage of exercising both regiments in desert movement.[194] Both were well to the rear of the front and the units facing Rommel were enjoined to avoid giving battle and withdrew as planned to Sidi Barrani. East of Sidi Barrani was the town of Mersa Matruh, a strategic prize. It was a defended port similar to Bardia and Tobruk, but with the logistic advantage of rail, road and sea access. It was here that the columns lined up for their first actions, cognisant of the fact that they were in a protective (passive) rather than aggressive role. They were brigaded with the 8th and 25th field regiments, RA, and two anti-tank batteries.[195]

The columns now endured a foretaste of digging in, digging in and digging in. It was a torrid 'welcome to the desert' and, by the time they returned to Australia some 18 months later, these two regiments had dug thousands of pits, command posts, slit trenches and doovers. Their rigorous training regimes in Palestine and Egypt were to stand them in good stead. CRA Brigadier Alan Ramsay was cast in the Morshead mould and a firm proponent of offensive defence. He soon took the initiative by despatching sniping guns to shell Halfaya at night and then withdrawing them at first light. Using single guns instead of the basic fire unit, a Troop of four guns, would have been anathema to him. He placated Eastick and Tinsley, who both wanted to move south-west and get some more action, until he found a role for them. Obviously Ramsay reasoned, as had his predecessor, that gunners generally prefer to give it rather than take it and thus agreed to the sniping gun concept. It should be noted that the 2/12th had already used this stratagem effectively in Tobruk, although the situations were hardly comparable.

Accordingly, on the night of 4/5 September, Eastick ordered a sniping gun forward from 13 Battery's A Troop. Sergeant 'Darky' McKinna was detachment commander and he and his crew moved west from the Troop position at Bir Tashida. They occupied a position and proceeded to dig the gun in, which they accomplished quickly in the soft sand. Nearby shallow scrapes indicated that its previous occupants (RA regiments) had not been as astute as they could have been. Telephone line was run from the OP at Point (Trig) 20, a feature occupied by the Scots Guards as escort for the OP party the day before. Captain C.L. Schrader began registering his zone at daybreak in conditions of poor visibility. The Germans retaliated and, given that this was a relatively quiet sector with little artillery fire, the OP signaller, Gunner J. Quilliam, was able to advise the gun position signaller of shells heading in his direction. He, in turn, called out a warning which dispersed the crew to slit trenches well clear of the gun. This quickly became the 2/7th's 'sniping gun drill' and comprised: fire, disperse, count enemy rounds bursting, take post.

The next day another gun crew from the same Troop went forward and occupied the same pit. Moments after Lieutenant P.F. Cleland ordered 'fire!', the gun was deluged by retaliatory fire, all so accurate as to preclude return fire. Cleland's BC agreed to move the detachment 50 yards to either flank, a wise precaution. The damaged gun sustained 20 hits and a round exploded the ammunition in the gun pit where men had sheltered earlier. Charge cases were found 120 yards away. In all, 204 rounds were fired at the gun although, somewhat miraculously, there were no casualties. In the fashion of the day, these events were described in terms of a cricketing score: 204 for 3. The gunners deduced that the gun had been observed by several German OPs and/or flash spotters.[196]

Concurrently, the 2/8th's CO, Walter Tinsley, detailed Captain Tulloch Roberts to bait the Germans on the coast by engaging opportunity targets and using dummy gun flashes to confuse their observers. Bombardier Campbell was in charge of this part of the enterprise, which Roberts positioned 350 yards to the left rear of his gun. Initially this ruse was successful, with enemy fire falling in the area. However, once Roberts had engaged a bathing party and motor transport on Halfaya Pass road, the enemy saw through this deception. Had Campbell's simulations been synchronised with the sniping gun, Roberts' ruse might have resulted in more ammunition being expended in their direction. Nonetheless, these sniping gun exercises had produced four significant benefits. First, they introduced gun crews and OP parties to real soldiering and exposed them to shellfire. Second, the exercises helped the counter-bombardment organisation to accumulate data on enemy artillery strength in terms of numbers of batteries and the calibre of guns. Third, it enabled counter-battery fire to be directed against hostile batteries, and fourth, the enemy was persuaded to expend much more ammunition thus placing additional strain on his extended supply line.[197]

The second essay by A Troop (2/7th) exposed the poor practice of occupying the same sniping gun position more than once. German hostile battery fire usually fell to within 10 minutes for line and a 10-yard bracket. This meticulous ranging was typical of both German and Italian gunners. Sniping guns were seen as useful exercises, but there were occasions when German patrols occupied or laid up near OPs and/or gun positions and waited until dawn to round up unsuspecting gunners in search of adventure.[198]

As described, the 2/7th and 2/8th were extended between Halfaya Pass and Mersa Matruh and, eventually, Headquarters Western Desert Force agreed to Ramsay's suggestion to extend the defence scheme at Matruh to deny the enemy observation of the garrison from the high ground. Also in the general area were several bulk forward supply depots, the westernmost known as F4. This depot held upwards of 4000 tons of fuel, ammunition, food and other consumables. Rommel, should he have succeeded in capturing it (before it could be destroyed) would have been a much happier man. In the last week of July both the 2/7th and 2/8th received orders to move south to the Sofafi area where there was a defensive position, the northernmost point of which was known as North Point while the southern end was coded Playground. Each was defended by British battalions—the 5th Coldstream Guards and the 9th Rifle Brigade. The 7th Armoured Brigade guarded the southern flank.

Towards the end of July, Captain G. Huggett's E Troop (2/7th) was sent to add its weight of shell to the defence of Siwa Oasis. This was 180 miles further south and no mere stereotypical desert waterhole. It was populated by Bedouins, stretched some 30 miles long (and is now a tourist destination). At that time the oasis had significant strategic value and Huggett and his detached troop enjoyed their experience. If Ramsay or Eastick were unhappy with the staff breaking up regiments into 'penny packets', there was more in store for them later on.

The regiment was now despatched across the trackless sands as a collection of mobile columns dispersed by day to avoid presenting the *Luftwaffe* with an attractive target and laagering by night. According to Goodhart, 'Nowhere in all our travels did we experience again quite the same thrill as belonging to these desert-worthy shows, feeling ourselves as a desert-worthy part of them.' Two columns were created for the move to North Point and Playground and Little Brother and Little Sister. The organisation to supply these columns centred on the night laager. Goodhart continues:

> Under the Egyptian stars, in a different patch of desert each night; with AFVs and guns nose-to-tail in long rows; with lanes between to be filled by echelon vehicles, with supplies; with Bren carriers and anti-tank guns around the perimeter; with a blackout absolute except for a smothered match flare to light a cigarette under a blanket; with the calm cold desert night humming with its own immensity; with the drone of enemy and our own planes overhead, unseeing and unseen; with hundreds of roosting vehicles, huddled in a silent flock; with the whole show up and away an hour before dawn—this was indeed mobile warfare de luxe, a romantic adventure, a big-game hunt, a tank-marauding expedition rather than a war. The night laager was protective as well as inconspicuous. The evening meal was cooked and eaten before the vehicle left its daylight position; breakfast was cooked on a primus or petrol fire after arrival at the daylight position. Every evening, at about 2100 hours, the lanes filled in quietly with echelon vehicles. Water, bread, meat, tinned foods, beer, mail, cigarettes, chocolates, changed hands in the space of minutes; so, of course, did petrol, oil and ammunition.[199]

In forward areas by day, officers stood behind drivers (depending on the vehicle type) for observation and navigation, the latter by vehicle odometer and compass. Red, yellow, and blue identification flags were carried on all vehicles that travelled in battery columns with 200 yards between vehicles and 300 yards between columns so as to avoid detection by aircraft. Rendezvous for halts during tactical moves and night laagers were given in a simple numerical code from a known point; for example, 1762300 meant 176 degrees bearing 2300 yards from Cairn 203. The driver-navigator, if delayed for any reason, only had to locate Cairn 203, then steer 176 degrees for 2300 yards.

The 2/7th boasted that its B Echelon never failed to deliver the goods. How it accomplished this feat is one of the great unknowns of the desert war. Much is written of generals' strategies and tactics and operations with humour, pathos, gallantry, attacks and withdrawals, but little is recorded of the vital function of a field regiment's B Echelon. It is a difficult task to draw up an establishment detailing the resources available, in this case, to the Regimental Quartermaster. The 2/7th's Quartermaster was Captain Harry Loveband, MM and bar. A regular soldier, he knew his way around the Army supply and maintenance system. He also had to rely on the skill of his staff in B Echelon and his vehicle drivers to deliver what was required. Loveband had a golden rule: 'Get the stuff up, even if you have to walk with it.' This was motivation in itself for the drivers to put in the extra effort to keep their vehicles desert-worthy. B Echelon was no sinecure, as vehicles designed to Army specifications to operate in a desert were prone to broken springs. This handicap was neutralised by careful maintenance and the regiments' green-fingered mechanics in their attached LAD.

During this three-month period the regiments had to endure two systems of resupply from ASC Supply Points. Orders promulgated by the senior ASC commander specified location (by grid reference) and time of operation—when the point was open and when it closed. Closing time was based on completion time for the supply points' restocking. At the Sidi Barrani area supply point, the ASC dumped supplies on the ground and, given the nature and prevalence of dust storms, food became contaminated and was often covered over. Each unit's allocation was 'ticketed' and, in an 'honour amongst thieves' convention, no unit appropriated supplies that were not its to take. Since all units knew the opening time at the supply point, a cross-country event nicknamed 'the ration derby' soon resulted. Upwards of 200 vehicles thus formed a dusty version of that famous race, and the newcomers, the 2/7th and 2/8th, tailed the field until they analysed the *modus operandi* of their rivals. Their B Echelon vehicles were faster than others in the 200-vehicle field and, eventually, the Australian regiments arrived first. There was an advantage in being the last to be re-equipped after all.

On the escarpment a different system reigned, as Goodhart describes:

> As at the night laager, we had our proper place, second row from the right. Each morning, the following day's rendezvous was arranged at a B Echelon conference, and supplies indented for. The RV was never in the same place twice, and hours were variable—sometimes as early as 0600; sometimes at late as 1300. Within the Regiment, supplies were split up with meticulous care, a ready reckoner being a prize possession of the 'Q' staff. Batteries sent their own 'Q' trucks forward to the night laagers, splitting up clothing, stores, ammo, water, mail, petrol, spare parts, springs, and canteen goods—first into 'Troop'; thence, through the TSM, to each staff vehicle or gun tractor.[200]

Almost every vehicle crew had a 'kitty' for canteen purchases and so the gunners lived well.

On 7 September Rommel issued orders for the *21st Panzer Division* and the *3rd Reconnaissance Battalion* to destroy a supply dump at Bir el Khieighat 16 miles south of Sollum. The attack on the dump was merely an excuse for a reconnaissance in force against British forces which extended from the coast to beyond Sofafi. These forces comprised the 22nd Guards Brigade on the coast, supported by the 2/7th and 2/8th Field Regiments, the 7th Armoured Division Support Group and the 4th South African Armoured Car Regiment, the southernmost unit. East of the latter were North Point (Sofafi) and Playground (Habata), both protected by extensive minefields to their west. On 14 September the motorcycles, armoured cars, tanks and guns of the *21st Panzer* swept south-east towards Habata. The British were ready for them and withdrew according to their tactical plan. When the panzers stopped to refuel near Sofafi, the Desert Air Force (DAF) pummelled them severely from above.[201]

On the coastal sector that morning, Lieutenant Colonel Eastick was visiting his 13 Battery when the Germans attacked the Scots Guards near Alam el Kidad. From B Troop's OP, Lieutenant E. Phillipson engaged lorried infantry at their most vulnerable phase of deployment, inflicting heavy casualties. Later that day, Major Ralph deployed his section (two guns) and Troops to conform to B Troop's movement. They made a tactical withdrawal with their guardsmen while the guns of another Troop fired concentrations on advancing German infantry. A Troop and a detachment from the 2/8th under the command of Captain Ken Mackay withdrew down the coast road, turned right at Sidi Barrani and took up positions at Samalus with Char column. 'Each column detached little rearguards ... to hold vital points which got ready to die for King, Egypt, Scotland, England and Australia.'[202] By this time the Germans had withdrawn with considerable losses. The 2/7th passed through the area two weeks later and Goodhart described the scene: 'Anyone who saw that graveyard with burnt out tanks and blown up tankers ... could hardly have known any doubt as to what stopped Rommel's recce in force.' It was the artillery's surrogate depth fire arm, the DAF.[203]

In their last weeks on the escarpment, an excuse was found for 14 Battery of the 2/7th to fire a predicted night shoot on an Italian fort and supply depot at Point (Trig) 207. The plan relied on the Troop occupying a position with minimal noise and using predicted fire to achieve surprise. The first essay by D Troop was compromised and E Troop, fresh from an engagement-free sojourn at Siwa, was given the task. The gunners 'surveyed in' the guns and were ready and undetected by midnight—H hour. In a 10-minute program the troop fired 210 rounds and was clear of the area without leaving any 'produce' 10 minutes later. Nothing succeeds like success, and so Brigadier Ramsay, in consultation with the 4th Indian Division, agreed to a re-run. This time, both E and F Troops were involved, firing 65 rpg at 3.00 am, a time of engagement that coincided with the arrival of aircraft (from the Fleet Air Arm) over the target. During the 15-minute task, aircraft observers could distinguish clearly between gun flashes (fan-shaped and in plain view) and shell bursts (circular) as they flew their 25-minute sortie over the target. This successful cooperative shoot excited the gunnery staff's interest in the tactical use of night bombers.[204]

However successful or fortunate the Australian field regiments might have been with their strong motivation and new equipment, using fire units in inconsequential roles and engagements was poor utilisation of artillery, particularly based on its significant contribution to land warfare. The overall experience of the 9th Division artillery had been one of piecemeal employment and 'ad hocery' and their sole contribution—apart from that of the 2/12th in Tobruk—had been inconsequential. The gunners of the 2/7th, for example, fired just enough rounds to 'wear in' their guns. It would be reasonable to assume that Brigadier Ramsay was keen for his two regiments with the Guards Brigade to see as much desert service as their brethren in Tobruk. Any comparison of their two quite distinctly different roles is nonetheless unfair to those in Egypt. If there was a lesson to be learned, it was that there was no substitute for hard work and operations that honed skills and inured men to the many hardships desert operations inflicted on them. Both regiments were fortunate that they were not on the receiving end of the kind of episodes related in the next chapter.

There are several vignettes worth recounting as they illustrate both the use and disregard of artillery once the 9th Division vacated Tobruk. The first concerns Lieutenant Colonel F.A. Burrows' 2/13th Battalion during operations involving the Tobruk garrison (then under Lieutenant General Scobie) and armour/infantry formations of the Eighth

Army fighting in the Tobruk corridor on 29—30 November. The Australian infantry took up positions north-east of El Duda and immediately came under fire from the 210-mm guns of the German Arko 104, *Generalmajor* Boettcher's group. These guns initially opened fire at long range on trucks carrying New Zealand infantry to Tobruk. The size of the shell bursts and earth vibration constituted a new experience for the battalion which had been sited in full view of German OPs. Burrows redeployed his battalion successfully, but not until after the Australian infantry had added the biggest gun to shell them to their impressive list. The action at El Duda, a commanding pimple north of Trig Capuzzo, involved a German attack by elements of the *15th Panzer Division* on the 1st Battalion, the Essex Regiment, and the tanks of the 32nd Army Tank Brigade.

As night fell, Burrows was advised that El Duda had fallen and was ordered to counter-attack. He allocated two companies to the attack and retained one to protect the gun positions of the 1st RHA near Belhamed. The attack by armour and infantry at 1.30 am was a complete success, surprise being achieved by the expedient of foregoing artillery support. Captured Germans were miffed by this apparent lack of battle ethics. The tally was 167 captured for the loss of seven wounded (two mortally).[205] During this battle the New Zealand Artillery suffered its worst calamity of the war when the 6th Field Regiment was overrun at Belhamed-Zafraan that morning by the *5th Panzer Division*. General Blamey remarked that 'No better example of lacking a sense of organisation and failing to recognise the importance of employing formations complete can be found in the use of the NZ Division in Crusader.'[206]

Both the 2/7th and 2/8th field regiments disengaged from the 22nd Guards Brigade and left for Egypt and Syria respectively, the latter to rejoin the division and the former despatched to the RA School of Artillery at Al Maza as the Depot Regiment. Regardless of how, in terms of doctrinal purity, they were deployed during this period, the gunners felt that they had matured as members of their regiment and had made a contribution. That other British Army regular and Territorial field regiments participated in a lion's share of the action should be judged against the good fortune of the Australians who were not to be exposed to hazardous engagements as a result of the incompetent application of an unsuitable doctrine. Thus they were available the following year when the Allied need was greater. As it happened, many RA units had been captured, reduced to a mere shell of their establishment, or packed off to the Far East, unhappily to meet the same fate as their brethren in Burma or Malaya/Singapore under General Percival's command (such as the Australian 8th Division).

During this period, the debate within the General Staff over the utility of the infantry division in desert operations gathered momentum. Enemy forces were almost entirely comprised of panzer/mobile formations while the British formations were predominantly infantry. Surely the best way to beat the enemy was to emulate him and his tactics and to rely on a greater reservoir of materiel to make up for shortfalls in doctrine, and not just artillery. The next chapter illustrates how these issues, which arose from operations over the next seven months, came to the fore as the mixture of success and failure did nothing to clear the air on the application of doctrine. This did not involve Australian artillery directly, although, while the British endeavoured to reach the 'right' conclusion, the General Staffs of the Commonwealth forces and their artillery advisers did not have the same enthusiasm for changing an organisation that had a signal success to its credit at Tobruk. Moreover, it is difficult to imagine General Blamey even entertaining the thought of reorganisation given the woeful losses British armoured forces inflicted on themselves and continued to do so after he left the desert. He did not win any friends in the British High Command by his intransigence when asked to break up his divisions, but he was absolutely right in saying 'no'. Indeed, his successor thoroughly agreed with him.

Chapter 10
The British Experience, 1941–42

No tank commander will go far wrong if he places his gun within hitting range of an enemy.
Lieutenant General W.H.E. ('Strafer') Gott

While the AIF units considered 1941–42 a busy year, the British Army, which included New Zealanders, South Africans and Indians, was to be far more active, enduring a series of trials, tribulations and triumphs as it confronted the combined German and Italian land and air forces. British command arrangements continued to be marked by a degree of fluidity which was not always a disadvantage given that it represented adaptability in the way the British extracted operational effectiveness from its front-line formations. Yet, as events unfolded, this fluidity became increasingly disruptive. Following Tobruk, I Australian Corps had handed over to Cyrenaica Command; by October, however, the Command had been reborn as Western Desert Force. When General Alan Cunningham took command, he presided over Western Army Headquarters which, on 27 September 1941, was rebadged as the Eighth Army. Western Desert Force then became XIII Corps. By contrast, the Axis forces under Rommel suffered only minor changes at levels well above his command that recognised certain sensitivities in egos rather than challenged the operational effectiveness of the forces themselves. In May 1941, as the siege of Tobruk unfolded, Rommel's 'special purpose' division, *ZBV*, became the *90th Light Division* and the *5th Light Division* became the *21st Panzer*. The former was commanded by *Generalmajor* Streich and the latter by *Generalmajor* Neumann-Silkow. These realignments to the PAA command structure rendered it even more formidable, and it was to retain this form with minor adjustments until the Libya campaign the following year. In October, Rommel assumed command of all Italian divisions east of the Cyrenaican 'Bulge' except General Gambara's *20th Mobile Corps* (*Ariente* and *Trieste Mobile Divisions*). *Generalleutnant* Cruwell then assumed command of the DAK and General Navarini's *21st Corps* (*Brescia, Pavia, Trento, Bologna* and *Savona divisions*). These latter divisions would provide the bulk of POWs in the battles to come.[207]

Following Tobruk and during the period October 1941 to July 1942, the British formations operated without their AIF component. Australian artillerymen, allowed the luxury of refitting and retraining in Syria, were able to absorb and apply lessons learned from their battles and experiences at Tobruk in 1941 and from fighting alongside the 122nd Guards Brigade. They also used the opportunity to test new techniques to improve their battleworthiness. The gunners would have been proud of their 7th Division colleagues' performance during their brief Syrian/Lebanon campaign in June under the command of Lieutenant General John Lavarack, a well-known member of the Regiment. His short campaign was fought, not across vast expanses of desert, but across cultivated and habited areas near or on the coastal plain and the mountainous interior. So effective was the campaign waged by Lavarack and the 7th that it effectively knocked Vichy France out of the war.

A review of 7th Division operations by HQRAA affirmed current doctrine with a qualifying comment on the necessity for OPOs to be better sited to bring effective neutralising fire to bear. In the opinion of Colonel Frank Berryman, the Division's senior staff officer, poor siting of OPOs often led to unnecessary infantry casualties. Other 'negatives' comprised factors peculiar to the campaign, although many of the other non-doctrinal but systemic improvements to operations could and would apply to present and future campaigns.[208] The 7th Division had been warned to return to Australia and, in October, the 9th began its move to Syria/Lebanon from Tobruk. In the desert theatre, the DAF (with three RAAF squadrons) continued to be reinforced and began contributing strategically to the disruption of its opponent's Achilles' heel—the logistic chain. The RN fleet, including RAN vessels, also targeted the now stretched Axis logistic chain.

The withdrawal of the 9th Division meant that Western Desert Force essentially operated without an Australian field force component. In the following months, from October 1941 to June 1942, British and other Commonwealth forces fought three major battles: Operation Crusader in November 1941; the Gazala battles of June 1942 and the First Battle of El Alamein in July 1942. This chapter describes these battles and concludes with a review of doctrine in the context of operational outcomes and the innovative practices that evolved during these fateful months.

This chapter is important in understanding the development of artillery doctrine and methods of employment (or lack of) before and subsequent to Operation Battleaxe. Methods of employment of artillery were designed to neutralise Rommel's advantages over his adversaries, not only in men and materiel, but also in the command and control of artillery. The 9th Division—in Syria with the Ninth Army—forged its artillery communication link with the British forces through the MGRA, then Aylmer Maxwell at GHQ Middle East, and his periodic reports on operations. Some of the reports covering this period make sorry reading given the violations of artillery doctrine leavened only by references to the very high standards maintained and bravery displayed by many members of the Regiment, from unit COs to gunners in their detachments. Maxwell had much to comment on, including innovative measures necessary to make artillery fire more effective and responsive in desert operations, but most particularly concerning his loss of guns. Just how the PAA forces were eroding British advantages in doctrine and equipment forms the subject of the next section of this chapter.

If their brief campaign in Flanders had taught the German commanders any of the necessities for effective armour-artillery-infantry cooperation, these were certainly lost on the British. One emergent factor was the variety of supporting weapons used to destroy infantry, anti-tank artillery or anything else aimed at the German forces. Prior to the development of an anti-tank gun superior to the 2-pounder, the panzer tank was able to move very close to British forward defended localities, prompting field guns to be brought into forward positions using both direct and indirect fire in support. Part of the panzer credo in the desert was to kill infantry and then, it was reasoned, success would follow (in Russia, the credo was to destroy equipment, i.e. tanks and guns). Streamlined staff work and battle drills enabled a regimental (or brigade) attack to be mounted in two to three hours.

A typical panzer attack employed a covering force of MK IV tanks, heavy (Spandau) machine-guns, PAK 35 (and later PAK 38) anti-tank guns and 75-mm field/assault guns approaching to within 2500 yards of a forward defended location. Behind this, a phalanx of 40 tanks across a front some 800 to 1000 yards wide led a lorried infantry battalion and supporting elements such as 88-mm and other PAK guns positioned on each flank in a formation known as the 'Box'. Behind came the field towed and self-propelled artillery. When the objective was taken, some tanks would go beyond it and take up defensive positions while the main body mopped up. These formations had the advantage of being able to react quickly if engaged on the move, rapidly adopting a position of all-round defence. In the advance, the tanks would move forward ahead of the Box in two echelons, each accompanied by field guns. If attacked, the tanks would be positioned on either side of the Box. If an attack threatened the Box, the assailed flank would withdraw to enable the tank guns to shoot frontally while the PAK 38 and 88-mm guns engaged in enfilade. The disengaged tanks, meanwhile, could move and take the attacker from the rear. If artillery from Arko 104 was available, these guns would add their considerable weight of fire on targets of opportunity. During some actions there were instances of fire shepherding a brigade group onto a panzer force leading to its subsequent destruction. *Generalmajor* Boettcher had trained his Arko to a high standard and he would soon take command of the *21st Panzer Division* for a short but memorable period.[209]

General Erwin Rommel, Commander, Deutsch Afrika Korps, with General Major Boettcher.
(Bundeswehr 183-B18239)

Operation Crusader was Lieutenant General Alan Cunningham's first and only battle as an army commander. Like most men of his age and experience, he was an infanteer and had not previously commanded armour. He owed his newly elevated status to his handling of a formation in Abyssinia fighting the Italians where his success paralleled that of O'Connor in the Western Desert. In November he planned an attack from Halfaya across the escarpment towards Tobruk to bring Rommel's tanks to battle where they could be destroyed by a numerically superior force of his tanks, with which the Western Desert

Army had been re-equipped. It was typical of the mindset of the General Staff that tank numbers were all that mattered in the desert war. Yet, as Bidwell points out, over 600 field and other guns had already been lost in 11 months of fighting. Cunningham's attack would see both XIII and XXX Corps advance north-west and assault the German-Italian forces concentrated around Gambut and Sidi Rezegh. For its part, the 70th British Division was at Tobruk and had been largely spared the organisational fragmentation inflicted on both formations assembled well ahead of their supply base. Brigade groups had been formed from divisions in an attempt to match the panzer formations and their well-handled field artillery and lethal anti-tank weaponry. Thus, supporting arms of artillery and engineers were under command and the means of uniting them if a concentrated artillery (or engineering) plan were needed to meet a tactical exigency, unrehearsed. This was potentially problematic and was critically dependent on sound communications for speed of reaction.

A division's role in all phases of war was to engage and hold an enemy so that a superior force could administer a *coup de grace*. One of the arguments that supported this theory was that the offensive power of an infantry battalion was significantly increased by its organic mortar and anti-tank platoon's guns. A division comprised an agglomeration of 'threes'—brigades, battalions, artillery regiments and field engineering companies. Under this arrangement it was possible to form battalion groups, and the more this level of autonomy was pushed down from the supremo, the further it violated tried and true practice of command, control and coordination of effort. At this time only the armoured division support group remained untouched organisationally. Set against this was the efficient working model of a German brigade of armour or infantry with its organic support. One consequence of Operation Battleaxe was that the Germans altered their artillery dispositions as their field artillery now had self-propelled guns. This enabled them to disperse their guns so as to make them a difficult target to neutralise because of their frontage as well as their mobility—a crucial advantage. Thus organised, battle was joined and raged at various levels of intensity from 17 November to 8 December. The detailed comings and goings of units and formations, the particulars of indecision in command etc. are not the focus of this discussion. Those that figure prominently in the artillery battle which was, doctrinally speaking, still in transition, will be considered as the chapter unfolds.

The 7th Armoured Division played a key role in the XXX Corps (left flank) attack from the south of Sidi Omar towards Gabr Saleh and Bir Gubi in the general direction of Sidi

Rezegh. The 7th possessed its own support group, now 36 guns (instead of 24) and three armoured brigades, the 4th, 7th and 22nd, with 16, eight and eight guns respectively—in all, 68 guns, all 25-pounders. Further west, the 1st South African Division was armed with ninety-six 25-pounders and 16 medium guns under command—a total of 112 guns. The New Zealand Division (of XIII Corps) boasted another twenty-four 25-pounders and a medium regiment. Counter-battery assets lacked range and, for reasons best known to the General Staff, two survey troops were included in the XIII Corps area close to the sea rather than with XXX Corps. As a result, the 7th Armoured artillery had to use relatively new and untried methods of establishing a regimental grid in the absence of accurate survey data. This was merely a temporary expedient and enabled a regiment or group of batteries to coordinate fire, for which there would be ample opportunities in the battles to follow. Even so, when every minute counted between accurately placed fire and 'ploughing the desert', it was, at best, just satisfactory.

The 7th Armoured Brigade and its support group occupied the Sidi Rezegh feature, withstanding determined attacks for two days before withdrawing on 23 November. On 21 November, the brigade's sixteen 25-pounders and four 2-pounder anti-tank guns were assailed by the *15th* and *21st Panzer Divisions* with artillery assets comprising forty 105-mm and twenty 150-mm guns, as well as sixty-three PAK 38s, twenty-one PAK 35s, 88s and LAA artillery.[210] Half-naked, weary, begrimed, shelled, battered from the air and machine-gunned by infantry, the brigade resisted attack after attack. The gunners earned the admiration of their foes, one of whom commented that the British artillery was the 'best trained and commanded element in the British army.' The 5th South African Brigade, sent to rescue the support group, was destroyed by the *5th Panzer Division*. Yet another tactical misadventure involving the 4th Indian Division and tanks occurred on 22 November when questionable artillery support, tanks late crossing the Start Line, poor counter-battery work and unlocated anti-tank guns all combined to turn a one-hour show into a two-day debacle.

However, many miles from GHQ, practical people were devoting serious thought to the situation. Oddly enough, it was the South Africans who formulated an organisational structure that worked, doubtless to ameliorate mounting political pressure at home concerning their losses in the desert under British generalship. The 1st South African Brigade took the organisation of a brigade group of all arms rather more seriously than its brothers-in-arms. The brigade was attacked at Taib el Esem on 25 November and had adopted a position on good tactical ground some 3000 yards wide. Each battalion had a frontage of about a mile and all faced outward, evenly distributed around the circumference.

Command, ordnance and medical assets were in the centre and each battalion had its own battery. Anti-tank, direct support field and machine-guns were sited with interlocking fields of fire. Engineer field companies laid mines on a priority basis. Once on the move, the brigade's attached armoured car reconnaissance squadron included artillery observers. An LAA troop was also employed and a mobile reserve created. This layout resisted two determined panzer attacks and the brigade retired with its perimeter intact having used all its ammunition. Impressively, this layout resembled the grouping that Rommel had created in the aftermath of Operation Battleaxe.[211]

Another engagement involved the 1st Field Regiment which was supporting the 4th Indian Division, whose CRA (Brigadier J.F. Adye) had issued an instruction regarding the engagement of tanks, the nub of which was that 'armour should not be engaged at extreme range so as to make the panzer alter their *schwerpunkt*, but to lure them into decisive range and destroy them.' This policy was put to the test four days later at Bir el Hurush when a panzer squadron confronted the division, initially at a range of 2000 yards, firing guns only from the front and flanks. The 1st Field waited until the armour closed to around 800 yards before engaging with HE. The balance sheet for the resulting action read:

Their cost:	7 tanks
Our cost:	5 x 25-pounders
	2 x 40-mm Bofors LAA
	18 KIA, 44 WIA, 4 missing[212]

Following this bruising encounter, panzers seldom took on a 25-pounder unit. In seeking to entice armour into field and anti-tank gun range, the British forced the Germans to modify their approach to defended localities, particularly those of infantry and anti-tank guns. The armour would swan around just outside anti-tank gun range until it forced the gunners to reveal their positions. The gunners would then knock out the tanks with their 75-mm guns or by observed field artillery fire.

Operation Crusader was a battle which General Frank Messervy called 'an epic of British doggedness and optimism in adversity'. As Barnett notes, the British had muddled through, despite having never come to grips with the intellectual challenges of modern mobile warfare.[213] It is often forgotten that Rommel's reverses at Crusader matched his defeat at El Alamein a year later. He had only 30 tanks left, and a third of his command was in ruins. It is also worth noting that Rommel was stronger at Crusader than Auchinleck and the reverse was true at First Alamein. Man for man, the Eighth Army had bested its opponents. Australian casualties all came from the 2/13th Battalion (39 killed,

36 wounded) which was withdrawn in cattle trucks to the bosom of its division on 18 December in Palestine. The bald statistics of the battle are:

	Eighth Army	PAA
* Total strength	118,000	119,000
	(40 per cent Commonwealth)	65,000 German
		54,000 Italian
* Killed in Action	2900	2300
* Wounded in Action	7300	6100
* Missing in Action	7500	29,000
(all causes)[214]		

Crusader was widely believed and reported to have constituted a reverse rather than a victory, an attitude which Carver ascribes to several fundamental reasons. First and most obviously, there was a widely held belief in the superior performance and design of German armour. This was, in fact, a myth, and would only be debunked in time and, as it eventuated, by the use of 'superior' American tanks. British armour was never concentrated, but rather employed in dispersed brigades commanded by inexperienced and poorly trained officers. There were flaws in the organisation of forces, compounded by poor coordination of effort and indecisiveness at the top. For the British, this problem proved endemic. Carver deplores the appallingly slow transmission of orders, exemplified by an occasion on which Lieutenant General Ritchie's morning appreciation of the situation was then 'finessed' (discussed and modified) with his corps (and sometimes divisional) commanders, leading to orders being issued at night for execution early the following morning. Rommel's PAA could effect a similar outcome in a matter of two or three hours, such was the quality of its staff and battle drills.

British field artillery fire was dispersed and medium guns used piecemeal, whereas German and Italian artillery was stronger in the heavier calibres. The PAA held anti-tank artillery 88-mm and (both) PAKs in greater ratios to infantry in its divisions than the British. Both were deployed well forward. While the superficial view was that the 88-mm gun was a great tank destroyer, in fact it was the PAKs that did far more damage to British armour. The PAK 38, using tungsten carbide solid shot, was lethal when used offensively.[215] Auchinleck correctly estimated that Rommel had been severely weakened by the ebb and flow of battle—his constant defence and continual need to attack. Auchinleck ordered the advance to Tobruk to continue, forcing Rommel to withdraw on 7/8 December. Tobruk was relieved and Cyrenaica was again in Allied hands, while Rommel's PAA, its logistic situation increasingly stressed and under considerable strain, retreated to its April position.

There was one shining example of the use of concentrated artillery fire—the recapture of Bardia by the 2nd South African Division and the 1st Army Tank Brigade on 27 November. They were supported by over one hundred 25-pounders and 40 medium guns and howitzers, good survey and an effective Counter-Bombardment Office (with flash spotting, sound ranging and air photos) and No.451 Squadron, RAAF. By a rare coincidence, it was almost a year since the 6th Division had won plaudits for its capture of the port.

The British took a long time to learn the lessons of all-arms cooperation taught by the German practitioners of modern warfare. Crusader provided these lessons in abundance. First, the brigade group triangular pattern of organisation was reviewed, as was that of the armoured division. Subsequently, in 1942, a CRA was authorised and the armoured division support group abolished. Thus reconstituted, an armoured division still suffered from an imbalance in arms and its commanders still lacked an appreciation of the employment of artillery. The infantry division retained its standard brigade configuration with a field regiment in support. Artillery field regiments were allocated to independent armoured brigades.

One argument against breaking up a division was that the formation of battlegroups of all arms (or groupings) meant that the position of CRA, hitherto vital for the coordination and control of divisional artillery resources, would be reduced to an administrative cipher attending to personnel matters, training and technical advice and whose expertise on artillery tactics would be subsumed (unwillingly) by regimental commanders within brigades or battlegroups. The gunners fought this outrageous notion tooth and nail, particularly the implication that, if a division did not have a CRA, a corps may not need a CCRA. It was also suggested to the horrified gunners that Corps Brigadier General Staff could act as the artillery guru. In the event, the revisionists were rebuffed with the argument that, while divisional coordination might not be necessary during desert operations, there was an equally strong chance that it might, in fact, prove essential. While there were occasions when a battle/brigade group performed well, the opponents of breaking up formations argued that this was simply 'a fancy name for the dissipation of effort', and that regiments pitted against well-trained and led divisions constituted a recipe for disaster. Battle/brigade groups with significantly less firepower available from field and anti-tank artillery simply could not stop an enemy division in its tracks. The British artillery did not have the means to deliver quality fire plans to influence tactical situations. There was also the problem of positioning these groups to enable them to be mutually supporting with artillery fire. This was possible in a static battle, but once manoeuvre became a factor, mutual support might well prove difficult to arrange, let alone deliver.

Operation Aberdeen in early June 1942 in which three brigades (33rd Tank, 10th Indian and 22nd Armoured) attacked the DAK at Sidi Mufta illustrated the consequences of a lack of mutual support. The Allied forces were swept aside for want of coordinated artillery fire. Griffith notes that doctrine was often formulated by armour and infantry regimental commanders, resulting in unnecessary confusion as regiments and formations were switched between higher commands (the churn effect), a common practice at this time.[216]

Another issue to be resolved concerned the advice provided by the BRA, CCRA and CRA to their commanders on the employment of field, medium, anti-tank and anti-aircraft artillery. Prior to the arrival of Montgomery in September, all these advisers found their operational influence restricted by the lack of necessary resources to fully staff higher artillery headquarters. Once the British retreat towards Egypt accelerated ahead of Rommel's PAA in June 1942, these doctrinal flaws were continually raised at GHQ. As a result, there was a reluctant and generally uncoordinated effort by Auchinleck's staff to progress doctrine and translate concepts to create a more 'panzer-like' divisional structure. As always, suggested changes to an infantry division and its deployment had their vociferous critics. Under General Auchinleck these changes were gradually adopted in line with his view of the tactical situation. First Alamein at Alam Halfa illustrates the implementation of these changes and forms the basis of the discussion in the following section.

Operation Aberdeen, General Ritchie's assault of June 1942, was mounted by XIII Corps. Bidwell describes this operation as an attempt to use massed artillery against *21 Panzer Division* when it was trapped behind the British line at Gazala, south-west of 'Knightsbridge', a track junction west of Hagiag' es Sidra. In short, the operation was a failure. 'Planning, command and control, reconnaissance and coordination were all faulty.' The artillery plan involved supporting the 10th Brigade and units of 7th Armoured by neutralising enemy assets. However, instead of identifiable 'Boxes', the British were confronted with 'a ring of mobile guns backed by tanks and indirect fire artillery offering no tangible target.' Ninety-two guns were available and OPs were now mounted in OP tanks, a significant improvement, but an innovation that of itself was unable to avert disaster. The *21st Panzer* made a nonsense of British attempts to produce a fire plan for a mobile battle and the 150th and 151st brigades were all but destroyed. The infantry had been given a role that had quickly descended into forlorn hope—they were tasked with trying to hold flat, rock-hard ground.

The ground, on which four 'Boxes' were deployed with inadequate dispersion was referred to in later accounts as the 'saucer'—which it was not. It was a long, gradual slope and all its occupants were exposed to observation from a feature at its western end. Armour was sent in to extricate the force, to no avail. In this lamentable episode—the defence of the Ualeb Box—the gunners of the 7th Medium Regiment's 25/26 Battery were formed into infantry companies only to run short of ammunition engaging targets a mere 2000 yards away. Their last melancholy action in the Box was to destroy their guns when their ammunition was spent.[217] Four field and sundry other regiments lost more than 96 pieces and 37 anti-tank guns. The 107th RHA (South Notts Hussars) was returning fire with vigour until it moved to direct fire to face the meat-grinding panzers. As just one part of the 22nd Armoured Brigade, the 107th's stand was epic. Alongside machine-gunners from the 1st RNF, the 107th fought to the last, with fusiliers, signallers and drivers joining the detachments as gunners fell. The detailed account of this action is horrific in its intensity. The motivation of men faced with eventual destruction, death and captivity released 'psychological forces' that are difficult to imagine in this day and age. Two COs fell, Lieutenant Colonel Seely (107th RHA) and Colonel de Graz of the support group. The 144th RHA, which had replaced the 2/7th in Egypt in October 1941, was completely destroyed.[218] Ironically, the Australian gunners of the 2/7th and 2/8th field regiments had earlier envied them for their prospects of gaining 'dinkum' desert experience.

A key factor in the German success was the amalgam of *21 Panzer* artillery and Arko 104. Their large-calibre longer range weapons were concentrated and used effectively, while the 7th and 64th medium regiments' guns simply engaged targets of opportunity. It was ironic that, on this day (after the battle was all but over), Auchinleck issued his instruction, 'He [the enemy] will be at once engaged with all available artillery of the division by the division or divisions nearest the threatened spot. Other divisions if not engaged will move at once ... and attack the enemy boldly and quickly, coordinated by corps commanders.' He also called for maximum mobility, greater alertness and quickness of decisions.[219] In three days Rommel, with his 60 tanks and 2500 German infantry of *21 Panzer* and *90 Light Division* did likewise. In all, around 6000 men were taken prisoner or lost. The *sequelae* of this dark episode in British military history did not end with the last shot fired near Knightsbridge and the Saucer, but with the disorderly retreat eastwards by the remaining units and sub-units. It was this spectacle (known as the 'Cairo Flap') that greeted the 9th Division as it moved forward on the coast road from Palestine to Ikingi Maryut and further west towards engagement with Rommel's forces. Such was the reaction to the overall Middle East situation that the Navy up-anchored and left Alexandria fleet base for the Red Sea and the RAF moved its tactical headquarters, hitherto co-located with Headquarters Eighth Army, for fear of being overrun. GHQ contemplated a withdrawal up the Nile to Port Sudan on the Red Sea.

It was Auchinleck's acting Deputy Chief of Staff, Major General Eric Dorman-Smith's suggestion that the battle for the Delta would hinge around El Alamein, for which he wrote a compelling appreciation.

In the Battle of First Alamein (Alam Halfa), Auchinleck regrouped his command, with the addition of some welcome reserves, to oppose Rommel's last, desperate thrusts at the Delta. At this point Auchinleck made several key decisions, the first to relieve Neil Ritchie of his Eighth Army command and assume the role himself on 25 June. Ritchie had succeeded Cunningham in November 1941 and replacing army commanders relatively frequently was hardly inspirational for the troops they were leading. Major General Dorman-Smith was to be a key influence on the battle that followed. Dorman-Smith was known as 'Chink' because of his slightly oriental features and was described as 'an Irishman having great charm, high intellect, a gift for wit and a short patience with the stupid.'[220] British forces comprised two under-strength corps (X Corps commanded by Lieutenant General Holmes and XIII under Lieutenant General 'Strafer' Gott) with 150 tanks covering Mersa Matruh and Minqar Quaim.

Auchinleck had summoned his reserves following the Gazala debacle and, on the day of his new artillery manifesto, the 9th Australian Infantry Division and 2nd New Zealand Division, fully trained and equipped, were despatched from Syria and Palestine en route to Egypt, a journey covered in very quick time and a triumph for British Line of Communications staff duties. The plan called for 'boxes' based on artillery disposed in depth (towards Cairo) with mobile forces to operate against PAA forces. When the battle began, a South African brigade held El Alamein on the coast with two battlegroups to the south and three defensive boxes extending southwards in a crescent shape. Two key ridges oriented roughly east-west—Ruweisat to the east and Miteiriya to the west—and inland from El Alamein were tactically important, as was a track junction Deir el Shein located midway. Both ridges tended to channel armoured forces to the northern and southern sides. A similar ridge further east of Ruweisat, Alam Halfa, was to become a crucial tactical feature in the latter stages of the battle.

Rommel planned two thrusts, one on each side of Miteiriya. The left thrust would envelop the El Alamein salient and the right thrust would take out the boxes, the southernmost of which was at Abu Dweiss. Launched on 1 July, the left struggled south of El Alamein and lost momentum, then stalled. The right received a bloody nose at Ruweisat and also faltered. From 2 to 5 July, the panzers made some progress eastwards, sufficient for Rommel to send his famed 'victory signal'. Auchinleck pushed his boxes north with armour to threaten Rommel's rear. Post-battle analysis suggested that, despite having intimate knowledge of enemy positions from 'Y Intercept Service' messages, there were problems with the execution of orders.[221] The third phase of the battle lasted from 8 to 11 July and involved a mobile defence on the south side of Ruweisat and, to the north, the 9th Division capturing the Tel el Eisa feature and eliminating the *Sabratha Division* from Rommel's order of battle. He then tried to envelop the British southern flank at Bab el Qattara, exhorting his desperately tired troops by assuring them the 'fleshpots of Cairo' were but hours away. They did not get far. Dorman-Smith suggested to Auchinleck that he target the Italian formations in a counter-attack which he proceeded to do. It was a winning strategy, forcing Rommel to split his formations to reinforce his ally's dwindling reserves of tanks, guns and troops. One after the other, the *Trieste, Parma, Brescia* and *Trento Divisions* were disabled. Throughout the battle heavy concentrations of artillery further reduced his armoured formation's tank and anti-tank gun strength. For the first time, corps counter-bombardment sections operated efficiently and the artillery was handled in accordance with Auchinleck's June instruction.

The results spoke for themselves, although success would not have been possible without the important contribution of the RAF. Diverted from their strategic bombing, RAF aircraft battered Rommel's supply line and troop concentrations so badly that even he suspected that he lacked the wherewithal to continue to the Delta. He was, however, beholden to Hitler's directives and would not have been permitted to contemplate a retreat even had such a prospect not been against his nature. Rommel would make one last attempt, an action that is described in the next chapter.[222]

However bleak the state of British arms portrayed thus far in this narrative, morale always seemed high. Contributing to this phenomenon was the intelligent application and analysis of operations at the regimental and lower level and the constant striving for increased efficiency so as to match, or better, the German and Italian methods. Ultimately, all effort was aimed at producing timely fire more effectively. In the desert, where trail dropping was associated more with a 'tank alert' than a normal deployment drill, one innovation involved positioning a gun with the Troop gun position officer while on the move. This 'pistol' gun could fire a ranging or position fixing round before the remaining guns arrived. To provide a suitable aiming post at the position, one side of a signals truck was painted with a black vertical stripe. Faster methods for producing barrages, smokescreens and engaging targets from the air were also introduced. Smoke could be laid within five minutes and shell bursts of ranging rounds used to fix the points of origin of the screen gave the OPO wind speed and direction. For barrages, standard layouts were adopted that halved the time previously required to produce gun programs from the receipt of orders and/or traces. It was fortunate that the desert was usually 'flat' and that one height to calculate angle of sight to the target would suit all guns on all 'lifts' (i.e. lines). Speed compromised accuracy, and so a convention was agreed on the distance between an infantry Start Line from the first line (500 yards). Better sideband radio telephony sets on the ground and in aircraft improved the efficiency of air cooperation squadrons whose pilots conducted shoots by adjusting fall of shot by corrections to line and range. Ladder ranging also increased the rapidity of engagement. For example, an OP would order four rounds at 6000 yards with 300-yard adds between rounds. This practice was usually discontinued once an OPO became familiar with his zone of observation. Observed fire was generally regarded as more accurate than predicted fire as the unreliability of meteor telegram data issued by the RAF Meteorologist (Flight Lieutenant Bradshaw attached to the 4th Survey Regiment) often resulted in inaccurate fire and wasted ammunition. This was not the fault of the meteorologist as the desert was notorious for its rapidly changing conditions. To overcome this problem, 'Datum Point' shoots established the current correction necessary to map line and range from weather factors. The Germans adopted the same practice, using airburst shells and cross observation from OPs and/or flash spotting bases in the battle area. Similar procedures were used to calibrate guns—so successfully that it was eventually mandated that units should not go into action without calibrated guns. All these improvements were aimed at gaining the upper hand in the artillery battle.

Bidwell notes in his summary of desert war tactics that, following Battleaxe, unorthodoxy reigned until First Alamein, not only in terms of dispersion, but also concerning concentration and other misuses and abuses of artillery principles. These went hand in hand with the methods and styles of various commanders, almost all of whom lacked experience at elevated command and in the tactical handling of armour and artillery. Senior artillery officers were unable or unwilling to force the issue with their colleagues in the correct application of artillery fire. There were, for example, three changes in RA advisers at corps level between August 1941 and July 1942. Fluid operations produced a 'churning' in command systems involving higher and lower headquarters which was highly unsatisfactory. CCRAs were almost powerless without communications and staff, suggesting that they were denied the use of their authority by their commanders. All these problems were to be swept away by two new brooms—one an infanteer and the other a gunner.[223]

Possibly the only mistake the Germans had made in their 18-month desert campaign lay in not providing their assaulting tanks and troops sufficient support for the Easter battle at Tobruk. They had taken every advantage and revealed many flaws in British methods, of which counter-bombardment was but one. This was recognised by the Artillery Staff, particularly the next BRA Eighth Army.[224]

Chapter 11
First Alamein and Alam Halfa

The fox knows much, but more he that catches him.

Proverb

The Western Desert Battle of El Alamein fought by the Eighth Army in October 1942 has justifiably been described as one of the most decisive battles in history. Indeed, noted British military historian Major General J.F.C. Fuller lists it alongside a mere six others in his analysis of the Second World War.[225] The battle has as its genesis the desperate stand by British and South African troops west of the El Alamein position from the third week in June when General Sir Claude Auchinleck was both commanding the Eighth Army and filling the appointment of GOCME.[226] At the same time, *Feldmarschal* Erwin Rommel was pushing his exhausted troops who were well below full strength in a final lunge for the Suez Canal. As the battle reached crisis point Auchinleck, prompted by his Deputy Chief of Staff's perceptive plans, was exhorting his command to use artillery effectively as outlined in his instruction of 25 June. At that point, Morshead's Australian 9th Infantry Division and the 2nd New Zealand Division were garrisoning Syria with the Ninth Army. On 25 June both received orders to move to the delta—and quickly.

Major General Eric Dorman-Smith, Auchinleck's deputy, had reached the realisation that El Alamein was the key to the defence of the Delta and made his appreciation of the situation a month later when the front had stabilised. This was an important document—probably the most important produced in that theatre at that time—for it ordained where, how and when the next battle would and should be fought. Auchinleck and his senior commanders readily agreed to Dorman-Smith's broad outline. Based on estimates of enemy strengths following the Gazala battles and the ever-increasing flow of materiel, troops and munitions to the Allied forces in the theatre, Dorman-Smith suggested that attacks be focused on the Italian formations in the PAA. Should these attacks prove successful, Rommel would be forced to undermine his own front to bolster his weaker partner. Auchinleck had launched the attacks, with impressive results. Dorman-Smith's next intervention came on 27 July in an appreciation that anticipated the extent to which Rommel would be reinforced. The date is again significant because, on that day, the 9th Division and the tanks of the 50th RTR were attacking Miteiriya Ridge incurring losses as a result of the usual litany of factors attendant on reverses, however small. The attack gave Dorman-Smith cause for reflection and he made a number of well-considered points:

- Although the Axis forces are strong enough in defensive action they are hardly strong enough to attempt conquest of the Delta except as a gamble under strong air cover.
- None of the Eighth Army formations is sufficiently well trained for offensive operations and all need either reinforcement or a quiet period to train.
- Therefore, as we are hardly fit to launch any attacks, our best course is the defensive combined with offensive gestures from time to time ... until we are strong enough to attack ... in mid-September at the earliest.
- Rommel will certainly attack before the end of August and, as the Eighth Army gains strength and depth, he will attempt to avoid it and seek success through manoeuvre. The Eighth Army will have to meet this on the southern flank ... and organise a strong mobile wing ... well trained in harassing defensive technique.
- Eventually we will have to renew the offensive ... which will mean a breakout through the positions at El Alamein. Newly arrived infantry and armour must be trained for this and the pursuit.[227]

On 30 July, following discussions with his senior staff officers, Auchinleck approved this approach in its final form. He favoured a set-piece attack that would exploit the strengths of the British Army and ordered Major General Gott (7th Armoured) and Lieutenant General Ramsden (XXX Corps) to prepare plans. The commander's focus was on the southern flank and Rommel's expected thrust south of the Alam Halfa ridge, with counter-strokes issuing south into his left flank.

As he formulated his plans and marshalled his forces to counter the thrust of his German foe, Auchinleck was also conscious of the influence of British politics on his command. His relationship with Winston Churchill was characterised by strained cordiality as a result of his subordinates' fiascos which were ultimately his responsibility. What Auchinleck did not know, however, was that the War Cabinet (after Churchill's Conservative Party handsomely lost a by-election) had agreed a political solution to what was generally regarded as a win-less Middle East despite the enormous quantities of materiel and men sent to the theatre. The upshot was that Churchill visited Cairo in August with the CIGS, General Alan Brooke, bearing an agreed blueprint for a wholesale clean-out of commands in HQME and the Eighth Army.

However, Churchill's blueprint was to suffer at the hands of fate. The proposed appointment of Gott under General Harold Alexander never eventuated as Gott was killed when *Luftwaffe* fighters intercepted and shot down a RAF transport in which he was travelling on 7 August. General Bernard Montgomery, who had been designated GOC First Army (Operation Torch) thus became GOC Eighth Army by default. Dorman-Smith (amongst others) was sent into obscurity amid feelings of frustration and disappointment. These sentiments soon turned to bitterness with Montgomery's claims that he personally had saved the Middle East from Rommel and planned Operation Lightfoot, the main assault on El Alamein.

The Australian forces remaining in theatre in Syria, Palestine and Egypt may well have breathed a sigh of relief that they were not destined to be uselessly sacrificed by less than competent generalship further west. Doubtless, following the shock 14 February 1942 capitulation of Singapore, their thoughts turned to their homeland rather than this dusty, dry and unlovely terrain.

During the Siege of Tobruk, the infantry of the 9th Division had fought as an entity rather than as separate forces divided by regimental affiliations. They quickly realised that a critical element in the successful defence of Tobruk—the inter-regimental, arms and service affiliations forged by battle—could be further forged in realistic training and exercises prior to the next offensive. The division concentrated first in Palestine, then Lebanon and Syria, where it was part of the Ninth Army garrisoning force protecting the northern approach to the Delta. Reinforcements arrived to make good the wastage of three months' continuous service and integrated with the desert-hardened veterans. Individual, sub-unit and unit training proceeded diligently under the watchful gaze of Morshead, his GSO1 and Chief Staff Officer, Colonel Henry Wells (Lloyd had returned to Australia), his artillery, engineer and signals commanders, infantry brigade commanders and their staffs.[228] Brigadier Alan Ramsay continued to serve as Morshead's CRA—his chief artillery adviser. Training took primacy in Morshead's wartime philosophy, typified by his comment to his commanders that 'We must train the men to give them confidence begotten of knowledge and experience.'

Brigades and their field regiments (and other units) were rotated in different areas in both countries. The 2/7th spent a month in Aleppo in February in support of the 20th Brigade, although it was primarily stationed inland from Tripoli at Zgharta and Bsarma. Officers were detached for tactical exercises with the brigade's battalions while the gunners improved defences in 'comfortable conditions'. The 2/8th dug tank traps and trenches for the infantry at Jebel Tourbal and living off field rations became the norm for all ranks for increasing periods. The 2/12th was ensconced around a Maronite monastery near Zgharta and batteries under canvas built sandbagged walls around their tents as shelter from the wintry weather. They constructed OPs and 'surveyed in' all probable targets that could be observed to the north, including the exact coordinates of bridges, crossroads and other features in the area. In April inter-regimental shooting competitions were organised by Brigadier Ramsay and his staff. The 2/7th's 57 Battery received high praise for the accuracy of its shooting and 14 Battery topped the anti-tank shooting at Tartus range. In late May, the 2/8th Field Regiment, affiliated with the 24th Brigade, conducted 'very serious exercises' (in the words of one participant) in the desert near Homs. To the north-east lay Forghlos and Palmyra, characterised by hot dry deserts, and all units suffered in the arduous conditions. The daily water ration was one gallon per day (to be shared with the cooks and drivers) in the hottest, thirstiest and most gruelling conditions. While the gunners gained some relief from their camouflage nets, the less protected infantry quickly became distressed by the conditions and many soldiers suffered heat exhaustion.

'Tactical exercises without troops' (TEWTs) were conducted for both artillery and infantry officers to allow each to develop an appreciation of the other's role. The 2/7th officers 'paired' with infantry officers from the 20th Brigade, mainly those from the 2/17th Battalion, although some were paired with men from the other battalions, the 2/13th and 2/15th.[229] The gunners conducted realistic course shooting east of Tel Kelakh and practised supported arms training. During this period the 2/7th Regimental Survey Officer, Lieutenant T.A. Rodriguez, enthusiastically trained his section in the finer points of flash spotting and located eight hostile batteries in 25 minutes, a feat which drew much praise from Ramsay. The training, which hitherto was based on the 'defence and withdrawal' tactical phases of war, concluded in late June with a demonstration of an attack—a portent of events. This initially involved battalion groups and then expanded into a brigade group (or 'box') with 600 vehicles deployed over an area 3500 yards wide and 6000 yards deep. A defensive 'box' saw battalions, companies and platoons of infantry sited so as to be mutually supporting with their weaponry, with artillery and engineer officers as advisers to brigade and battalion commanders. Anti-tank troops of guns were deployed along the most likely enemy approaches. Wire entanglements and minefields were used to channel tanks and troops onto the 'killing ground'. Wireless links and line communications were established according to requirements. To the rear were the gun positions, other arms and services etc. to support the 'box' concept. This was nothing if not realistic, except for the fact that the 'box' signified a defensive rather than an offensive outlook. The training climax came with a simulated attack by the 24th Brigade on 'German' lorried infantry, followed by a supporting arms demonstration of artillery defensive fire. A strafing run was staged by 451 Squadron, RAAF.[230]

News from home of Japanese military successes and the threat to Australia created morale problems within some units. These were managed by judicious leadership and focused training—and the welcome onset of warmer weather. One brigadier, when quizzed as to whether the division would return to Australia or remain in the desert, replied, 'Go west, I think.' His reply was given little credence. It is worth noting that Morshead's division was regarded favourably by the Lebanese and Syrian populace for its general conduct and demeanour, not only when the men were on leave, but also when they were interacting with locals on garrison duties.

Notwithstanding these extensive training exercises, the Army's broader commitments still had to be met through the transfer and posting of all ranks. 'Normality' had to be maintained even in abnormal times, lest this vital intelligence, like small pieces of a jigsaw puzzle, provide the enemy some knowledge to his advantage. Sometimes this had a 'lighter' side. In the 2/7th, an OP Ack, Bombardier K.G. Wilson (a promising Adelaide King's Counsel) was promoted overnight to major in the Legal Corps. Three officers of the regiment found themselves transferred out to return to Australia to impart their battle lore to others. And, as if to remind soldiers of the ever-present dangers of campaigning, even as garrison troops, fatalities from accidents and illness continued to occur. The 2/12th, for example, lost three men during this period.[231]

The announcement in June that Tobruk had capitulated had an enormous impact on the men of the division. However, their initially bitter reaction gradually yielded to a feeling of resolve 'to become a cohesive fighting force without frills—just to stop the German advance was not enough; we knew we had to push him back.'[232] Exercises and training over, the division concentrated around the Homs-Jdaide area and, on 26 June, amid much 'pseudo-secrecy', with the 26th Brigade leading, the division boarded its transports and moved through the Baalbek Valley to Gaza, Kantara, and Cairo, travelling non-stop for three days and nights, and finally halting near Al Maza. The divisional convoy was an extraordinary 50 miles long and represented a triumph for the Q Movements staff in Cairo which planned the move at short notice.

At Al Maza (the site of the School of Artillery and RA Depot) the gun regiments spent some days checking and re-checking their guns and equipment, drawing additional batteries, D3 and D8 signal cable, replacing vehicles and generally getting themselves ship-shape. Gunner Mountjoy, an enterprising driver from the 2/8th, found a 6-wheel Bedford lorry abandoned by its previous Eighth Army driver. On the back of the lorry Mountjoy discovered six motorcycles for which he planned a variety of uses. His Troop Sergeant Major, WO2 C. Smart, was delighted with Mountjoy's presence of mind. This initiative extended to the gunners' accumulating a wide variety of small arms and automatic weapons, much of it German and Italian, for their personal use.[233] The situation highlighted the extraordinary scrounging ability of the Australian soldier. Later, when the order came to load first line ammunition, all ranks knew that their training in the Levant was soon to be put to the test.[234]

Rommel's German and Italian troops had achieved impressive victories over the British and Commonwealth forces since he had taken command in February 1941.[235] He had previously confronted Montgomery, then a divisional commander in France in the BEF's retreat to Dunkirk, and now had 18 months' experience in desert warfare—a considerable advantage. Montgomery was not without some form of desert experience, having served in Palestine as a battalion commander before the war. His predecessors had been completely unable to counter Rommel's tactics, mainly due to a superb intelligence system that delivered Rommel's opponent's plans to his staff and to exceptional battle drills based on sound doctrine. However, this was about to change, thanks largely to a brilliant episode of minor infantry tactics by the 9th Division's 2/24th Battalion. Thus, in September, Montgomery inherited an army with superior materiel and intelligence functions than those of his predecessors. He was to use it brilliantly from the time he took over as commander.

On 30 June the 2/7th led the division's gun group west towards the division's area at the western end of Lake Maryut and passed a rabble heading towards the Delta, described by the 2/7th historian as the 'ugliest sight in war—chaos past caring.'[236] The 2/7th, which was attached to the 26th Brigade, reached its positions on 4 July and began to dig in. The men were promptly strafed by the *Luftwaffe*, although without casualties. The other regiments deployed to support their brigades as they arrived over the next three days. On 9 July Morshead's division arrived at El Shammama on the coast under command of XXX Corps

(Lieutenant General Willoughby Norrie). The 24th Brigade, with three Troops (D, E and F) of the 2/7th and 2/12th in support, took up positions on Ruweisat Ridge (Trig 93). The 26th Brigade (with A, B and C Troops of the 2/7th) were sited close to the coast. The 20th Brigade was in reserve. XXX Corps artillery units in support of the division were under strength; Norrie only had the 7th Medium and South African Artillery (SAA) field regiments that could be brought to bear on a target. General Auchinleck was to replace Norrie with Major General W.H.C. Ramsden (of the 50th Division) within days, such was the fluid state of affairs in the Delta High Command at that time.[237] Morshead and Wells were unhappy with these changes, considering that they would not be enduring and that 'something had to give'.

The appointment of Ramsden was yet another case of relatively inexperienced senior British generals trying to cope with tactical situations that went far beyond their competence to address. None had Morshead's grasp of command, much less his experience, and this was obvious to the troops. Yet Morshead had not exercised command in mobile warfare—only a successful withdrawal and siege. If the performance of Ramsden's division was any guide, he would not survive at corps level either. It was no surprise that he clashed with Morshead over the role of the 9th Division which he proposed to break up—a proposal its commander refused to countenance. The division itself lost no time in making its presence felt.

On 7 July, the 2/43rd Battalion raided Trig 71, west of Ruweisat Ridge, inflicting casualties and damage and capturing prisoners. The division's first big show was an attack on Tel el Eisa on 10 July, and Brigadier Ramsay had, for the first time, his three 25-pounder regiments, the 7th Medium (recently reconstituted after losses at Gazala) and SAA that could be brought to bear for the assault. The 2/7th was deployed astride the road east of Tel el Eisa, the importance of which was obvious—it was 'vital ground' in army terminology. If the Eighth Army were to defend El Alamein and eventually mount a major offensive, Tel el Eisa had to be occupied. Trig 24, which ran east-west and was around a mile and half long, dominated the area as far as the eye could see. The 2/7th was in 'direct support' of the 26th Brigade, and the command post officer of the 14th Battery (Lieutenant Roger Fitzhardinge) was given a gun area astride the road and under direct observation from the feature to his west. Night occupation of a gun area after a daylight reconnaissance under observation from the enemy was a good test of training drills. Well before dawn, the 2/7th reported 'regiment ready'.

Supported by the divisional artillery, SAA and 7th Medium, the 26th Brigade attacked along the coast towards Trig 33; the 2/8th was in direct support of the 2/24th Battalion and its gunfire was much appreciated by the infantry. The Italians holding Trig 23 (the first objective) found themselves captive by breakfast time. On Trig 33, the 2/24th Battalion captured Unit 621, a German Radio Intercept Unit, whose ardent commander was killed, while most of his men were taken prisoner. This infantry company (supported by the battalion's carrier platoon) success was to have extraordinarily far-reaching consequences for the Allied cause. The battalion then swung south to take the Tel el Eisa feature which remained a scene of bitter fighting for several days, with tank battles and heavy artillery fire from both sides as the Germans and Italians attempted to recapture it. The usual scenario involved the appearance of armoured cars and tanks, followed by the enemy infantry forming up to attack plus the supporting artillery preparation. From their high ground the OPs could see clearly and ordered regimental concentrations (Mike targets) to be fired—first 10 rounds of gunfire, then 20, then 'Fire until ordered to stop'. For many OPOs, this was their first experience of a typical desert battle with, among other sensations, German GP 88s firing airburst shells at their parties and infantry localities. While most of the field artillery support was for the infantry, counter-bombardment targets were engaged actively by medium guns. The *Sabratha Division* artillery, one and a half battalions of *Bersaglieri* and a battalion of the *Trieste* were all destroyed. One Italian recorded that, 'allied artillery assailed the heavy artillery of 52nd Group 152/37 and 33rd Group 149/40 [*Bersaglieri* and *Sabratha*]. This amounted to a disaster.'[238]

This attack was repulsed, only to be resumed early the next morning. A phalanx of German armour and lorried infantry advanced on 'the cutting' at Tel el Eisa. At Trig 24, Captain J. Elder (OPO) with OP Ack Gunner Alan Kinghorn, Signallers E. Atkinson and R. Ware and Driver B. Fenton, established their OP under continuous fire. Elder and three other officers sent forward were all wounded. Gunner Kinghorn was awarded the Distinguished Conduct Medal (DCM) for his initiative in taking over and directing the 16 Battery fire with great accuracy. On 12 July, the 2/8th was supporting the 2/23rd Battalion on 'The Hill of Jesus', as Tel el Eisa was known, when another ferocious attack was launched at 6.00 pm out of the setting sun, a typical panzer feature. A DAK regiment advanced from the 'Leg of Mutton' feature. Shellfire had cut line communications between the OP and the batteries and maintenance signalmen worked frantically to reinstate communications. By this time the forward elements of the enemy were only 600 yards away. The battalion's history records the scene: 'The OPO brought down a regiment's fire in front of the enemy. They hesitated. The slaughter began. Up and down the line of advance the guns played ... Even to the hardened defenders it was a sickening sight, and even at the height of the battle men found time to feel sympathy for a courageous enemy.'[239] Captain Tulloch Roberts

(2/8th) won the MC in this series of actions. Australian casualties were moderate, while many prisoners and guns (mostly Italian) were taken. Tel el Eisa—the key to the defence of El Alamein and the route to the Delta—was never again occupied by Axis forces. The artillerymen also noted that the initiative in the air had passed to the DAF, which made a very welcome change.

As previously described, at that time 'Jock columns' were a feature of British tactics, and comprised a mixed group of arms with a specific mission of short duration. Major John Day, BC 57 Battery (2/7th) commanded a raiding force comprising a squadron of 12 Stuart tanks of the 6th RTR; four 25-pounders and four armoured cars from British units; the 57th Battery's eight 25-pounders; a platoon of the 2/2nd Machine Gun Battalion; and a platoon of four Bren Gun Carriers from the 2/23rd Battalion. Its objective was Miteiriya Ridge and the DAK's communications. In the early morning of 11 July, the column swiftly overran the Italian infantry (*Bersaglieri*) positions. As day broke, the DAK reacted swiftly and DAYCOL (as Day's Jock column was known) was bombarded by artillery fire from Italian 75-pounders. Thereafter an 'artillery duel' developed with 57 Battery's OPs established on the gun position just 2000 yards away from the enemy. They targeted the enemy gun flashes and silenced all eight guns, but not before taking some gunner casualties of their own. A sally towards the ridge by the tanks and 6-pounder anti-tank guns (of the RHA) resulted in the destruction of three Italian tanks. By the time the sun was high in the sky, Day realised that the column, which had partly achieved its objective, had become an obvious target and should withdraw. It was overlooked by high ground, was being continually shelled, and was a long way from its given objective. Day showed great courage and determination in extricating his force from a sticky situation throughout the afternoon, during which it was also dive-bombed. The infantry mopped up, prisoners were marshalled and, at 4.00 pm, the column 'leap-frogged' its way through surrounding minefields bringing its 1034 prisoners with it. The 57th Battery suffered three killed and five wounded. Day was Mentioned in Despatches and Gunner T.P. Hill was awarded the MM in the same action for leading a troop of guns to safety through a minefield.[240]

Australian field regiments had not had the same exposure to German panzer tactics in attack or counter-attack as their British brethren. There was a stronger likelihood of the division's facing an attack or a counter-attack at this stage of the battle rather than conducting offensive action to support formal attacks (before Operation Bulimba was conceived) or encountering action that would see both attack and defence phases. A key to the enemy's success was the cooperation between tanks, artillery and infantry. To dominate and disrupt this battle drill required accurate and timely defensive fire from artillery—both field and medium. To counter enemy (field) artillery fire on the defended locality required effective counter-bombardment methodology and having guns available for this, preferably mediums. To neutralise enemy observers and armour, smokescreens were laid down—weather permitting. Tanks that continued to advance through the smokescreen could be engaged at favourable ranges by anti-tank guns. High explosive percussion or airburst shells placed beyond the line of tanks separated the infantry from their armour and reduced their ability to cooperate. The force was at its most vulnerable when the infantry was debussing.[241] This 'steel curtain' of regimental defensive fire 450 yards long and around 100 yards or more deep could be moved around by the OPOs to maximise effect. To effectively neutralise PAA tactics, the 9th Division preferred night operations, something their opponents were reluctant to undertake as they used the period between dusk and dawn to repair equipment, recuperate and so on. The enemy infantry was also afraid of the Australian infantry, believing that 'after dark they did not take prisoners'.

Once battle had been joined on the coastal sector, it produced its share of excitement for C Sub-Section, F Troop, 58 Battery of the 2/8th, commanded by Sergeant Vernon Garrard. On 12 July, the battery deployed on the spur of a ridge near Trig 26, parallel to the road at Tel el Eisa. It was attacked by Stukas and bombs damaged Garrard's gun which suffered a shattered wheel and range cone, and a seized breech block, while the limber on its side was peppered by shrapnel. As the LAD recovered the gun for repair, the battery was shelled by 'Alamein Annie', a 210-mm naval gun from Arko 104 firing from close to the Sidi Abd el Rahman mosque. The next day, while engaged on a counter-bombardment shoot, several 88 rounds passed through the camouflage net above Garrard's gun. After a busy day in support of the infantry, Garrard's men faced a rare PAA dusk attack. The tanks of the RTR were ordered to withdraw from hull-down positions and panzers ploughed through the infantry. As they were 'crested', the gunners could not use open sights. Instead they loaded HE with 119 fuse and Charge 1 with some success. Tracers from the tanks' machine-guns passed overhead. When the tanks crested the ridge, the gun position officer gave the 'death or glory' order: 'Gun control; down camouflage nets; AP shot; engage own targets.' The tanks hesitated, then withdrew and enemy shelling stopped. There were only a handful of guns between the panzers and the coast. It was, recorded one witness, very timely relief.[242]

During July, HQRAA allotted the 2/7th to fire concentrations in close support of the infantry by observed shooting and counter-bombardment shoots. This was a reflection of the expertise gained while the 2/7th was Depot Regiment at the School of Artillery and on which the Commandant had commented favourably. Invariably, these targets involved RHQ and their technical gunners' skills, as well as BHQ command post staff. As Goodhart notes, 'It was work [using traces, tasks for tricky predicted shoots] which required speed and accuracy ... work carried out, most of it, in the fuggy atmosphere of dugouts, blacked out with army blankets across "inner" and "outer" doors: work that required not only rapid calculations, but relied on highly efficient and well-oiled communications as well ...'[243]

The month of July would see all gun regiments endure every facet of hostile enemy action—bombing, strafing, air and ground observed shoots, hostile battery fire, mines, booby traps, anti-tank, small arms and machine-gun fire in the course of their duties. The 2/12th historian reckoned 17 July to be the worst: 'a bitter day for both sides'.[244] A 2/7th gun from B Troop received a direct hit on its breech and the entire crew was killed or wounded. However, by dint of its leadership and training, the 9th Division coped well and thwarted Rommel's attacks. The regiments fired linear targets, concentrations, Stonks and smokescreens in support of their infantry. It was serious campaigning and exhausting for all concerned. Enemy artillery chose to inflict particularly severe damage on the division's infantry rather than engaging in concentrated counter-bombardment shoots. On Miteiriya Ridge on 23 July, another battle began, and the 2/7th OP, Captain Bill Ligertwood, distinguished himself when attached to the 2/28th Battalion which, four days later, much to the mortification of the division, was overwhelmed and many men taken into captivity. Ammunition expenditure was prodigious. The 2/12th fired more than 2000 rounds a day in early July. On 17 July it fired 9643 in support of the 24th Brigade's attack and withdrawal. The 2/7th fired 20,129 in five days.[245]

There were numerous occasions, and these were to be repeated, when members of the OP party in their Bren Gun Carrier became prime targets for enemy field, anti-tank and 88-mm guns. In all regimental accounts there are constant references to OP carriers and their occupants becoming casualties or having to improvise to perform their role—or both. Most decorations would go to this group in the battle. One which made a big impact on the 2/7th was the loss of Captain Arthur Fielding whose party of two had been involved in the 2/28th Battalion debacle and who was killed in his Bren Gun Carrier. Gunner Athol Manning, the carrier driver, won his MM for showing initiative and defying enemy fire and mines to deliver Fielding's message of the battalion's plight to Brigade Headquarters, unfortunately too late to save them. The breakdown in artillery communications was cited as the most crucial reason for the failure of the operation.[246] Some measure of the intensity of the battle is apparent in the casualty rate—note that, for slightly different periods, killed in action (KIA) includes those who died of wounds:

> 2/7th: 7 KIA, 14 WIA
> 2/8th: 10 KIA, 65 WIA (10—31 July)
> 2/12th: 4 KIA, 13 WIA

In other regiments, both field and anti-tank, there were numerous examples of heroism, probably the most extraordinary (and posthumous) that of Sergeant A. McIlrick, MM, of the 2/3rd Anti-Tank Regiment, who single-handedly destroyed six tanks with 19 rounds of 6-pounder shot.[247] The anti-aircraft gunners of the 2/4th LAA Regiment were deployed to protect (with some success) the gun regiments and headquarters which came under heavy Stuka attack on 10 July and throughout the month.

July was an important month for Ramsay, comprising the first big test for his complete artillery. He had spent an anxious month nursing his three regiments through continuous operations. His report covering the salient points of supporting infantry and armour dwelt on the efficiency of smokescreens to blind the southern flank of Trig 33 from 9 to 11 July and the importance of medium and SAA field guns in counter-bombardment roles. There were times when the Axis gunners fired from positions that had not been fixed accurately enough for counter-bombardment fire to be effective. They also moved their guns frequently and, in so doing, focused the minds of the embryonic and, at this time, generally inexperienced counter-bombardment staff.

During the mid-month engagements, the infantry was happy to report that 'very heavy and accurate fire' on lorried infantry helped operations considerably. There was, however, one exception. On 15/16 July, the 2/23rd Battalion was critical of artillery support at Tel el Eisa. While the battalion noted that the minimal damage to guns from enemy aircraft was due to the effective siting of the LAA guns in the gun area, the comment was made that the anti-tank gunners tended to suffer casualties when they were pushing their guns forward during the consolidation phase. They were also less effective engaging the leading elements of enemy armoured/infantry groups than field guns. On one occasion an OP directed his fall of shot to 'herd' four armoured vehicles into anti-tank gun range, where they were destroyed. He had maximum support from the South African gunners and the medium regiments. The line signallers had difficulties with US Army cable on issue and 78 miles of D3 was required to make good losses. There were also a number of casualties: 12 officers and 120 other ranks were killed, wounded or died of their wounds.

Replacements from artillery training establishments amounted to seven officers and 193 other ranks, as well as many more for anti-tank units which suffered the heaviest losses during this period. The lessons Ramsay distilled concentrated on the tactical handling of his force to best support the infantry. Foremost was the accurate location of enemy guns. When this was not possible, a barrage was ideal to cover the infantry approach. During a 2/28th Battalion attack, enemy gunfire 'closed' the minefield gap, threatening to cut the battalion's withdrawal. The 2/28th also felt the sting of the German 88s firing airburst and, as if to add insult to injury, the relationship between the infantry commander and his artillery adviser proved less than ideal. Both headquarters should have been cooperating closely; however, the CO was forced to fight his regiment as an entity.

There was no shortage of distraction, as personnel from RA field regiments new to the desert were sent forward to 'get the form' from their experienced colleagues. One Major Williams, a counter-bombardment specialist from the US Army, visited headquarters and, on 29 July, as part of the relentless attempt to improve cooperation between air and ground, 451 (Army Cooperation) Squadron, RAAF, staged a demonstration. The War Diary recorded: 'Very successful day. Many lessons obtained. Captain Grimston, RA, offered excellent advice.' Sustained operations during the month took their toll on the guns, and Ordnance (DADOS) arranged for four guns per week to be taken to the rear for overhaul. At the same time, Routine Orders promulgated the ruling that going AWOL (absent without leave) was a court martial offence. As July came to a close, General Auchinleck noted that 'the division was a more self-confident division.'

August was a period of static defence for the division, although this proved a misleading term for the field gunners. Regiments engaged hostile batteries which inevitably retaliated. Because German flash spotters were so expert, the regiments moved guns out into sniping positions each night to 'harass around the clock'. Flashless propellant of Canadian manufacture was not reliably so, with only a third proving truly 'flashless'. To confuse flash spotters, simulated gun flashes were lit either side of the sniping guns. These were none too reliable either![248] When the front became static (and before the next offensive) gun regiments were redeployed frequently. It was an exhausting business: 'Often one battery was on the move and the other two remained in action. As soon as all three were dug in, another new move commenced. To the gunners it appeared to be bastardry on a large scale.' Well may it have been, but the main reason was protection from enemy counter-bombardment. The 2/12th gun crews recorded that, in the six weeks to 22 August, they changed position 22 times. Each new position required the gunners to dig a 360 degree gun pit—the gun was mounted on a circular platform for ease of movement—rather than a 'key-hole' type pit allowing a 40 degree 'switch' left or right of the centre of arc (about a third of the size). Little wonder they submitted a requisition for 8000 sandbags! This was also a period when the regiments were to feel the force of Ramsay's leadership. The 2/7th experimented with Matilda tanks as OPs, but the conventional wisdom amongst OPOs was that a Jeep (4 x 4 ¼ tonner) was superior because it offered a smaller target, while armoured vehicles attracted enemy fire. OPOs joined the 23rd Armoured Brigade in exercises early in the month while Ramsay instituted a policy of using his field guns to demolish enemy defensive works by day and then engage their working parties with harassing fire at night.

Throughout August Morshead maintained the pressure on his opponents. The 2/8th supported the 20th Brigade on 14 August and the guns also supported fighting patrols. Ramsay went to Tel el Eisa and watched the registration of high frequency tasks on the Rahman track. The 9th Division's next major effort was launched on 23 August as part of XXX Corps operations. Its role was to strike at Axis supply routes south of Tel el Eisa. At the same time, enemy attacks were expected from the south, and command post staffs were busy completing gun programs in support. Operation Bulimba had limited objectives and the 20th Brigade's 2/15th Battalion was the force employed. The gunners had 12,000 rounds of HE, 1560 rounds of Smoke and 1530 of Supercharge dumped at their positions. They provided one FOO and three OPOs per regiment with Major Johnston appointed Liaison Officer at brigade headquarters. The infantry was supported by the tanks of the 40th RTR while the 2/13th Field Company Engineers gapped minefields. The Germans reacted vigorously notwithstanding that 16 defensive fire tasks, mostly linear targets, were spread around the bridgehead to protect the battalion. The 2/7th looked to the west, the 2/8th to the south and the 2/12th to the north. Enemy casualties were around 300 while the 2/15th Battalion casualties amounted to 173.

Following this and earlier battles, all the gunners (and they were not alone) were left feeling somewhat jaded by their continuous exertions. Until the front stabilised to the High Command's satisfaction—and there was no 'relief in the line' in sight—sound leadership was essential to prevent morale sagging. One enterprising command post officer from the 2/7th who, like most of his battery, was keen on horse racing, organised a 'race meeting' for a Saturday afternoon. The troops could nominate a horse, complete with lineage, which they did with enthusiasm; names such as 'Victory', by Rommel out of Africa, and 'Misery', by Pub out of Beer, were typical of the titles that peppered the field. Using a dice and the sandy floor of the command post, the 'horses' moved around the 'track' when no targets were being engaged, and the 'results' were passed by the battery exchange to the Troops. There were some amusing incidents including one that found its way into regimental lore and concerned a gunner who had problems looking smartly turned out, preferring to do the minimum possible. One Saturday, the command post officer came across him polishing his boots, brass and attiring himself in his best dress. 'Turning over a new leaf, are you, Smithers?' the command post officer asked. 'No, Sir,' replied the soldier, 'I'm going to the races this afternoon.' There were also race meetings in Cairo, although it is assumed that the key figures in HQME were too busy with other matters to attend.

Auchinleck placed battlegroups south from the coast, and one of Morshead's first actions was to tour the front and inspect the dispositions with which he had to conform on his southern flank. He did not like what he saw. Mobile columns were still part of the tactical scenario which must have added further to his unease. These were structured around a battalion with two batteries of field artillery. The 50th Division, Ramsden's former command, the 10th Infantry and the 5th Indian divisions were also fragmented. Brigadier Meade Dennis, CCRA XXX Corps, found his ability to influence the battle undermined by a lack of staff and resources, particularly in the area of communications.

During the battle an inexperienced British armoured brigade (the 23rd) charged its opponents' tanks without any planned fire support for its adventure. In similar fashion, the 15th Indian Brigade was destroyed at Deir el Shein when isolated by panzers without armoured or artillery support. Even the experienced New Zealanders suffered a reverse when they outran their artillery support and took a bloody nose. Other features of the battle included divisional commanders who 'interpreted' their instructions as orders had not been issued. It was a week of mismanaged opportunities. One of the two bright spots was that the SAA and the 7th Medium delivered a salutary warning of what lay ahead when they brought the *90th Light Division* to an abrupt halt with an Uncle (divisional) Target on 7 July. The other cheerful note was that, without intelligence from his now defunct eavesdropping wireless unit (U621), Rommel's staff did not have a clear picture of Auchinleck's dispositions. Until he could replace that expertise, Rommel would suffer a crippling disadvantage.

Preparations for countering Rommel at Alam el Halfa and launching an offensive from El Alamein occupied the High Command and, in August, the Eighth Army found itself with a new commander. One of General Montgomery's first decisions had been to relocate Air Marshal Tedder and his HQ DAF close to the Eighth Army Headquarters. Then began the exodus of senior officers. Lieutenant General Ramsden, of whose abilities Morshead thought very little, relinquished command of XXX Corps to Lieutenant General Oliver Leese a few weeks later in mid-September. Auchinleck vacated his post (and was sent to India) and Morshead wrote him a letter praising his achievements as a commander in extraordinarily difficult circumstances.[249]

Montgomery set about revitalising the Eighth Army and succeeded brilliantly. Morshead considered him 'a breath of fresh air', relishing Montgomery's philosophy which he outlined in discussions (summarised below):

- The line (El Alamein) will be held.
- Formations not to be broken up—no more 'Jock columns', 'Battle Groups'. Divisions will fight as divisions.
- The word 'box' is not to be used.
- 'Consolidation' is not to be used (it implies a handicap to momentum, and 'reorganise' is better).
- No more 'outposts'—they will be called Forward Defended Localities (FDLs).

In late August Rommel planned an attack to roll up the British left flank and then strike north-east to the Delta. One account of this period states that Rommel's plan was partly based on an intelligence deception executed by skilled cartographers in HQME. They produced a topographic map that falsely showed 'hard going' in an area south of Alam el Halfa Ridge, which was new territory for the panzers. The map was placed in a derelict lorry and found by PAA troops. In fact the area was not hard going, but soft sand and therefore unsuitable for tanks.[250] Attacking at night what he thought was a lightly defended and mined area, Rommel's armour ran into extensive minefields (there were 180,000 mines laid), lost momentum and was heavily engaged by Corps artillery. The next day, at 6.00 pm, he swung north-east and encountered tanks and anti-tank guns dug in on Trig 32.

Montgomery had brought with him as his BRA Brigadier Sidney Kirkman, and now gave Kirkham 'his' plan for the defence of the ridge. Kirkman thus became the first senior artillery adviser to have the unequivocal backing of his army commander. When the SAA induced a state of panic in the *90th Light Division* troops in July—a highlight of First Alamein—it indicated quite clearly that their adversaries were contemptuous of British artillery methods. The SAA had taken maximum advantage of this.

The current situation was, however, entirely different, and was exacerbated by Montgomery's having relocated his DAF Headquarters to an area in close proximity to his own. At Alam el Halfa, the artillery and DAF bombers again pounded the enemy throughout the night and Montgomery's line held comfortably. Rommel's troops, though wearing and plagued by shortages of all kinds (a result of DAF successful interdiction attacks on his supply routes), were driven by a desire to get to Cairo. However, Montgomery countered Rommel's tactics by fighting on the ground of his own choosing, using day and night tactical bombing so that Rommel's formations lost momentum. He used artillery concentrations to break up attacks as they formed. Montgomery noted that, at Alam el Halfa, 'artillery and armour were used in concentrations, and had been so positioned that the enemy armoured thrusts were dealt with quickly and efficiently. It had been unnecessary to conform to Rommel's thrusts ... Army and Air Force worked to a combined plan.'

Ammunition expenditure and gunner casualties for Operation Bulimba and the Battle of Alam Halfa totalled:

2/7th: 20,129 HE	2/8th: 20,519	2/12th: approx 20,000
11 KIA, 19 WIA	1 KIA, 9 WIA	2 KIA, 11 WIA

From July to September, all regiments acquitted themselves splendidly in supporting the Eighth Army at the El Alamein position. An important part of their contribution was the omnipresent counter-bombardment obligation which effectively countered the enemy's guns and howitzers. With XXX Corps was the 1st Composite Battery of the 4th Survey Regiment, RA, yet again supporting the Australian artillery with its survey, sound ranging and flash spotting. The battery's A Troop laid out four flash spotting bases (between Ruweisat Ridge and the sea, south and west of El Alamein twice, and for the Alam el Halfa Battle, towards the Qattara Depression), locating 125 hostile batteries in this period. The sound ranging Troop entered the fray in late August, establishing bases in the northern salient, then at El Mreir and finally at Tel el Eisa. During this period they recorded an impressive 165 hostile battery locations. The PAA had gained helpful stocks of field guns and ammunition for 25-pounder and 4.5-inch guns when Tobruk fell in June and were utilising these with great effect. Intelligence reports at this time indicated that British medium guns had been seen in a convoy on the coast road heading east. *Generalleutnant* Krause, Rommel's artillery commander, now had more guns to match the munitions and lost little time in using them against their former owners.[251]

It was therefore no surprise to Lieutenant Colonel H.C. 'Toc' Elton, DSO, CO 7th Medium, to find himself on the receiving end of high class hostile battery artillery fire. His Intelligence Officer at RHQ, Lieutenant Alan Benson, soon located their positions on air photos. He set up a bombard using three OPs to concentrate their fire on each gun in turn, firing 10 rpg from both batteries, completely silencing the hostile batteries.

General Major Krause commanded the DAK artillery in the Western Desert.
(Bundeswehr 183-J12335)

Air reconnaissance subsequently showed that each weapon had been hit. As part of an active counter-bombardment policy in force at this time, the 7th Medium moved one of its guns forward to range on an enemy flash spotting tower. Once again, the Intelligence Officer played a leading role. He set up two artillery boards, secured fresh meteor data from the 4th Survey Regiment and had five OPs linked to him by telephone. Using airburst fuse 210 HE, they had a verified bracket in five rounds. The sixth destroyed the tower. All regiments fired 'bombards'—gunner jargon for engaging hostile batteries. Lieutenant 'Tim' Rodriguez, the 2/7th Regimental Survey Officer, laid out a flash spotting base and the quality of response to calls for counter-bombardment fire drew praise from the Corps CBO, Lieutenant Colonel M. 'Stag' Yates.[252] All regiments, including the mediums, had also been on the receiving end of frequent counter-bombardment artillery fire and bombing raids from the *Luftwaffe* and had suffered casualties.

However, it was not long before the glaringly obvious shortcomings of British armoured employment with infantry during desert campaigns simply had to be redressed. There were many lessons to be learned, particularly from Operation Bulimba and at Alam Halfa. The 2/7th historian wrote of this period:

> Long before the end of August, it was realised ... that interlocking keys of cooperation were essential, as between infantry, artillery and armour. Such keys would have to be cut to a precision tool scale of efficiency not previously envisaged, if the combination lock of Axis minefields, defensive firepower and rapidly counter-attacking armour was to be faced.[253]

Any reading of the accounts of the Eighth Army over the tempestuous period following Gazala will highlight the difficulties senior commanders faced in trying to combat Rommel's forces. As far as the 9th Division was concerned, an enduring theme was the stark realism of its commander and his GSO1 in assessing the merits, or otherwise, of the plans of operations issued by XXX Corps in which they were tasked to take part. On 3 July Morshead refused to break up his division by detaching a brigade to another headquarters. Eighteen days later, he was involved in a serious dispute with his corps commander, Ramsden, over a plan which he thought was, to say the least, impractical and likely to result in unaffordable casualties. These were two important events given that, by September, Morshead's division was still intact but his confidence in British armoured doctrine (but not their bravery) had reached rock bottom. Accounts of the Eighth Army's forays against PAA forces over time make depressing reading, particularly the Gazala debacle. An underlying theme is one of the squandering of lives, particularly those of the Commonwealth armies of Australia, New Zealand, South Africa and India, because of poor leadership. One is left with the impression that Morshead's refusal was a cathartic event in HQME higher command circles. He was, in the jargon of today, making a significant statement that Major Generals Bernard Freyberg (New Zealand) and Daniel Piennaar (South Africa) had thus far not seen fit to make with sufficient force. The subject of competency for higher command from the Australian point of view has been well covered by Maughan and by Blamey's biographers. Even Bidwell speculates that Morshead was another Monash and should have been given a corps. But within the British Army at this time, and coinciding with Montgomery's assumption of command, the issue did not arise. Generals and brigadiers may have been sacked, but divisions fought as divisions thereafter and the higher staffs 'got the message'.[254]

Some time would elapse before British Army doctrine on armour-infantry cooperation progressed. If anything was gained from the withdrawal to El Alamein it was that the CO, OPOs and BCs saw at close quarters the terrain over which they would be fighting their regiments in October and November, had they known of Montgomery's plans. In operations to October, divisional artillery casualties amounted to 110. Infantry casualties of almost 2000 caused concern as Morshead's ability to call on reinforcements was limited compared to the British, Indian and (at the time) New Zealand armies.[255] After Alam Halfa, Montgomery decided that the Eighth Army would stand fast and defend the El Alamein line prior to launching the decisive attack. Offensive operations ceased and both sides dug in and laid extensive minefields in length and depth. The PAA also constructed well-protected gun positions that were to be a feature of the counter-bombardment engagements to come. The remarkable *modus operandi* of the PAA formations, command and intelligence functions and artillery resources are the subject of the next chapter.

CHAPTER 12
THE OTHER SIDE OF THE HILL: ENEMY COMMAND, TACTICS AND RESOUCES

The whole art of war consists in getting at what is on the other side of the hill.
The Duke of Wellington

Rommel wrote extensively on his strategy and tactics and gradually refined his doctrine for desert warfare over a period of 18 months. His central assertion was that armour was the core on which every aspect of the battle turns. Concentrating force, disrupting his opponent's supply lines, speed of movement, maintenance of momentum, and concealment of intentions were the primary objectives Rommel identified. Technically, a tank with a bigger gun would always beat that with a lesser armament as it could engage at a greater distance. The same principle applies to artillery, which must have greater range and be capable of carrying its ammunition in large quantities. However, what Rommel never alluded to in his writing was the enormous advantage he gained from his comprehensive intelligence-gathering network.[256] Rommel's results lent him an aura of invincibility and, although he came close to being beaten several times, he remained undefeated until October 1942. It is worth examining some examples of British intelligence and operations research at HQME that gradually saw the Eighth Army transform its mode of warfare and gain the upper hand.

There were many factors that contributed to the successes enjoyed by DAK armoured formations, not least of which was the efficient employment of all types of artillery in the armoured and light (i.e. light armoured) divisions. Italian armour was inferior to British and thus less of a consideration when formulating tactical plans. There were fundamental

A On the move

Reconnaissance elements proceed in direction of *schwerpunkt* given by higher command. Their objective has been located by wireless intercept. Recce element 1–2 miles ahead of tank group. Lorried infantry in trucks and half tracks. Anti tank and aircraft guns in travelling position. Field guns/SP may be supported by medium and heavy guns well to the rear. British forward elements (armoured cars with artillery OPs) report Panzer advance.

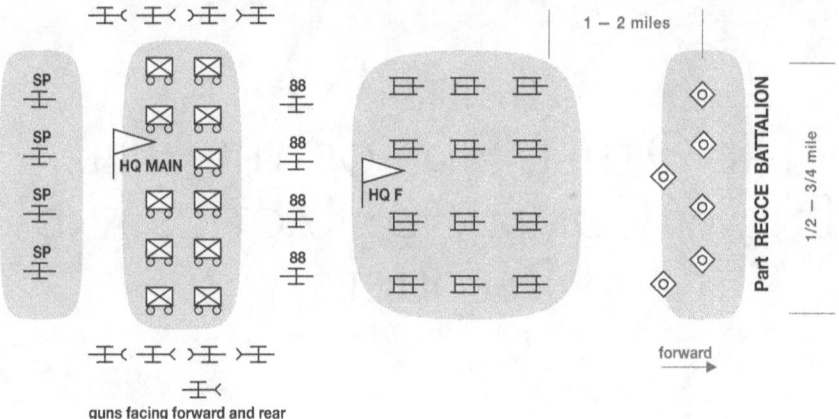

guns facing forward and rear

B Encounter battle

Recce group probe limit (flanks) of British defended area. Tanks and commander comes forward. Tanks and 88s engage anti tank guns; field artillery engages infantry FDLs and field artillery. Heavy machine guns (on tanks) when in range suppress defences. Lorried infantry dismount and form up. Anti tank and anti aircraft guns deploy facing outward. Field/SP artillery take up positions.

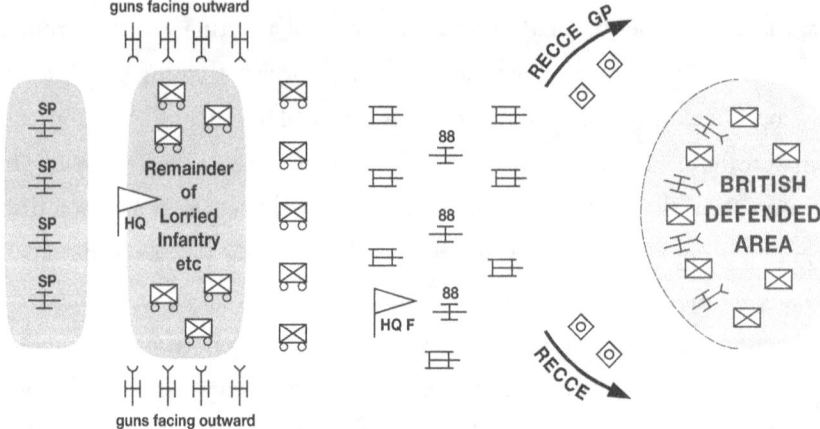

Symbol	Meaning	Symbol	Meaning
SP	Field gun (self propelled)		Medium tank
88	'88' GP gun		Armoured car and/or motorcycle
	Anti aircraft gun		Lorried infantry
	Anti tank gun		Infantry

C Assault phase

After British anti tank guns are destroyed tanks move to flanks and attack infantry, HQs and service units. Tanks and 88s go forward to engage field/medium artillery. British positions overrun.

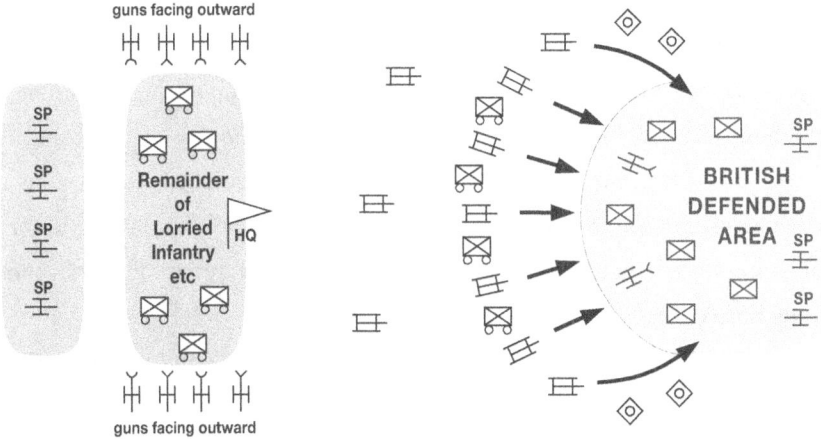

D A version of Panzer mobile defence

British tanks (and armoured cars) meet Recce Group and tank formation. They withdraw to allow 88s and anti tank guns to engage tanks. Lorried infantry and field artillery in rear are guarded by anti tank and anti aircraft guns and field medium/heavy artillery. German tanks split and withdraw to either side of infantry/services assembly, positioned for possible counter-attack.

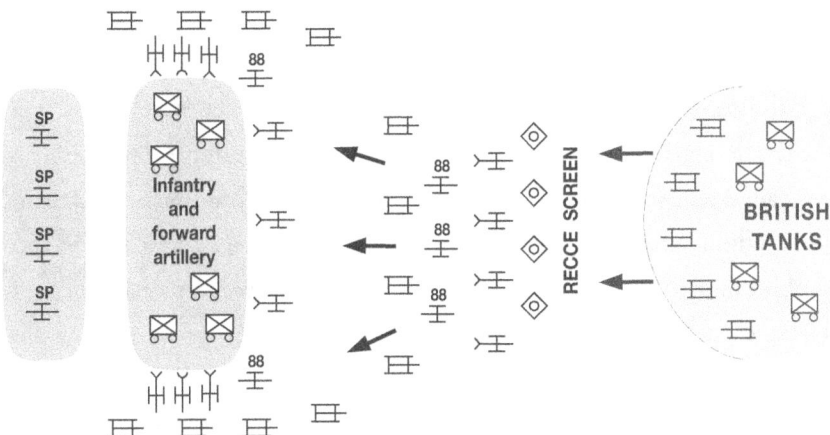

SP	Field gun (self propelled)			Medium tank
88	'88' GP gun		◇	Armoured car and/or motorcycle
	Anti aircraft gun			Lorried infantry
	Anti tank gun			Infantry

(doctrinal) differences between the British approach to armoured warfare and that of the German-Italian forces. Analysis of the DAK *modus operandi* identifies the effectiveness of the reconnaissance battalion as one of the contributing factors to its success. Polished battle drills showed the following panzers possible approaches and reported enemy strength, defensive layout and resistance to regimental and divisional commanders. Communication with higher command was excellent. A reconnaissance battalion using light and heavy armoured cars, half-tracks, motorcycles, mortars, light and medium machine-guns, light infantry artillery and anti-tank weapons was usually the first to be observed by the opposing forward troops and OPs. In battle at a distance, the armoured cars resembled tanks and calls for neutralising fire were rarely heard for 'armoured cars'.[257]

One advantage the Eighth Army enjoyed, however, was its ability to anticipate the thrust line (*schwerpunkt*) of German armour. It was well known that the DAK exploited any captured British equipment—guns and weapons of all kinds, tanks and vehicles, ammunition and maps. The latter were particularly useful since the DAK did not have the same resources to map North Africa as the British. In directing his corps and divisional commanders, Rommel would indicate a bearing for his *schwerpunkt* and leave the details to his commanders. The British had captured a device known as the '*strosslinie*' which was made from thick talc and consisted of two axes (one horizontal—'A' and the other angled at 90 degrees to it—a vertical 'B') each graduated from 0 to 12. The former slid up/down the latter. To indicate the direction of the day's thrust, the DAK commander placed the B (vertical) axis over the thrust line and the A scale adjusted over the map feature/objective. This gave two numeric values that were converted into a simple code for transmission. Late in 1941, the staff of the 4th Indian Division broke the German three-letter codes and, to speed up the process of breaking the *strosslinie* code, they arranged for their divisional artillery to fire a regimental target each morning. The German OPs and command posts sprang to life with wireless messages, while the Corps Special Wireless Section listened to enemy fire orders and matched the *strosslinie* codes to the fall of shot given by the flash spotters and OPs. While this procedure often resulted in counter-bombardment fire, for the expenditure of 50 rounds of ammunition, Corps Headquarters gained valuable data on the dispositions of the DAK units, the bearing of the thrust line and the day's code.[258] Unhappily, the data gained was often squandered.

However, the biggest coup of all was the 2/24th Battalion's capture of the German Radio Intelligence Company U621 of the DAK Signal Corps at Tel el Eisa in July. Anticipating a rapid advance, the company had deployed close to the FDLs instead of further to the rear in safety. Its zealous commander (*Hauptman* Seebohm) was mortally wounded during a spirited action and most of his company taken prisoner. Equipment and records yielded a veritable bonanza of data. Rommel was furious when told of this loss, for it removed from the battlefield his prime source of information on which to base his plans. During interrogation, one of Seebohm's subordinates (*Leutnant* Herz) gave his captors valuable information on British wireless security—or lack of it—on the Eighth Army's command nets. These lapses allowed Rommel to anticipate almost all of his opponent's moves. British brigadiers, generals and their staffs used 'codespeak'—low-level nicknames and code words—to refer to locations and commanders. The battalion's captive informed intelligence officers that this was easily broken by his operators. Herz averred that his operators' linguistic skills were more important than their ability as cryptographers. Most spoke perfect English acquired when working in London's West End hotels! They were at a disadvantage, however, when British commanders with Indian Army backgrounds spoke fluent Urdu communicated 'in clear' when exceptional circumstances required.[259]

U621 analysed battlefield traffic and reported its conclusions rapidly to DAK Headquarters. The unit monitored and deciphered the signals traffic of hostile or neutral governments, embassies and all naval, army and air headquarters. Analysis of a large quantity of captured records in Cairo revealed just how much the German High Commands in Berlin, Rome and in Africa knew about British forces—commanders' personalities, order of battle, plans, resources, personnel and vehicle states, etc. Moreover, this information enabled Rommel's spy centre in Cairo to be captured intact. This accomplished, the British were able to send false and misleading information on its wireless link to DAK Headquarters which was assumed to be genuine.[260] It is difficult to imagine a more important single minor action in World War II with broader strategic ramifications than the success of the 2/24th Battalion in capturing U621, an outcome undoubtedly made more complete by the quality of Brigadier Ramsay's artillery preparations for the battle. U621's successor was far less expert and, consequently, DAK operations across North Africa, Sicily and Italy never enjoyed the same intelligence advantage after July. Another significant benefit arose from the amended British approach to radio traffic and voice/code procedures which was tightened considerably, wireless silence enforced on reserve formations, 'padded out' messages used with dummy traffic and so on.

For their part, the British had established Ultra special intelligence links with the UK. Alexander and Montgomery had been alerted to Rommel's order of battle and plan prior to the battle of Alam Halfa. In the battle zone itself, by far the most important input came from the Corps Special Wireless and Wireless Intelligence Sections. In September/October they set up their tents in the centre of the front, near the flash spotting towers of the 4th Survey Regiment's No. 2 Composite Battery. They plotted all the headquarters, fixed intersections on gun position command posts and were reading tank-to-tank wireless traffic. This led (for security reasons) to DAK armoured squadrons signalling one another using flares during manoeuvring before or during the battle, which the flash spotters were able to report instantly to artillery intelligence channels.[261] The Eighth Army was 'reading' DAK encrypted wireless traffic—daily assessments of tank strength, gun availability and ammunition state, fuel state and deliveries, all of which contributed to the final form of the Eighth Army plan.

But the reverse was also true. The DAK knew the Eighth Army order of battle and was reading its daily manoeuvre and map codes (Slidex and Unicode) and administrative traffic between El Alamein and Helwan where GHQ was now sited. Pressure from London on Montgomery to bring his attack date forward led to staff 'contingency' arrangements. In this instance, the revelation of D Day came from a most arcane source—GHQ. Alexander had moved it from Cairo, requiring eight days' notice to provide stretcher-bearers. The Eighth Army's first message was on 8 October which led to the DAK 'standing to' on 15 October. When that date came and went, another message on the same day confirmed the DAK's analysis that D Day, and the full moon, would be on 23 October. In addition, as a result of *Leutnant* Herz's disclosures of security lapses, the Eighth Army's methodical Chief Signals Officer ordered days of 'wireless silence' on 4, 14 and 22 October. Because of the extensive buried line/cable network, this did not cause any insuperable problems; however, by the third wireless silence date, German suspicions had been aroused as to a probable D Day. Perhaps the greatest advantage that signals intelligence gained for the Eighth Army lay in decoding the traffic between the DAK and the *Luftwaffe*. Liaison arrangements between the two—vital to the land battle—were decrypted from 'Scorpion' Enigma. It was thus a simple matter for the DAF to sortie its fighters and fighter/bombers to destroy enemy aircraft on the ground—the most economical in terms of manpower and aircraft—rather than in the sky. As a result, *Luftwaffe* efforts became uncoordinated and generally ineffective over time.[262]

Rommel noted that 'deception' was a key part of desert warfare and both sides used ruses and stratagems to mislead and confuse. The historiography of many Western Desert battles includes examples of mistaken identity of vehicles, the principal cause of 'surprise', self-delusion and disabling and/or fatal 'hesitation'. The PAA used this more often as a result of having captured thousands of vehicles of all types, as well as guns. They also repaired immobilised vehicles and turned them into 'runners'. Sometimes they painted black crosses on the doors of lorries which meant that, from a distance, they resembled ambulances. Rommel used aero engines mounted on trucks to simulate the assembly of his armoured forces by creating dust clouds. However, the British were fast learners in this field and had many surprises in store for the PAA once the October battle commenced.[263]

The German artillery facing the Eighth Army was organised along similar lines to the British insofar as overall command was vested in *Generaleutnant* Krause as adviser to Rommel. Artillery was organised into gun regiments, Arko 104, and supporting artillery functions of flash spotting, sound ranging and intelligence *(Beobachtungs Abteilungen)* and survey and mapping *(Vermessung und Karten Abteilungen)*. This organisation absorbed a command arrangement in which both panzer divisions were under DAK command and the *90th Light* and *164th Infantry* (later to become a Light) *divisions* reported through Rommel, as did the three Italian corps. Initially, however, German divisional artilleries facing the Eighth Army comprised:

Generalmajor Gustav von Varst
15th Panzer Division: 33 artillery regiments; 3 battalions

Generalmajor Heinz von Radow
21st Panzer Division: 155 artillery regiments; 2 battalions

Generalmajor Theodor Graf von Sponeck
90th Light Division: 190 artillery regiments; 2 battalions

Generalmajor Karl-Hans Lungerhausen
164th Light Division: 220 artillery regiments; 2 battalions

Arko 104 remained a formidable threat and now comprised the *1st* and *2nd African Artillery Regiments*, the *221st Artillery Regiment (Motorised)*, the *408th Heavy Artillery Regiment (Motorised)* and five artillery battalions, the *364th, 408th, 528th, 533rd* and *902nd*. The *19th Flak Division* consisted of the *606th, 612th* and *617th Anti-Aircraft Battalions* of 88-mm GP guns, half of which were used as flak, the remainder as anti-tank guns. The *Luftwaffe* (under *X Fliegerkorps* command) provided the *102nd* and *135th Anti-Aircraft Regiments*; *II Battalion, 25 Anti-Aircraft Regiment*; *I Battalion, 33 Anti-Aircraft Regiment*; and the *Ramcke Parachute Brigade* under *Generalmajor* Hermann Ramcke (three battalions and one artillery battalion).[264]

The panzer division artillery field regiment consisted of an RHQ, two battalions (sometimes three) each of twelve 105-mm guns and one battalion of eight 150-mm howitzers, and a strength of 89 officers and 2156 other ranks. Within the division, a panzer grenadier brigade fielded six 150-mm gun/howitzers, either self-propelled or towed, and fifty-eight 81-mm mortars. The latter were used very skilfully and were difficult to locate by any means—technical or human. The infantry and light divisions had the same field artillery regiment organisation. However, DAK infantry divisions also had organic assault artillery of 75-mm guns with infantry regiments—their titles prefixed with 'IG'—while the *90th Light* had Russian 76.2-mm guns. In summary, each division had its own preference for a particular balance of weapons and the numbers provided here are indicative only.

If British/Commonwealth armour and infantry held any grudges against their generals, they should have been directed at armoured theoretician J.F.C. Fuller (retired major general) who, by now, had recanted his 1926 statement concerning the primacy of the gunner.[265] His latest hypothesis was that infantry and artillery were less important on the battlefield than armour. Guderian believed there was a use for field artillery in infantry divisions, although the light division at Alamein had four infantry battalions, two anti-tank battalions and one field artillery battalion. British and American units had a balance of anti-tank, field and heavy artillery at corps level, not divisional. Panzer and light divisions and other formations under Rommel's command were all mechanised, the only German formation so equipped. As Gudmundsson notes, 'infantry and field artillery had almost disappeared' very much as a consequence of Fuller's ideas.[266] Both arms had been reduced relative to mobile elements, armoured and others. As his example Gudmundsson takes the *90th Light* during the period May–June 1942 when it consisted of seven mobile battalions—four infantry, two anti-tank and one field artillery. Von Sponeck, its commander, described his battalions as 'entire arsenals' which were 'expected to serve either as conventional motorised infantry or anti-tank groups'. Light and heavy machine-guns, mortars and infantry guns completed the establishment. The latter ordnance, which was to discomfit Australian infantry later in the year, was the 76.2-mm field gun of Russian design (and manufacture), an all-purpose weapon with field and anti-tank capability. Six of the seven *90th Light* battalions were equipped with these. In this division, field artillery battalions were the only ones that provided indirect fire. As mentioned, DAK divisions used captured British ordnance almost as a matter of course to supplement or replace their own.

The other division in the DAK's order of battle was the *164th Infantry/Light Division* (so described by all historians except Gudmundsson). He noted that *Generalmajor* Lungerhausen's division, a reinforcement from Crete, was converted to a motorised formation 'richly supplied

The Other Side of the Hill: Enemy Command, Tactics and Resouces

Brigadier Meade Dennis, RA, Corps Commander, Royal Artillery (CCRA) XXX Corps, at left. He is shaking hands with Major General Frank Messervy, formerly GOC 4th Indian Division, who had just escaped form captivity.
(IWM E 12872)

with anti-tank guns as was 90th Light ... Its divisional artillery consisted of a paltry five batteries—three of 105-mm howitzers and two of 75-mm mountain guns.'[267] It can be safely assumed that Lungerhausen was quick to utilise von Sponeck's desert battlefield knowledge and match the organisation of his mentor. As has been noted, the *90th Light* caused the Eighth Army much grief, particularly at Gazala, before both the *90th* and the *164th* were destroyed by the 9th Division in October. Thus, Arko 104 assets were the backbone of offensive support.

Chapter 3 described how a panzer attack was launched or a defence was conducted. In either phase of war, control and coordination of all artillery resources—assault, field, medium and heavy (Arko) were combined with the extensive use of numerous machine-guns, including panzer-mounted machine-guns. In a typical engagement of panzer forces bearing in the direction of enemy FDLs or their gun line, whichever happened to be out in front (the location obtained, for example, by their *strosslinie* as described earlier), it took only six minutes for the tanks with their mounted machine-guns to move from opening fire at 3000 yards to close to 500 yards—the engagement range for 2 and 25-pounders. The arrival of their bullets on the gun shield (an entrée) was also a harbinger of assault by field and medium artillery (main course) or, if Arko heavy ordnance was used, a second helping. Dessert comprised 81-mm mortar fire which added to the weight of shell. The German mortar crews were, by reputation, very competent. In the October battles, their bombs caused more casualties to infantry than field artillery.

By far the most dangerous artillery piece was the 88-mm Flak GP gun that fired fixed ammunition at a very high muzzle velocity (3000 feet per second). Originally designed as naval anti-aircraft artillery in 1914, an improved mark made its debut in land-air warfare in 1917. Its utility in a ground role was soon recognised.[268] Its solid shot was highly destructive to tanks, fortifications and machine-gun posts. Its airburst shell was lethal and its signature sound heard prior to its gunfire reaching its mark. It was also very accurate in both roles. The gun and its half-tracked tractor had a high profile and were easily spotted and soon engaged by observed fire. The Germans used the 88 more as a sniping gun. It was easily distinguished in air photos as it was always deployed with a heavy machine-gun and mortar sections. The 88s came to the battle as part of the division's anti-aircraft command which was organised as two heavy batteries of 88s (16 guns) and two light batteries of sixteen 3.7 or 20-mm anti-aircraft guns.

Within Italian divisions, the scale of artillery support was somewhat reduced.[269] Italian corps artillery was well represented in battle, and was deployed in the following formations:

General Edouardo Nebbia
X Corps: 49th Artillery Battalion
           147th Artillery Battalion
           17 Pavia Division: 26th Artillery Regiment
           27 Brescia Division: 55th Artillery Regiment

General Guiseppe di Stefanis
XX Corps: 357th Light Artillery Regiment
           132 Ariete Division: 132nd Artillery Regiment
           161st SP Artillery Battalion (75LI8)
           133 Littorio Semovente Battalion (SP)
           3 Littorio Battalion, 3rd Mobile Artillery Regiment
           101 Trieste Battalion, 21st Artillery Regiment
           185 Folgore Battalion, 185th Artillery Regiment

General Enea Navarini
XXI Corps: 8th Artillery Regiment, 331st, 52nd and 131st battalions
           25 Bologna Division, 205th Artillery Regiment
           102 Trento Division, 46th Artillery Regiment

Italian (Libyan) divisional artillery regiments deployed twenty-four 65-mm howitzers, twelve 75-mm guns and twelve 75-mm howitzers. Two of these guns dated from prior to the First World War, while two were produced soon after. Within the Italian motorised/armoured divisions, a mechanised artillery regiment comprised twenty-four 75-mm guns, eight 90-mm guns, eighteen 105-mm howitzers and twenty 75-mm self-propelled guns. A self-propelled and shielded anti-tank gun, type *Semovente Modello* 41M 90/53, was also used, and relied on an accompanying vehicle to carry its ammunition, as only six rounds could be held on the chassis.

Italian medium artillery comprised a substantial weak link in Krause's resources as its range was less than both British medium guns and thus the guns suffered severely during counter-bombardment battles. However, it was generally recognised that, depending on circumstances, division and unit, an Italian gunner was a far better soldier than his infantry counterpart.

In reports of the battles from July to October, there are few mentions of DAK field artillery firing any other shell type than HE, although coloured smoke was sometimes used to mark objectives, flanks of advance, etc. The lethality of DAK ordnance was comparable to that of its opponents, and infantry dispositions (e.g. distances between platoons and companies) were less predicated on this factor than on the characteristics of infantry weapons. References to illuminating shell shoots are rare in command post logs and counter-bombardment reports.

Nonetheless, DAK/PAA artillery resources were a considerable force to be reckoned with, and Montgomery and his subordinates were under no illusion that artillery domination would be first priority in future plans. One account quantifies the number of guns as 500, although establishment figures indicate that a figure of almost 1000 may be more accurate. It may well be that recent losses had considerably reduced the number of serviceable weapons available. As will be seen, strenuous efforts were made by all the counter-bombardment agencies and the DAF to accurately determine the number and calibre of DAK artillery. On the day of the battle, not all the enemy artillery within range of the Eighth Army guns had been located prior to Zero Hour. Yet another aspect of the firepower facing the Eighth Army's armoured forces was the menace of anti-tank artillery as, in terms of sheer numbers, it formed a higher ratio to other artillery than it did in British divisions.[270]

These formidable resources for use against the Eighth Army were 'commanded' by *Generalleutnant* Krause, who was more an adviser to Rommel. How effectively or otherwise Krause used his artillery would be a vital factor in the success of Montgomery's tactics.

Would he pursue an active or passive counter-bombardment policy? Would he use mortars and guns against Allied infantry and rely on his 88s and anti-tank artillery to neutralise the armour? Or would he use his heavy artillery against headquarters? This was the question facing the Eighth Army's Brigadier Kirkman and his staff. In the event, Montgomery told Kirkman to 'make the best fire plan [i.e. use of artillery] you can.' Kirkman did.[271]

It is interesting to note that artillery historian Gudmundsson wrote that German artillery headquarters and the command posts of artillery units organic to or supporting the *78th Division* (on the Russian front) were connected by teletype machines for the transmission of firing data of targets to their front. While no mention of such an interesting adaptation has been found relating to DAK artillery history, Maughan's account of (Second) Alamein contains several references to coordinated enemy artillery fire from 25 October onwards from the south-west and north-east of the Australian salient, peaking on 1 November. Could this have been pure coincidence?[272]

Mention has already been made of Germans using captured British equipment. In the lead-up to the battle, aerial photographs showed several Troop positions of captured 25-pounder and 4.5-inch guns, booty from previous actions in Cyrenaica (and possibly Europe). Under the semi-static conditions of September, the 46th Survey Company, Royal Engineers, produced battle maps from almost daily aerial photography to give 'as accurate a picture of his dispositions that camouflage would allow'. They showed infantry-defended localities, headquarters, minefields etc., with such accuracy that fire planners could fire concentrations on these rather than rely on a barrage. The CBO noted that field guns (75-mm) and 88s were within 2500 yards of the FDLs. The 25-pounders and 105-mm guns were 6000 yards and the 150-mm medium equipment 8000 yards from the FDLs. To establish the location of hostile batteries often required the 9th and neighbouring 51st divisions to follow a deliberately deceptive tactical 'stunt' to which the enemy would react with retaliatory fire. The sound rangers and flash spotters, forewarned, were able to gather more data and compare 'locreps' (as location reports were called) with the latest air photos or reconnaissance reports. In this way, the fixation of hostile batteries could be revised for status, accuracy, calibre and number. This included enemy anti-aircraft artillery which was still very much intact and would pose problems for DAF fliers during the battle.

The first two weeks of October saw the Australian gun regiments withdrawn from the line in rotation to 'rest areas' where they engaged in further training, maintenance and generally enjoyed a less stressful life. Under its commander, Brigadier Alan Ramsay, the divisional artillery had developed into a highly efficient organisation since its concentration almost a year before. Ramsay's field regiments and those of the Royal Artillery in XXX Corps, which would become the backbone of the artillery effort in the forthcoming battle, are the subject of the next chapter.

CHAPTER 13
THE 9TH DIVISION ARTILLERY PREPARES

The armourers, accomplishing the knights with busy hammers closing rivets up, give dreadful note of preparation.

Shakespeare
King Henry V

For the almost 12,000 all ranks who comprised the 9th Division, the pace of operations continued unabated: training courses, digging, guarding, communicating, conferring, building dummy trucks and stores proceeded relentlessly. Morshead told his unit commanders to 'weed out' those who could not be relied upon if they had not already done so (as distinct from those 'left out of battle' as a matter of policy to 'make good' unit casualties). Routine Orders, the Army's means of advising personnel matters (promotions, demotions, postings etc. and disciplinary measures) and other procedural/systemic items were produced regularly in units and by superior headquarters and circulated throughout the division. This was a necessary procedure designed to ensure conformity in method and reduce misunderstanding, the foundation of 'good order and discipline'. Within the division men took ill (jaundice was prevalent), were killed, wounded and involved in accidents. It was illustrative of Morshead's leadership that, two weeks before the battle, several divisional courts martial handed down their verdicts on cases of desertion, failure to obey a lawful command, and absence without leave by other ranks. These offenders all received maximum sentences (18 months' hard labour and dishonourable discharge). An officer convicted of embezzlement of mess funds was cashiered. These were minor but important aspects of a division's morale and regimental COs ensured that details of these cases were promulgated within units.[273] Soldiers moving between B Echelon and rear areas saw a detention centre

Brigadier Alan Ramsay, Commander Royal Artillery, 9th Australian Division. This photo was taken in Syria when the division was garrisoning Syria/Lebanon, prior to July 1942.
(AWM 024075)

built near the Coast Road for those 'sent to the rear' for whatever reason: no Cairo comforts for them.

RAA regiments were now the second most numerous in the 9th Infantry Division and comprised headquarters, three field regiments and other regimental units under command, including heavy and light anti-aircraft, anti-tank and survey.[274] Together they provided most of the offensive power of the division—its efficient use would be vital in the forthcoming battle if victory was to be secured. Organic support would come from sub-units and detachments from other corps, namely Royal Australian Corps of Signals (RA Sigs) and EME, which were attached permanently to HQRAA or RHQ. Specialist officers from other corps, with or without other ranks, such as Medical, Ordnance and Chaplains, were included in establishments as 'attached' personnel.[275] By October, Brigadier Ramsay's staff (establishment six officers and 23 other ranks including regimental secondments) at HQRAA 9th Division comprised:

Majors James Irwin and Hylton (Bill) Williams, former and incumbent BMRA respectively of 9th Australian Division. Irwin was the first ex-Militia officer appointed a BMRA in the war.
(AWM 024076)

Brigade Major, RAA:	Major H.E. Williams	(2/12th Field Regiment)
BM RAA (L):[276]	Captain H.C. Kaiser	(2/8th Field Regiment)
Staff Captain:	Captain W.C. Radford	(2/12th Field Regiment)
Staff Captain:	T/Captain W.C. Stevens	(2/7th Field Regiment)

Ramsay relied enormously on his BM RAA, Major Hylton 'Bill' Williams, another World War I veteran and former 15th Field Brigade captain whose first AIF appointment was Staff Captain HQRAA I Australian Corps. He had transferred to Ramsay's headquarters and, in May 1941, gained his majority and became BM (L) to Major Jim Irwin, formerly second-in-command of the 2/7th. In June 1942 Irwin returned to Australia and Williams moved into this key role for the El Alamein battles.

Williams, possessor of a fine military pedigree, was a stickler for detail and accuracy in the operation orders he crafted with Ramsay.[277] By 20 October they had already compiled many of these, models of precision and clarity. Ramsay provided the framework and his BMRA filled in the detail—which Ramsay then checked meticulously. The process was streamlined to provide increased efficiency during battle to the point that the XXX Corps CBSO noted with some annoyance that HQRAA 9th Division 'did not use the proper channels for the request, interpretation and distribution of air photos'. Williams found it more efficient to bypass 'the system'.[278]

Other key staff were all experienced ex-Militia regimental officers who had staff experience at regimental level. Temporary Captain W.C. Stevens, for example, who was responsible for ammunition supply, had been Adjutant of the 2/7th. These officers were supported 24 hours a day by the attached Signal Troop signallers, drivers and clerks with draughtsman skills to craft traces and overlays. All Ramsay's field regiments were now organised into three batteries, primarily effected by taking one Troop from each of the two original batteries and creating a third. The 2/8th now comprised 15, 16 and 58 batteries. These were commanded by Majors F.D Stevens, M.R. Ralph and A. Salter. The 2/12th was reorganised into 23, 24 and 62 batteries. For the coming battle, battery commanders would be Majors Milledge (23 Battery), Maurice Feitel (24) and Archie Carter (62). Adjutant Captain W.M. Henty was appointed in July 1941 and the Quartermaster, Captain L.R. Seabert, was an experienced officer, having served since October 1940.

After three months in the line, the gun regiments had been relieved in rotation for training and affiliation exercises to prepare them for the forthcoming battle. The training was conducted with infantry and, where possible, armour, in an area to the rear of the Tel el Eisa—El Alamein defensive position. These rigorous exercises spared no discomfort or eventuality, the brunt of which was naturally borne by the heavily laden infantry. Minefields and defences were constructed to reflect their employment under Montgomery's plan. The aim was to ensure that all soldiers were battle-hardened and fit ('confidence borne of experience', as Morshead had commented) including reinforcements recently 'marched in' to make good losses. One commentator noted dryly that these exertions were directed at the advance and attack phases of war, unlike some of the Syrian training. Equipment shortages were rectified, regiments brought up to establishment strength, some officers reposted and troops enjoyed good food and amenities.[279] Similarly, three months on operations had revealed that some soldiers simply could not cope with strafing, bombing and hostile battery fire. These cases were handled sympathetically by the Regimental Medical Officer (RMO) in consultation with the officers concerned. Men were no longer sent to B Echelon as no improvement in their condition had been observed. Captain Sibree, RMO of the 2/7th, noted only one case that was bad enough for evacuation, believing that it was preferable for the men to remain at their job 'until they become moderately useful or break down altogether'. In September the first of a series of six-day leave periods was granted to all ranks to visit Cairo and Alexandria where it was possible to buy bottled beer. This was a very welcome interlude given the exertions of the previous nine weeks.

The emphasis on morale and training was a hallmark of the style of the new Eighth Army Commander, General Sir Bernard Montgomery. He set about providing the leadership necessary to enable his forces (land, sea and air) to defeat the Axis armies and restore the fortunes of the Allied cause. He removed officers he did not trust to do his bidding and his staff began work on his plans while he visited his command—some 200,000 soldiers—in his various styles of headgear which included an Australian slouch hat that he wore in a distinctive but not prepossessing way. The Eighth Army was boosted considerably by the flood of men, ammunition and equipment pouring into the theatre, all part of an overall strategic plan.[280] Montgomery made a point of addressing all officers above the rank of lieutenant colonel on whom would devolve the responsibility for putting their unit and troops 'in the picture', a practice he insisted be followed. Soon after he arrived he would ask private soldiers 'What is your role in ... ?' If they did not know, the unit commander was 'bowler hatted'. The message was soon disseminated amongst the officers. A regimental historian reported that, at one of his conferences to general officers, Montgomery had finished his exposition of his plan of attack and other details, and then asked, 'Any questions?' A newly arrived (in the theatre) brigadier, obviously intent on drawing the general's attention to himself asked, 'Have you thought of such and such?' Monty turned to him and said, 'Brigadier, what is your substantive rank?' He replied, 'Major, Sir.' Monty said, 'Brigadier, you will resume your substantive rank tomorrow. Any questions?' Needless to say there were none.[281]

In the static period prior to the battle, the 2/7th, again in support of the 20th Infantry Brigade, was located in the Tel el Eisa area, west of El Alamein station, as were the headquarters and the other two regiments. The first fortnight in October was relatively quiet, prompting a unit historian to note, 'Never had it been so quiet.' OPOs undertook observed shooting at enemy batteries when permitted as part of an overall counter-bombardment plan. Otherwise they observed bombards fired at hostile batteries in fixed or sniping positions. These were often 'engaged', with fall of shot deliberately placed some hundreds of yards away from their known location to induce the enemy to remain rather than move to another position. This saved the locators the trouble of finding them again if they moved.

On 6 October General Montgomery issued his plan to his corps and divisional commanders and their senior staff officers. Montgomery did not believe that his armoured

Lieutenant Colonel Thomas Eastick, Commanding Officer of the 2/7th Field Regiment.

Lieutenant Colonel Tinsley. Commanding Officer of the 2/8th Field Regiment.

commanders were capable of fighting a battle of manoeuvre; thus his plan was to 'stage manage' the battle using his infantry. In brief, his rationale was:

> ... to aim first at the methodical destruction of the infantry divisions holding the enemy's defensive system. This would be accomplished by means of a 'crumbling' process, carefully organised from a series of firm bases: an operation within the capabilities of my troops. For success, the method depended on holding off the enemy's armour while the 'crumbling' manoeuvre was carried out. It was also vital that the 'break in' battle, designed to gain a foothold in the enemy's defences should achieve complete success, so that they might be attacked from the flank and rear and supply routes in the forward area could be cut.[282]

The Eighth Army plan for XXX Corps required a four-division attack commencing at 10.00 pm which would reach its final objective by 3.00 am on 24 October. The corps front would extend 10 miles to the south where, simultaneously, XIII Corps would attack. Opposite the 9th Division, the most northerly corps was *XXI Corps* (German-Italian) comprising two regiments of the *7th Bersaglieri* and the German *164th Light Divisions* and two battalions of *Ramcke Brigade*. The *102nd Trento* and *25th Bologna divisions* were opposite the 51st Highlanders, 2nd New Zealand, 1st South African and 4th Indian divisions. Then followed *X Corps* (*27th Brescia*, *185th Folgore* and *17th Pavia divisions*) and two battalions of the *Ramcke*, while the *33rd Reconnaissance Battalion* and *Kiehl Group* (mobile forces) occupied the southernmost positions. Rommel's reserves in the north, the *15th Armoured* (panzer) and *133rd Littorio divisions*, were sited behind *XXI Corps* and the *21st Armoured* (panzer) with the *132nd Ariete Division* behind *X Corps* in the south. The *90th Light Division* (German) and the *101st Trieste Division* and two other battalion-sized units were in reserve near and south of Ghazal. The corps boundary was south of the Indian Division and north of the Greek Brigade running east-west. All the Italian divisions were motorised. Ultra signals intelligence decrypts revealed that both the German High Command in Berlin and Rommel's headquarters were quite confident that they would defeat Montgomery's Eighth Army. Axis forces totalled 180,000, less than half of which were German. The Eighth Army numbered 220,000.[283]

Montgomery had given Morshead's 9th Division the most important task in his plan for the battle. Never again in the Australian experience would there be a massed, concentrated fire plan involving the fire of 360 guns from 15 regiments as there was during the October battles. The 9th Division advance was on a 3000-yard frontage to a depth of 9000 yards, with RED line (Phase I) beyond the enemy's forward defences 5000 yards from the Start Line. The 26th Brigade occupied the right flank with the 20th Brigade on the left. The 24th Brigade in reserve was to hold the northernmost coastal sector, conduct a diversionary attack, and mine

2/7th Field Regiment Survey Section. Its OC is Lieutenant T. A. Rodriguez, who was awarded the Military Cross when an OPO at El Alamein. This section with their RA colleagues did flash spotting from towers behind the front line.
(AWM PO 2558.003)

the northernmost flank towards Trig 24 westwards before and during the attack. The ground the brigades would cover was well known to the Australian troops:

> The area of the forward defence lines consisted of undulating, hard, stony, gradually rising desert broken by occasional swells and low ridges. The only vegetation consisted of low, scrubby, prickly bushes, commonly known as camel thorn. The area was under observation from the Sidi Abd el Rahman Mosque.[284]

The plan would see X Corps pass through the bridgehead with an armoured brigade supporting the infantry. Phase II, the final objective, was the BLUE line. The corridor of advance was divided into segments laterally from the Start Line to the second/final objective and vertically into three report lines after the Start Line which were named after divisional personalities. Three battalions, the 2/15th, 2/17th and 2/24th, were to assemble on the Start Line and, once Zero Hour ticked over, begin their advance. Engineer mine clearance teams, using techniques developed especially for this task, would accompany them to clear the front of extensive (in numbers and depth) minefields. The FOO parties would 'marry up' with their allocated battalion or company commanders as ordered.

Thus October and November 1942 represented a high water mark for Australian artillery. El Alamein would mark the last time an Australian division's organic artillery would (or could) be deployed to function as a whole with adequate armour, engineer, signals and service resources—headquarters, field, anti-tank and anti-aircraft—in a major operation crucial to the outcome of a world war.

A brilliantly conceived 'cover plan', Operations Bertram and Slender, formed a key part of Operation Lightfoot. Montgomery wrote that the plan was:

> ... aimed at misleading the enemy about the direction of our main thrust ... The basis of visual deception was the preservation of a constant density of vehicles ... so that the enemy would be denied inferences made from changes in day to day air photos. By means of pooled transport resources, construction of dummy lorries, the layout and density ... for the assault was laid out on 1 October. Guns and quads were handled in the same way. Rear areas ... from whence the units came were maintained at their full vehicle quota by erecting dummies as the real transport moved out. Dumps were concealed by elaborate camouflage and stacking stores to resemble vehicles. Slit trenches were dug one month beforehand ...
> Meanwhile ... a dummy pipeline was started late in September, and work progress timed to indicate completion in November: dummy dumps were also made working to a similar date. HQ Eighth Armoured Division's wireless network was used to assist the notion that armoured forces were moving to the southern flank.[285]

Having virtually driven the *Luftwaffe* from the skies over the front during the first weeks of October, the DAF had taken the offensive. A total of 500 fighters and 200 bombers pounded enemy communications, headquarters, stores and depots, tanks in reserve laagers, vehicles and strongpoints, in support of the artillery plan. At Zero Hour they were to turn their attention to areas closer to the front where enemy armour was harboured and to gun positions and headquarters.[286]

Arrangements had been made by XXX Corps for 'affiliation exercises' between newly arrived British formations and the Australians. The 9th Division hosted the 51st (Highland) Division. Over several days, exchanges took place between key personnel of both divisions' three field regiments, the 126th (2/12th), 127th (2/7th) and 128th (2/8th), after which comments recorded related to the Australians' comparatively relaxed approach to discipline and its possible effect on the Scots![287] Over the next two weeks, the COs and Adjutants attended a number of conferences at HQRAA to finalise details. At this time, Lieutenant

Colonel Goodwin was CO of the Artillery Training Regiment and Major Houston was promoted temporary lieutenant colonel and CO of the 2/12th. On 20 October they visited their batteries and lectured all the officers so that they could put their gunners 'in the picture' and explain the plan to all those available.[288]

However, the early release of the orders meant that much preparatory work had to be completed by D Day, with accommodation made for the inevitable changes that arose from the 'fine tuning' which flowed from such conferences. The amount of technical work performed in RHQ, battery and Troop command posts was prodigious. Not only did the staff help dig their command posts at night, they also worked on the gun programs by day. Using slide rules, range tables and logarithms and, as the occasion demanded, drivers to check their arithmetic, the staff worked for 70 hours without a break. In the 2/12th's 24 Battery, a colleague noted, 'Howard Smith leaning against the dirt wall of the post—his legs buckled beneath him and he slid, ever so slowly, down to a squatting position—he was sound asleep before he crumpled.' Such scenes were to be repeated many times over the next 13 days. In the allocated battery areas, the gunners, drivers and signallers dug gun pits and command posts by night three days prior to Zero Hour. In the 2/12th, as with the other regiments, the historian noted, 'The men commenced digging new positions. They had never prepared positions so far forward before and were ordered to work silently. To dig a pit in silence was never within the normal capability of Australian gunners. Spades hit rock and oaths blistered the air—the men's innate sense of humour could not be kept quiet for long.'[289] The pits were some 30 to 35 yards apart and staggered alternately some 10 to 15 yards behind or ahead of the pivot (right hand and No.1 gun).

Telephone cables were laid deep under main traffic arteries. All vehicle movements were controlled according to a Vehicle Movement Table. Camouflage discipline had to be perfect. Movement of troops in daylight was severely restricted as the divisional area was overlooked by German OPs from the mosque at Sidi Abd el Rahman. The Australian gun area was directly behind the FDLs held by the battalions of the 24th Brigade. Never had the infantry been so close to the gun line which was a mere 2000 yards away. They were later to complain about the noise! They were, no doubt, unaware of the enormous amount of work the artillery staff in division and corps headquarters had devoted to supporting them with concentrated firepower.

Chapter 14
Artillery Command, Resources and Counter Bombardment

Do unto the enemy before he does it to us.

Gunners' 'golden rule'

Notwithstanding the fact that the arrival of 300 Sherman tanks added considerably to the Eighth Army's armoured strength, boosting it to almost four times that boasted by Rommel, Montgomery's battle plan resembled one more suited to the last great conflict. His plan comprised an assault against entrenched positions in depth with stronger artillery followed by a breach in the defences to enable armour to be used to advantage. Thus the Eighth Army artillery plan bore a striking resemblance to one of the 1918 'genre'—if not in guns, then in length.

The men who were to direct these resources came from various backgrounds. They enjoyed some similarities in that four had served in World War I while two had not. Both Brigadier Kirkman (Eighth Army) and the CCRA XXX Corps, Brigadier Meade Dennis, had served in World War I, so for them the artillery plan with its heavy reliance on counter-bombardment was less than novel. In the absence of a Headquarters 5 AGRA (5 Army Group Royal Artillery—then being progressed through staff channels), the counter-bombardment tasks of one field and three medium regiments were directed by the CCMA, Lieutenant Colonel H.S. Hunt, DSO, MBE and his Corps Counter Bombardment Officer (CCBO), Lieutenant Colonel M. Yates. The office was staffed by a BMRA, Major P. Carey, a Counter-Bombardment Staff Troop of two detachments and an Air Section, captains and lieutenants, clerical personnel and 'office' other ranks totalling 28 all ranks.

The role of this specialist unit was to determine the location, strength and grouping of enemy guns by type, calibre, arc of fire and areas habitually shelled. Also on the task list were enemy movements, concentrations, locations of OPs, headquarters and command posts. An APIS was situated close to Corps Headquarters and the CCBO where the real strength of the APIS function soon became apparent. Better cameras with longer focal lengths and vertical format enabled the precise identification of types of gun position (roving, sniping), calibre, etc. Counter-bombardment methodology had not yet reached peak efficiency given its users' relative inexperience compared to that of its DAK adversaries and also because of the paucity of counter-bombardment fire opportunities over the previous three months. Another contributing factor was the loss of officers from medium regiments at Gazala who could have made the transition to CBO staff work with ease. Other reasons were not hard to find either. When the BEF withdrew from France, the losses of experienced officers and gunners had to be made good for units and headquarters sent to the Middle East. When the Japanese entered the war, further experienced personnel were sent east. This development, coupled with losses in Greece, Crete, Gazala and Tobruk left the RA stretched to the limit for experienced personnel, and this affected all the medium regiments and higher formation artillery staff appointments.[290] It was unfortunate that this impacted most heavily on Corps artillery staffing and operations, where a great deal of experience of enemy artillery methods was necessary for it to fulfil its role effectively.

Apart from Brigadier Ramsay, other CRAs included those of the 51st (Highland) Division, Brigadier G.M. Elliott, RA, a former commander of the 7th Medium Regiment, and Brigadier C.E. (Steve) Weir, CRA 2nd New Zealand Division, who succeeded Brigadier Miles after the Belhamed fiasco of November 1941. A highly respected artilleryman, Weir had graduated in 1927 from the Royal Military Academy at Woolwich and progressed slowly and surely through the ranks to regimental command in 1940. At 37, he was younger than his peers, had an enquiring mind and a forceful personality. Brigadier F. Theron, who had been CRA 1st South African Division for some time, was appointed major general and his country's representative at HQME. He was succeeded by Colonel C.L. deW duToit and, in the fullness of time, by Colonel Frederick Theron, former CO 7th South African Field Regiment and, at 27 years of age, the youngest CRA in the theatre.[291] The CRA exercised tactical command of his divisional artillery and, if the tactical situation demanded, he could use another division's guns to augment his own, liaising directly with his opposite number. This flexible arrangement was to work well during the battle.

Within XXX Corps, the field regiments of the Royal New Zealand Artillery (RNZA), the field regiments, units and sub-units of the RA and the SAA supported the armour and infantry battle. The whole front was supported in observation/locating and survey functions by the 4th (Durham) Survey Regiment, RA (TA).[292] These were formidable assets and had prompted Rommel to comment with some prescience on the way the battle would evolve. He had, after all, 18 months' experience of his adversaries from which estimate their military worth. He noted that the 'full value of the excellent Australian and New Zealand infantry would be realised and the British artillery would have its effect.'

Army and Corps artillery arrangements were vital to Montgomery's plans, both for direct support of the infantry and armour and for counter-bombardment work. The key units in support of the Australians were three that had served with them previously on various occasions. It is worth describing their background and experiences as both were called on to give their all to support Morshead's formation in the weeks that followed. The 7th Medium Regiment, RA, was the basis, supplemented later by the 64th and 69th medium regiments, of what was to become 5 AGRA, formed later in 1942.

The 7th Medium was a regular British regiment of two batteries. The 25th/26th 'Battleaxe Company' Battery began the war in Palestine, having come from India at the end of 1938. The troopship bearing the battery unloaded its stores, but not the gunners' personal gear which completed the voyage to the UK without its owners. This minor administrative misfortune seemed, as events transpired, somewhat portentous. The regiment held a practice camp for two weeks in Palestine in 1939. The 27th/28th Battery now occupied El Alamein, so a few 'old hands' knew the area well. Reconstituted as D and E Troops which, with RHQ, had fought in the rearguard action at Gazala, and 107 Battery (formerly 107th RHA, South Notts Hussars, TA) A and B Troops, the regiment was equipped with 16 new 4.5-inch Mk II guns with split trails. These guns had arrived in crates without assembly instructions, range tables or other essential data in what was yet another setback for the gunners.

The CO, Lieutenant Colonel H.C. 'Toc' Elton, DSO, who was regarded as an inspirational leader, soon had the regiment operational. Elton, a former RHA officer, was a 'live wire' CO and, in September, while reconnoitering OPs in his Jeep with his driver/operator, attracted shellfire which destroyed the vehicle. The men escaped unhurt and headed east on foot until they discovered an 'abandoned' 15 cwt truck. Elton and his driver climbed gleefully aboard, managed to turn the engine over, and set out for RHQ. They had not travelled far when there was a furious hammering on the cabin roof. They stopped and checked the back of the vehicle, to be greeted by an annoyed Australian driver who had been asleep in the rear and wanted to know, in no uncertain terms, what they were doing with his truck! Elton explained, and the truck, dodging minefields, eventually reached safety and the 7th Medium's RHQ. Elton shouted the driver a whisky in the Officers' Mess and the 9th Division recovered its truck.

The Mediums were glad to be back 'amongst friends' since their travails and disappointments with the 6th Division in Greece and Crete. This was the regiment's first association with HQRAA 9th Division. The regimental historian noted that 'They were grand fighting soldiers and very easy to get on with. In a short time very cordial relations were established between the officers and men of the regiment and those units they were supporting.' The British/Australian Army convention of long-term regimental affiliations was regarded as positive and a boost to more efficient operations.[293]

When operations slowed, the new 2/8th RMO, Captain R.A. Douglas, was allocated personnel from the nearby 7th Medium for routine sick parades, a morning Army ritual in all units. Unaware of the background of the regiment, Douglas greeted the bombardier in charge with, 'You're 7th Medium, are you?' The bombardier drew himself to his full height and saluted. He said, 'Sir, we are not mediums, we are Royal Horse Artillery!' After all they had been through, it was illustrative of the regimental spirit abroad and gained the admiration of the Australian gunners.[294]

At El Alamein in September/October, the 7th Medium was ensconced in the northern sector of the front; in the interim, however, 107 Battery had exchanged its 4.5-inch guns for 5.5-inch. Both ordnance were identical except for calibre, and a 'mixed' regiment was to prove itself an advantage during the battle. The 4.5-inch guns had superior range for counter-bombardment tasks and fire 'in depth', whereas 5.5-inch projectiles were 40 pounds heavier with a maximum range that was 2400 yards shorter. Operations analysis by artillery staffs on the employment of medium regiments in the desert had confirmed that the most damaging enemy action against medium regiments was not counter-bombardment but came from the *Luftwaffe*'s dive-bombing Stuka Ju87s. This continued until the revelation from a *Luftwaffe* prisoner under interrogation that the high-roofed Matador tractors cast such a long shadow that guns on the move or dispersed in wagon lines were readily spotted from the air, thereby inviting attack. The CO tried several times to have tractor profiles lowered by removing the roofs, at least when in static positions, but to no avail. To overcome the problem of vulnerability to air attack, a Troop of four guns was laid out in a diamond or square formation, not a semi-circle, with an interval of around 50 yards between each. Thus the German pilot was presented with a target that appeared to be not worth the risk. This placed a greater strain on the gun position officer in terms of control and communications, but nonetheless reduced casualties. When the regiment moved into the northern sector in early October, it changed its gun positions frequently to banish the 'getting settled in' mindset and fired occasional bombards to CCBO directions. The regimental historian noted that, during the battle, its RHQ and HQRAA 'shared a fine series of concrete dug-outs—a most convenient arrangement'— which it was to prove beyond any doubt.[295]

The 64th (London) Medium Regiment, RA (TA) comprising 211th and 212th batteries was raised at Islington, a north-eastern suburb of London in April 1939. The 53rd Medium, a TA unit, provided cadres and trained in several locations before boarding the *Otranto* in November 1940 (with units of the 18th Brigade, AIF). Six weeks later, it was at Mena and the Pyramids. The 212th Battery was despatched to Eritrea while RHQ and 211th Battery, armed with 60-pounders, joined Lieutenant General O'Connor's XIII Corps after the capture of Bardia and took up positions outside Tobruk. They added their weight of shell to the 6th Division's attack on the fortress on 21 January 1941. They were then attached to the 7th Armoured Division for the advance to Agedabia and, when Rommel bared his fangs, RHQ and 211th Battery returned to the Delta without firing a shot for deployment to Greece then Crete. Their first CO, Lieutenant Colonel R.L. Syer, was killed at Lamia Pass in Greece by a bomb splinter and the reduced unit suffered further casualties until withdrawn to Crete where it augmented a pitiful forlorn hope of artillery support for the mixed British, New Zealand and Australian troops until evacuated. Later, 212th Battery supported the 7th Australian Division in the Syrian campaign to good effect, despite being plagued by shortages of equipment and trained personnel.

In the autumn, the unit garrisoned Syria and Lebanon and it was not until April 1942 that it was re-equipped with 4.5-inch Mk II guns for 211th Battery and 5.5-inch guns for 212th. Both batteries were engaged in operations in Cyrenaica and Egypt including the fiasco at Mersa Matruh (engaging panzers over open sights). They lost two guns, had three damaged and suffered casualties at Gazala in June before lining up behind Ramsay's artillery for the second time for the attack on Miteiriya Ridge on 27 July. Their expenditure of 138 rpg was the highest for the month, and they then joined the Eighth Army before El Alamein to fight yet another day. Despite problems with unserviceable guns, 211th Battery fired 460 rounds on counter-bombardment tasks a week before the battle. This drew accurate retaliatory fire from heavy German 170-mm guns. They were well dug in and suffered only a single casualty.

In comparison, life for 212th Battery was relatively quiet. The medium regiments were involved in the same sort of preparatory measures as the 9th Division gunners. They calibrated their guns, dumped 400 rpg and, during the night of 20/21 October, occupied their positions. At 8.30 pm, the 7th Medium fired two airburst rounds at different ranges to check the accuracy of the 'Correction of the Moment'. These shoots (at a Datum Point) were observed by the flash spotters of the 4th Survey in their towers and confirmed the predicted fire values for field and medium guns. At Zero Hour (less time of flight) with two other medium regiments (the 64th and 69th), the 7th Medium was the first to fire.[296]

A newcomer to the Eighth Army's order of battle was the 66th Mortar Company, Royal Engineers, comprising three sections each of six 4.2-inch mortars. This was a brand new unit with new ordnance and was delivered to the Middle East with its HE and chemical (White Phosphorus) ammunition. One section was to support the 24th Brigade's feint attack while the others deployed southwards.[297]

The 4th Survey Regiment, RA (TA), was raised in Gateshead, Tyne and Wear, County Durham. It provided battlefield survey data (fixation and orientation), flash spotting and sound ranging troops to locate enemy artillery. Its Meteorological 'Met' Section provided meteorological data for all Eighth Army units. The 4th Survey's misfortunes during the war were similar to those of the 7th Medium and Australia's 6th Division in Cyrenaica, Greece and Crete, and it too had suffered losses in men and equipment. When the establishment for a survey regiment was approved in August 1940, it was based on the 'Phoney War'. Its RHQ had a survey section attached and three batteries, one each of flash spotting, survey and sound ranging. Each battery comprised a headquarters and two Troops. In mobile operations, this allowed a corps to allocate a Troop per division. Survey Troops performed the usual survey tasks but could also produce 'known ranges' for anti-aircraft guns and assist in calibration tasks. Flash spotters could deploy a short base to control airburst ranging, a useful technique in the desert. While the 1940 establishment worked well enough, it was not good enough for the desert. In fact, Lieutenant Colonel R.O. Cherry (2/1st Survey Regiment, RAA) had already moved towards a composite organisation before the RA regularised it. It was recognised in mid-1941 and the resulting doctrine (in Artillery Training Pamphlet Vol. III, No.11) promulgated. This changed the composition of battery groups in both the advance and withdrawal phases of war. The gunners adapted well to the new form.

Thus, when the 4th Survey was reconstituted in Palestine and subsequently trained in the Levant, instead of being a functionally organised regiment, its three-battery establishment was modified to two composite batteries embodying three functions. No.1 supported the 9th Division during the siege of Tobruk for 177 days; No.2 Composite Battery relieved No.1, only to go into captivity when the fortress capitulated in June 1942. At El Alamein, No.1 Composite Battery's OC, Major Kellett, established himself in an old concrete dugout close to the coast road near the XXX Corps CBO where the headquarters of A Troop Flash Spotting and R Troop Sound Ranging joined him. Y Troop Survey eventually set up its computing centre near the flash spotters a week before the battle. Nearby, Captain Edwards, RA, established his counter-bombardment computing centre manned by his group of specialists. They were all located in the 9th Division's area, where the AIF gunners regarded them as 'old friends'.[298]

The 9th Division attack was on a westward axis for 6000 yards from the Start Line and was some 3300 yards wide. This meant that the divisional artillery, located at the northern end of the front, would not be able to cover the more southerly hostile batteries. The 2/7th gun area was initially 1000 yards behind the other two regiments, around 2500 yards from the infantry Start Line. The remainder of the Corps artillery was deployed so that its arcs of fire at a planning range of 11,000 yards (for 25-pounders) and 18,000 yards (for 5.5-inch guns) covered the objective (Blue Line) and the latter extended beyond it. Thus, available to the 9th Division were batteries from:

51st Highland:	126th, 127th and 128th Field Regiments, RA
2nd New Zealand:	4th, 5th and 6th Field Regiments, NZA
1st South African:	4th, 15th and 75th Field Regiments, SAA

The 126th Field Regiment, RA, an Army field regiment, had six Troops available for counter-bombardment tasks. Its six Troops were attached to the 2/8th (U, V, W) and the 2/12th (X, Y, Z) by a system of command known as 'Link Troop Control'. In effect, each Australian battery had an RA Troop in addition to its own, so a battery command post oversaw the operation of three Troop command posts. The RA regiment did not deploy OPs.

The strength of the field artillery deployed by the Eighth Army, XXX Corps and the 9th Division amounted to:

	Eighth Army	XXX Corps	9th Div
25-pdr field	832	288	96
4.5-inch medium	32	24	-
5.5-inch medium	20	16	-
105-mm	24	-	-
Anti-tank 6-pdr	112		
Anti-tank 2-pdr	1256	756	54

This number varied during the battle, and there is much debate over how many guns were tasked for the various stages of the battle's first and second phases.[299]

Thus, relative to the corps frontage, a greater proportion of the field artillery was under command of XXX Corps, which was stronger in medium artillery. This reflected the importance of the counter-bombardment program, particularly against the enemy medium artillery as the 2/8th was earning a reputation for its counter-bombardment gunnery. Ramsay had attended a preliminary conference on the operation on 16 October and had been notified that he would command the division should Morshead become a casualty. Any of the divisional artilleries of the 51st Highland, 2nd New Zealand and 1st South African that were in range were also tasked to support the 9th Division, subject to the advice of the CCRA. Three field regiments (18 Troops) from XIII Corps artillery were also allotted to XXX Corps. XXX Corps medium artillery, used mainly for counter-bombardment tasks, was under the control of the CCMA. The CCBO was located in the 9th Division area and, through its telephone exchange, was linked to the survey regiment's battery headquarters. The three British medium regiments were:

7th Medium Regiment	(27/28 Battery):	eight 4.5-inch guns*
	(107 Battery):	eight 5.5-inch guns
64th Medium Regiment	(211 Battery):	eight 4.5-inch guns*
	(212 Battery):	eight 5.5-inch guns
69th Medium Regiment	(221 Battery):	eight 4.5-inch guns*
	(222 Battery):	eight 4.5-inch guns *

*all 4.5-inch guns were Mk II

In the artillery planning for the battle, a convention was adopted whereby the Troop (of four guns) would be the basic fire unit. That is, higher headquarters would decide whether it was a one, two or three (or more) Troop target and the designated regiment would then decide which Troops to use. Flexibility was the key criterion.

Other artillery units and sub-units on the Corps order of battle were located behind the FDLs to support the field artillery. In addition to the 4th Survey Regiment, RA, the 46th Survey Company, RE, produced battle maps with the location of enemy guns, posts, headquarters, minefields, telephone cabling, ditches and earthworks, wire entanglements etc. with sufficient accuracy for artillery task tables (fire plans) to be compiled. The Meteorological Section of the 4th Survey provided meteorological data several times daily to all gunner units. Anti-aircraft artillery (heavy regiments of 3.7-inch anti-aircraft guns and light regiments with 40-mm Bofors guns) was deployed to protect gun areas, headquarters and other vulnerable points. Anti-tank regiments/batteries/companies were deployed with the infantry brigades.

The CCBO was a key figure in the battle, and it is worth reviewing some of the factors with which he and his office had to contend in relation to artillery siting, command and deception plans—both friendly and enemy. These included gun positions, whether main, temporary, roving or sniping, and the extent to which they should be developed with real or dummy weapons. Manpower resources figured prominently in these considerations, as well as in relation to the overall deception plan developed by Army Headquarters. Deception plans had to be realistic and thoughtfully prepared and included the laying of line and command posts, dummy or otherwise. The problems facing a CCBO and his office also included the necessary accumulation of a vast database on the counter-bombardment activities of the enemy gunners. Managing such a database required experience and specialisation, both of which were, in general terms, in short supply in the Middle East at this time. Quite apart from the simple question of whether to destroy or neutralise, and/or maintain an active or passive policy, the type of retaliatory fire that would be most effective in the current circumstances also had to be considered. For example, a comparatively light concentration of eight rounds per minute or the equivalent might only bring temporary relief while the fire continued. On the other hand, short concentrations of no fewer than three to one (i.e., three batteries engaging one hostile battery) could provide temporary neutralisation. Sustained fire achieving an average of one round every three minutes could also halt the firing of the hostile battery.

Counter-bombardment fire which aimed to suppress hostile batteries for periods of up to an hour demanded high ammunition consumption. Much heavier concentrations were required if the aim was to cause severe disruption and/or damage. Airburst shells rather than ground burst were found to be more effective during counter-bombardment fire that involved round-for-round exchanges with hostile batteries. As counter-bombardment preparations had been made for an attack on well-defended and plentiful enemy artillery, the scope and timing options for the CCBO were considerable. The lesson of Ramsay's onslaught on Italian artillery at Tel el Eisa had not been lost, if only for the fact that the heavier Italian guns lacked the range of the British guns. The offensive counter-bombardment capability of the DAF was also a useful consideration. With air superiority now assured, the CCBO had no hesitation in planning the use of prearranged bombing by light bombers, rocket and strafing attacks by fighter aircraft or having a 'cab rank' system of omnipresent air cover (combat air support) on call. Key postings at Corps level comprised the G Ops (Air) and ACBO (Air), who worked closely with the APIS Section.[300]

Enemy artillery had not been idle during the quieter periods and, as well as many 'fixed', well dug-in positions, there were numerous roving and sniping positions facing the Eighth Army. The 5 AGRA historian noted that the enemy made excellent use of folds and depressions in the terrain to secrete his guns. As well as the minor stratagem of deliberately 'aiming off', the CBO would order 'no retaliation'. However, when moving to the 'active' policy, the CCBO would look for a ratio of 50:1 for a bombard. Depending on the accuracy of location, the CCBO would order the guns to concentrate their fire on a pinpoint target (instead of the convention of guns firing on parallel lines). This was known as 'Ball of Fire'. Where accuracy was less assured, a 'Murder' would be ordered comprising a frontage of 60 yards or guns 'half-concentrated' (i.e., concentration at twice the range).[301] British bombards were ordered so that the fall of shot on a hostile battery was synchronised and followed a different pattern to the German. While the Germans would fire four guns or more at a high rate for an hour (for example), British practice was to repeat the bombard at random times throughout the day and/or night. Prisoner interrogation revealed that the enemy gunners 'hated it'.

Only three medium regiments were available to the artillery planners, two of which were deployed in the XXX Corps area. The 7th and 64th were 1200 yards east of the 2/7th, while the other (the 69th) was south of El Alamein station in the South African divisional area, 10,000 yards south south-west. A feature of the disposition of the Corps artillery was that all hostile batteries were in reach of the medium guns. There were 45 hostile batteries facing XXX Corps, 33 field and 12 medium/heavy calibres. Most of the enemy artillery north of the northing (PURPLE) grid line 2840 could range onto three of the four assaulting divisions, while some 23 were out of range of the 9th Division artillery. The enemy artillery was very effectively deployed, tending not to reflect the deceptive measures undertaken by the Eighth Army, but rather recognising that the northern flank was the most likely area for Montgomery to launch his attack.[302]

This plan, based on all sources of artillery intelligence, originated in its entirety at Headquarters Corps Artillery, Brigadier Kirkman's headquarters close to the DAF Headquarters. The DAF contribution to this plan is beyond the scope of this history, but it was substantial, involving around 500 aircraft, some of which were used as heavy artillery. The medium bombers were restricted to bombing from altitude by the German anti-aircraft defences and the formidable *Luftwaffe* fighters, although the destructive effects sought by the DAF were secondary to the impact on morale and the disruption to the German routine. On the night of the assault, heavy bombing of enemy gun areas was scheduled.[303] The noise overhead was also used to mask the noise made by the mass

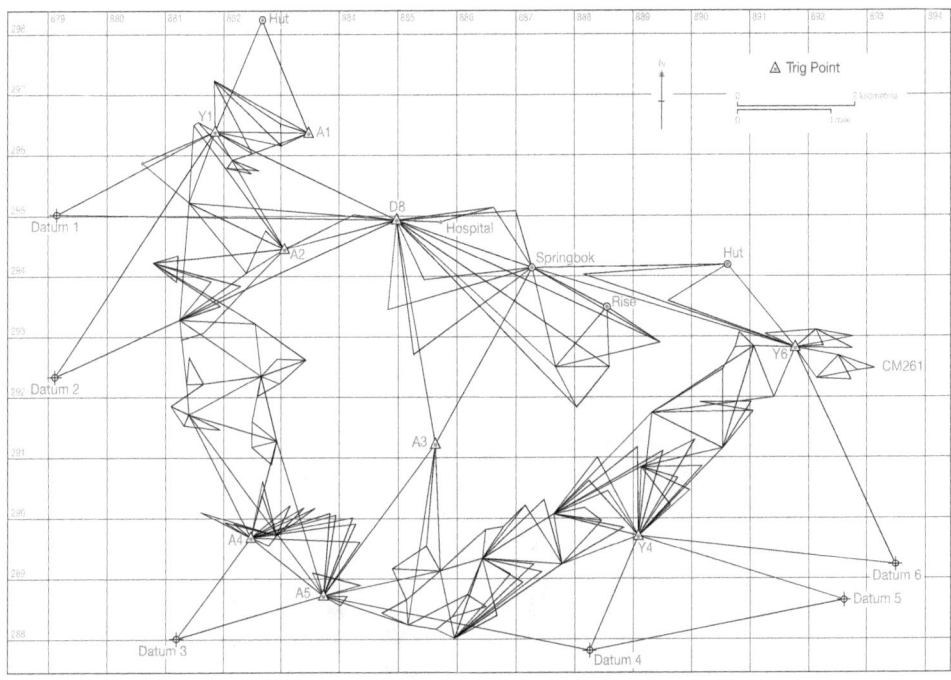

THEATRE GRID

4th Survey Regiment, RA established Theatre Grid for Eighth Army, the form of which replicates its curved front line, and is a safe distance from enemy lines. Its rearward extension is toward heavy and medium artillery positions. Datum point coordinates enabled more accurate gunnery from several gun positions.

(Source, Army History Unit)

movement of the Eighth Army vehicles rolling towards the gun and infantry areas. All these activities had to conform to the Army counter-bombardment plan authorised by Brigadier Kirkman. He later wrote:

> Before the attack the enemy was believed to have in action 200 field guns, 40 medium guns and 14 heavy guns which could bear on the front of the main attack. It was necessary that the positions of these guns be known, and that nothing should be done which might cause any of them to move at the last moment. Consequently, on the northern part of the front, for over two weeks prior to the attack, a silent CB policy was adopted, except for certain destructive shoots against a particular active battery, and neutralisation shoots against roving guns. Since it was impossible to neutralise all the enemy guns during the attack, the policy of employing all available on CB fire immediately prior to Zero Hour was adopted.[304]

Captain Frank Hamer, OC 'Y' Survey Troop, 4th Survey Regiment, RA, takes a break near the shore of the Mediterranean.
(Private Collection)

Thus the first 15 minutes (Z-20 to Z-5) were allocated to counter-bombardment fire on enemy positions. The accurate location of enemy guns was the work of the flash spotting and sound ranging sections of the 4th Survey Regiment, RA, combined with air photographs and operational reports from RAF photo reconnaissance and reconnaissance squadrons. Some elements of the artillery plan could not be finalised until the last aerial photography cover was flown by the RAF on 20 October. The processing time for air photographs from sortie to interpretation at Corps Artillery Headquarters APIS had been considerably reduced and, in a matter of hours, the latest results were ready for dissemination through the artillery intelligence channels. The 46th Survey Company, RE,

4th Survey Regiment, RA surveyors establishing Bearing Pickets. They are using a theodolite to establish the precise location of a Bearing Picket (BP) in a featureless desert. BPs were essential for anyone with a map of the area to accurately site a gun, headquarter, OP or rendezvous.
(Major Frank Hamer collection).

produced gridded maps of the battle area with enemy artillery positions, used, disused or suspect, clearly annotated. Facing the 9th Division were 33 of the 45 enemy hostile batteries. On 20/21 October, Corps hostile battery List No.18 added 15 new and unfixed hostile batteries. All but one (B grade location) was Z grade—its position accurate to within 25 yards. Two were suspect, two had been upgraded to fixed, three were fixed and four deleted. There were nine anti-aircraft gun sites located, all Z or A grade. The counter-bombardment operations report of 22 October had the last word: 'AP [aerial photos] show little change in the enemy main dispositions apart from the move of 6796NP Group to 6794UX. The principal changes have been filling the gaps and the addition of single troops to strengthen his defence in depth over the whole front from N295-285. There has been a decrease in the number of heavy guns.'[305]

Flash spotting bases and sound ranging bases had covered the whole front from late August. Connecting microphone bases and towers had required a total of 300 miles of D8 cable. There were duplicate communications between the sound ranging troop, CCMA and medium regiments, primarily for meteorological information. Adopting a static position 30 miles long enabled the battery to fulfill its potential for the first time. The unreliable Austin 8 cwt vans had been left in Egypt, exchanged for the ubiquitous Jeep, which was regarded as a far better vehicle.

All the gun regiments were placed on permanent grid (based on Egyptian PURPLE grid—metric). On the day before the battle, all the bearing pickets were checked.[306] A very important aspect of flash spotting was the accuracy of fixation of tower OPs. Their special range-finding theodolites were surveyed to less than one metre in accuracy (as a general rule), given that an error of one metre between widely separated OPs would result in a 'best' accuracy of more than 30 metres—B Grade—which was not good enough. The north FS base (No. 5), some 15,000 metres long and installed by A Troop on 1 October (until 26 October), ran south-east along the high ground, turned south and could locate guns south-west of Trig 26. Towers were essential to allow the gunners to observe west; the first towers were 60 feet high, and one was later extended to 80 feet. These were constructed of tubular scaffolding and erected by the RE. Prior to the battle the flash spotters' 'bread and butter' were destructive shoots and calibration. The base made it possible to calibrate all the guns of the 9th Australian, 51st (Highlands) and 1st South African Divisions. A Troop extended a base (8000 yards—No. 6) further south in the central sector of the line for six days, and this was handed over to its sister Troop (B). On 26 October the north base (towers Ack, Beer, Charlie and Don), which was extended by other towers, Pip, London and Nuts, was manned by RHQ surveyors from the 2/7th and 2/8th Field Regiments. Bases 5, 6, 7, and 8 specifically covered the 9th Division's operations when it struck north and north-west in the latter phases of the battle from 1 November. When the battle opened, the towers were close to units, approximately 5000–7000 yards from the enemy front line and located at:

> Tower/Post A: near the 7th Medium Regiment, RA
> Tower/Post B: by the main road, near Survey Battery HQ
> Tower/Post C: south of El Alamein Station 8000 yards
> Tower/Post D: near the 69th Medium Regiment, RA
> Tower/Post F: near the 4th South African Field Regiment
> Tower/Post E: within the 4th Indian Division area
> Tower/Post V: on Ruweisat Ridge

The base established by B Troop (No. 2) was further south and ran from north-east to south-west, and was specifically sited to cover the breakthrough corridor of Operation Supercharge. This operation was supported by the 9th Division artillery. The operation order for the 2/7th

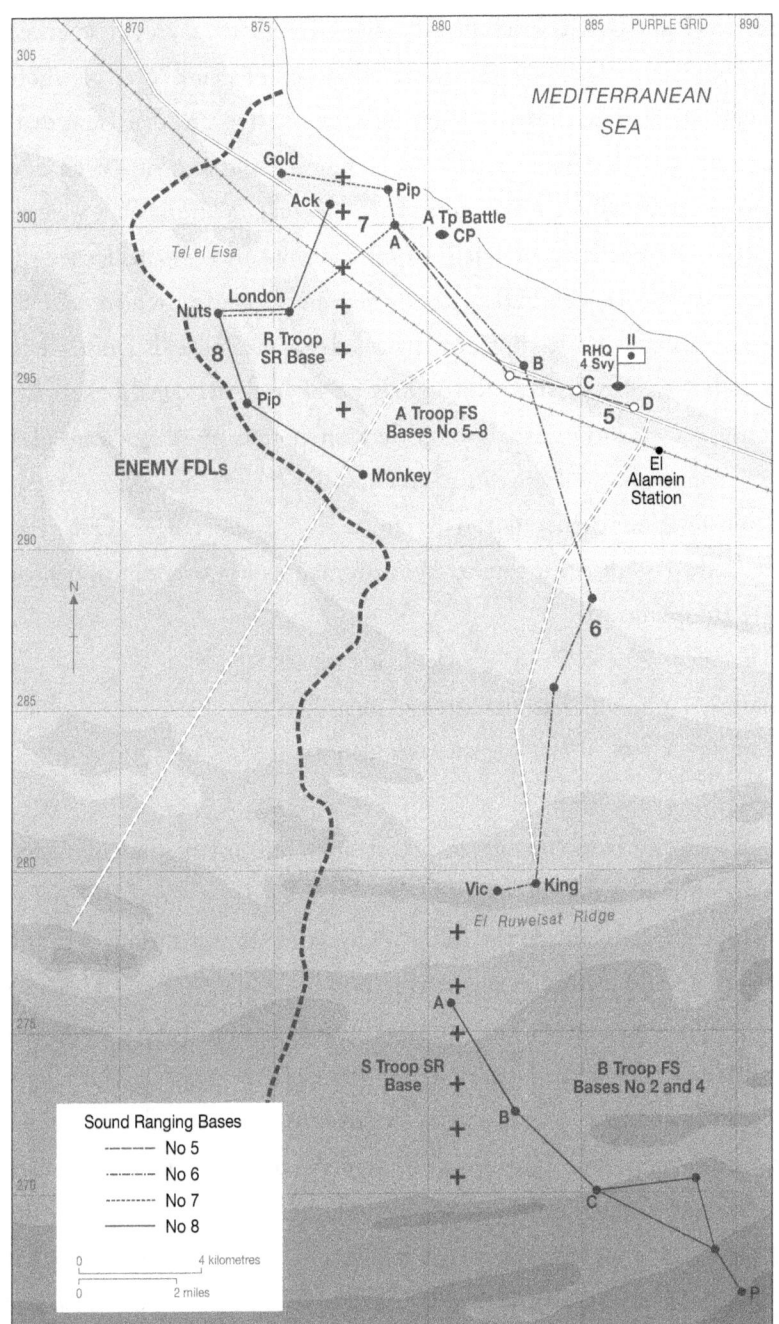

MAP FLASH SPOTTING AND SOUND RANGING BASES

4th Survey Regiment deployed its full resources of flash spotting and sound ranging assets to locate enemy artillery and provide comprehensive battlefield intelligence from their pipework towers.

(Source, Army History Unit)

noted that towers B and D had car headlights mounted on them for use as reference objects by the guns (and beacons for fixation by everyone). This was not an unalloyed benefit to the flash spotters or their guardians, as will become apparent. The flash spotters had views that surpassed those of the Germans in the Sidi Abd El Rahman mosque, and were sited at an elevation that allowed them to see 20,000 metres or more depending on conditions. They were able to feed observations on enemy defences of much tactical value into artillery intelligence, thus leading to accurate delineation of enemy FDLs. Flares, signals and ground movement were all reported. Fixation of petrol flares lit by friendly patrols signalled gaps in minefields. During the battle they gave accurate reports of progress based on the location of successive signal rockets. At the height of the battle, this was a priority task. However, the flash spotters' most significant advantage was that they were continuously effective no matter how much battle noise surrounded them. On the other hand, it was noise that nullified the efforts of the sound rangers. One measure of the effectiveness of the Eighth Army counter-bombardment policy over time is contained in the flash spotting Troop's records:

Base 5:	55 hostile batteries identified
Base 6:	24 hostile batteries identified
Base 7:	14 hostile batteries identified
Base 8:	2 hostile batteries identified

R Sound Ranging Troop first deployed a straight line north-south of the regular base with six microphones and a sub-base of 1200 yards, later extended to 1500 yards with advance posts 1500 yards inside the FDLs. A longer base was impractical as it worked satisfactorily to 14,000 yards, but accuracy fell off beyond that. After a brief spell in the New Zealand Division's sector, the Troop returned to its northerly base. In the week before the battle it recorded 83 locations: Z: 48; A: 12; B: 5; C: 3; D: 15. Seventeen of these were confirmed active, with particular attention paid to calibre.[307] This was vital work, particularly as poor conditions prevented the use of air reconnaissance prior to the battle. Successful ranging shoots were conducted, although flash spotting proved faster. The intensity of the fighting on the 9th Division's front and the incessant artillery fire, anti-tank battles, aircraft overhead, bombing and tank engine noise negated the advantages of sound ranging.

There were many valuable lessons distilled from 14 days of hectic fighting, most not germane to this account. However, without the expertise of the 4th Survey's Territorials, its inspired leadership and dedication at all levels, the casualty lists at El Alamein would have been far longer.

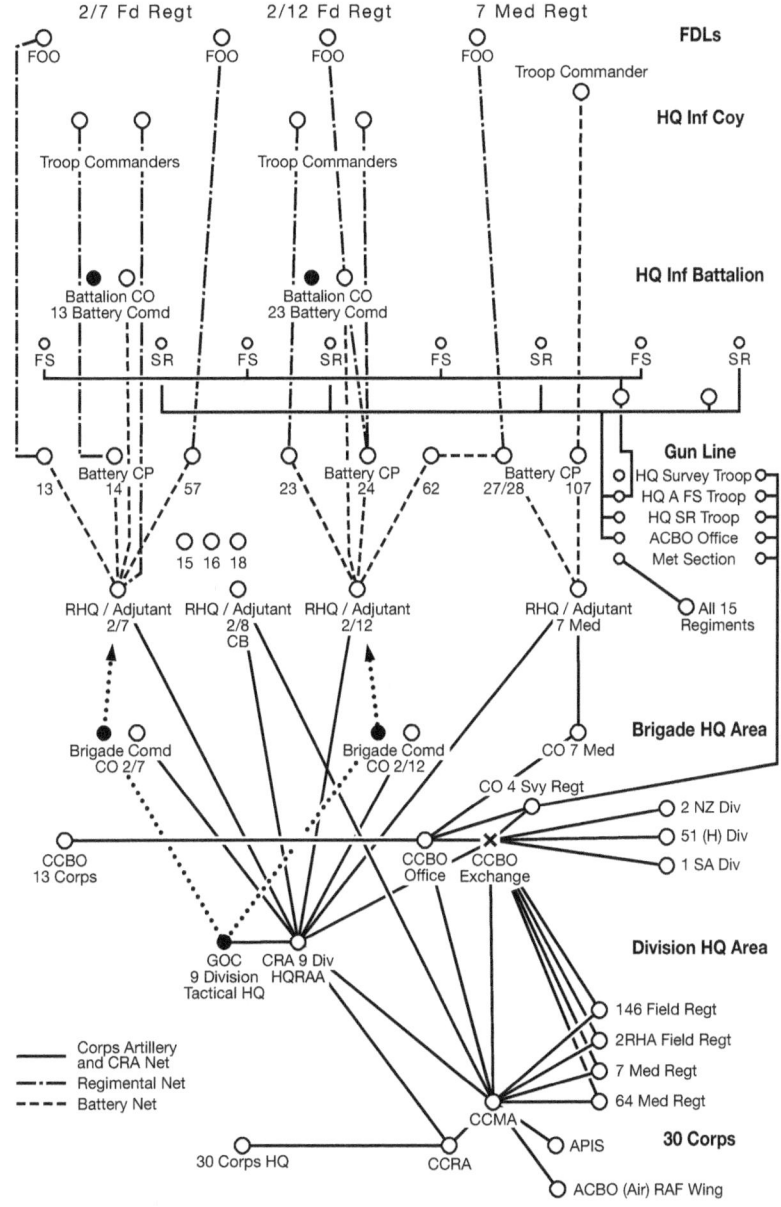

CORPS and DIVISIONAL ARTILLERY COMMUNICATION NETWORKS
This diagram shows the extensive wireless and line communication networks supporting the 9th Division, from Observation Post Officers to their batteries, then regiments, then division and on to the Corps Artillery Commander. Lateral connections to infantry and armoured command centres provided command and control during the battle to commanders at all levels.
(Source, Army History Unit)

CHAPTER 15
A REGIMENT PREPARES—
THE 2/7 ON 23 OCTOBER 1942

*The business of gunner COs is to train their regiment,
and then train the infantry brigadiers how to use them properly.*

Field Marshal Viscount Montgomery of El Alamein

The 2/7th Artillery Regiment RAA was to have some of the most onerous artillery tasks allocated to it during the 13-day battle. While many of the tasks applied equally in a general sense to the other two regiments, the 2/8th had a slightly different role to play in the artillery plan. Much had happened since the 2/7th was raised, had sailed and trained in November 1940. The 2/7th's gunners had gained plaudits from several formation commanders, officers and other ranks of supported units and formations for the quality of their fire.

The CO of the 2/7th, Lieutenant Colonel Tom Eastick, was an engineer by profession and methodical in his ways.[308] He had been commissioned in April 1922 and, eight years later, was a battery commander with the 13th AFA Brigade in Adelaide. Eastick was not a technical gunner, preferring to leave the finer arts to his subordinates, provided they did it 'by the book'. His message to the regiment when it came to El Alamein in July reflected this—and his conservative view of the world. He neither drank nor smoked nor was he frivolous in any sense. He seldom interfered with his subordinates and his strength was his concern for his other ranks. His nickname was 'February', bestowed on him because those malefactors opting for CO's punishment were invariably 'fined 5 pounds and 28 days' CB'.[309] Only 18 of his original 36 officers staffed his regiment when they left Egypt and, prior to the battle, further changes occurred when some of his 'desert-wise' officers departed for other important postings. One such was Adjutant Captain Bill Stevens who went to HQRAA as Staff Captain. Eastick's key subordinates at RHQ were now his second-in-command Major Roy Johnston, (Adjutant) Captain Archie Forwood, (Quartermaster) Captain Harry Loveband, MM and bar, (Survey Officer) Lieutenant Tom Lutz, and Lieutenant Ken Whyte (Signals Officer). Loveband had been awarded a bar to his MM in World War I and was the only Permanent Forces officer in the regiment. Forwood, Lutz and Whyte were all Militia men who had volunteered and had been commissioned in late 1940.

The Battery Commanders, Majors Alf Rogers (13 Battery), Norman Munro (14) and John Day (57) were supported by their command post officers, Lieutenants Henry Newland, Roger Fitzhardinge and 'Tim' Rodriguez. Alf Rogers, a dentist by profession, was commissioned in the Militia in 1940, promoted major in the 3rd Field Brigade and had then dropped rank to become an original in 14 Battery. Munro had been Deputy Assistant Adjutant General in Western Command in 1940 before resuming regimental service. Day had interrupted his Militia service during the Great Depression and soldiered on in Queensland before returning to the 13th AFA. He was commissioned in the Militia in 1940, his captaincy confirmed in May of the same year.

The command post officers all had extensive experience as junior battery officers before progressing to the pinnacle of technical gunnery required of that demanding position. All were commissioned in the Militia before being seconded to the AIF, their substantive ranks as lieutenants gazetted in 1940. They came from many different callings—accountants, commerce, industry, shipping. One was a banker, one a surveyor and another a mining engineer. Unsurprisingly, they exhibited a broad range of approaches to their commands—astute, direct, pleasant and polite, formal, 'hands on'—but, above all, capable and diligent and, more importantly, inspirational. Each Troop commander, a captain, was deployed as an OPO or FOO, although subalterns were also assigned to these postings. They normally 'lived' with the infantry company commanders of the battalion they were supporting directly. This continued the association of the battalion CO with a battery commander as his artillery adviser (an association which began at army level). The infantry commander's confidence in his artillery adviser was established during battalion and brigade exercises. The real test came when under fire.

At RHQ, Lieutenant Colonel Tom Eastick was reckoned to be a good judge of character by those who served with him. He held daily conferences with his headquarters staff and battery commanders, his Quartermaster, Signals and Survey Officers, RMO and LAD OC to ensure that the regiment was up to the mark. Each battery commander covered similar topics concerning his battery, although operational activities tended to predominate. Battery commanders frequently inspected their Troop commanders and command posts, ensuring they were well organised, trained and that reinforcements were settling in. They relied on their Battery Captain (BK) for provisioning and administering the battery. BKs worked closely with Quartermaster Loveband. The latter now used a more standard supply system compared to that employed on mobile operations. Ammunition, rations, mail, stores, petrol and water for each battery were distributed with minimal handling in a process that worked well. At battery level, each officer's appointment was filled and there were also some supernumeraries. When the regiment assumed its position in the coastal sector of the front, it comprised 33 officers (with three others attached) and 663 other ranks.[310]

As previously described, gun pits, cabling and command posts were completed two nights prior to Zero Hour. At last light on 22 October, the guns moved forward to their prepared positions, each battery's vehicles a column stretching for over a mile once they vacated their old positions. The divisional gun area was a skewed 'square' approximately 6500 yards by 5500 yards. The disposition of Troops of four guns, Troop and battery headquarters meant that a distance of roughly 300 to 400 yards separated these centres—sometimes more, sometimes less. HQRAA and the 7th Medium shared dugouts and the RHQs were located a short distance away. The guns were sited well forward and positioned to fire to the limits of their range from the sea to the northern edge of the 51st Division near Trig 30. Both the 7th and 64th Medium were located to the rear of the Australians in the 'premature zone', as one gunner remarked. Once in position, there was to be no registration of targets by shooting (ie., there would be silent registration), only predicted fire, with all targets engaged using only Charge 3 for greater accuracy. Batteries were to be prepared to move from first light on 23 October, travelling to the forward areas. Y Troop of the 1st Composite Survey Battery, RA, was to make a complete check of all bearing pickets prior to D Day. Ammunition allocation for the field regiments was 600 rpg, plus 190 rpg under divisional arrangements, 160 'last line' and 130 rpg 'on wheels'. Medium regiments were allocated 400 or 500 rpg depending on calibre.

Support for a full-scale operation which was expected to extend from seven to 12 days exerted enormous pressure on all ranks, particularly the officers. The arrival of recent reinforcements helped to some extent, but those replaced took with them much desert and battle wisdom. Officers acting as OPOs and FOOs were highly trained in the ways of the infantry companies and battalions they were to directly support. The more experienced captains (Troop commanders) were replaced by lieutenants of lesser experience. A night attack under a near full moon was an ordeal requiring strong nerves, not always replicated on exercises and this was to be a crucial difference. Within the division two experienced Troop commanders from the 2/12th, Captains Nelligan and Calder (FOO with the 2/24th Battalion), were authorised to call directly to HQRAA for the divisional artillery to fire Stonks.[311] Lieutenant Alan Mason (2/12th) was FOO with the 2/48th Battalion and Lieutenant Peter Turner was his relief. The 2/7th's CO would be located at Headquarters 20th Infantry Brigade as adviser to Brigadier Wrigley, its temporary commander. Major Johnson was Artillery Liaison Officer. The 2/7th observing officers were deployed with the infantry battalions as follows:

2/13th Battalion:	Captain A.G. Jones (OPO)
	Captain W.L. Ligertwood (FOO)
2/15th Battalion:	Lieutenant K.W. Everett (OPO)
	Captain W.J.L. Cole (OPO)
2/17th Battalion:	Lieutenant T.A. Rodriguez (OPO)
	Captain R.H. Rungie was 'in reserve'

Similar arrangements applied in the other two divisional regiments with Captain R.M. Jones from the 2/12th appointed Liaison Officer to the 26th Brigade.

Unit historians would later regard Operation Order No. 19 of 21 October as the most significant document in their short history.[312] The order began with the 'big picture' and referred to the latest intelligence summaries detailing the tactical dispositions of the enemy and his anticipated intentions distilled from numerous sources. Next came the maps necessary to fight the battle on the current and intended front/areas of operations. The advance of 3000 yards westwards also applied to the 9th, 51st Highland, 2nd New Zealand and 1st South African divisions. The divisions were to capture and hold an area so that X Corps armour could exploit westwards. The attack would comprise two phases with a one-hour pause between them. A composite force of divisional armour (cavalry), anti-tank guns, machine-guns and pioneer (infantry) was to guard the northern flank.

The 2/7th Field Regiment was to 'support the attack of the 9th Australian Division and support it against counter-attack. On capture of final objective ... [it] will support the 20th Infantry Brigade.' The 2/7th would lay zones of fire 'within limits of range, from the sea to the southern boundary of the 51st Highland Division.' Observation was set as west of the 872 Easting grid line and north of the 296 Northing grid line. OPs were to be established at first light the following morning. The 57th Battery would support the 2/7th with one OPO while another OPO would remain at 20th Brigade Headquarters until relieved by the second-in-command, Major Johnston. Targets could not be registered, and revised Meteor Telegrams were due at the batteries at 7.30 and 11.00 pm and at 6.00 am for predicted fire. Liaison Officers and observation parties moving forward with the infantry companies were urged to pass information quickly to the rear by wireless and line. It had to be 'accurate, detailed and up-to-date' to be of any value. To simplify communication of target data (map coordinates), code words for concentrations and area targets were issued to observers and command posts.

The fire plan was outlined in detail:

Z-20 to Z-5:	all XXX Corps artillery under command of CCMA for CB
Z-2 to Z+7:	all field artillery reverts to enemy FDLs
Z+7 to Z+300:	2/8th fires CB program

(9th Division would share targets with the 7th Medium and one battery of the 64th Medium Regiment)

Z+7 to Z+300: all field artillery reverts to command of divisional CRAs; Corps medium artillery to CCMA.

Brigadier Ramsay's tactical headquarters was sited in the 2/12th's area with the guns in close proximity to the tactical headquarters of his regimental commanders. The 7th Medium and 4th Survey's tactical headquarters were virtually 'next door' to Ramsay. Not far away was Morshead's tactical headquarters and the CCRA, Brigadier Meade Dennis. RHQ survey parties were to 'tee in' the pivot gun once the positions had been occupied at 9.00 pm, the 'time to be ready' on 23 October. Their orders also directed that the wounded were to be evacuated through battery aid posts to the RAP, with further evacuation to the 2/8th and 2/11th field ambulances advanced dressing and medical dressing stations (ADS and MDS), and to the 2/3rd Casualty Clearing Station. Those requiring more complex surgery ended up at the 6th or 7th Australian General Hospital where surgical teams operated. They were supported by Nos. 14 and 20 dental units, whose expertise was also sought for patients with facial injuries. Medical vehicles were to patrol tracks to the rear of the advance to collect walking wounded. Captured enemy medical personnel were to be permitted to continue working. The location of the POW cage was also designated, as were the locations of ration, petrol and ammunition points to the rear. Communications would be vital, particularly wireless, as line (underground) was secure and safe from shellfire and bombing. Wireless silence was enforced from 10.00 am on D Day, code words for objectives and defensive fire

(SOS) tasks listed, the latter to be called for by a Very Light sending up 'a large number of yellow stars'. The infantry had carrier pigeons, but only for use in an emergency.[313]

The staff work involved in moving and placing hundreds of units and detachments of all kinds was immense. Merely moving the assaulting battalions was a complex task and Divisional Headquarters excelled itself in that regard. Moving and siting the artillery and calculating the space required for dispersed guns was no less complex because of the importance of accurate survey. However, to move the whole phalanx of an army corps to its battle positions required an enormous amount of coordination, restrictions and their policing to reduce movement to absolute essentials. In the fortnight prior to D Day, XXX Corps had no fewer than 52 assembly areas for vehicles, stores, ammunition, petrol, oil and lubricant (POL) in the Corps administration area. Vehicles had to move constantly since the DAF would occasionally permit a *Luftwaffe* reconnaissance plane to overfly the front.[314] For example, if 34 vehicles moved out of Areas 24 and 28, 34 had to come in. The decoy pipeline being built in a south-westerly direction at a given rate was designed to mislead the Axis forces as to the time and direction of Montgomery's attack.

Speed by day was 15 mph at a density of 20 vehicles per mile and, by night, 10 and 40 respectively. Tanks and guns were limited to 6 mph but allowed 8 mph on the coast road. Thus a field battery gun group on the move—some 40 vehicles—extended almost half a mile. RHQ vehicles were more dispersed on the ground, but its column at night was about the same. The ubiquitous Jeeps, a recent addition to vehicle strength on artillery establishments, were to be treated as 'motorcycles' for route planning and traffic control purposes. All tracks and roads were to be kept clear and free, stopping on roads was not permitted and halts were to be clear of the road. Thus, on the night of 22/23 October when the regiments occupied their forward positions, the 9th Division artillery moved out of Areas 33 and 34 at 7.15 pm, blocks of vehicles either 19, 46 or 57 in all had to move to the gun area and be back in their areas by 6.00 (2/7th), 4.10 (2/8th) and 4.00 am (2/12th) the next morning. This was made possible by the efficiency of the RASC and Military Police control posts.

The Divisional Provost Corps, never regarded as the most popular of military personnel, regulated traffic in its own divisional area. Artillery drivers had the onerous task of bringing up ammunition, food and stores in a terrain devoid of landmarks. Each main route from the rear was marked discretely and unit locations designated by petrol drums painted a distinctive colour/s bearing the unit 'code number' and an arrow placed at track junctions and turn-ins. For example, artillery colours were red over blue and the 2/8th number was '43', while HQRAA was '40'. Drums had holes punched in the letters, numerals and an arrow so that a hurricane lantern placed inside at night would allow drivers to distinguish their unit, battery or Troop drum as a turning point. Divisional transport under Headquarters ASC control comprised four transport companies, 10, 11 and 12, and 101 (General), which busied themselves bringing forward vital supplies of all kinds from bulk holdings, ammunition and POL points etc., and from their divisional area to B Echelon areas and returning to the rear with salvage etc.

The Administration Area (RED Grid) was south of the main road to 905 Northing grid line and between 454 and 459 Easting grid line. All vehicles had to be self-contained for rations, water and fuel. The inspirational message from the Corps Commander was read to every driver. HQRAA's Administration Instruction specified one gallon of water per man per day, with bulk held in containers in an underground cistern near El Alamein railway station. Each man would have three days' hard rations, and one hot meal per day would be served at night from B Echelon to the gun lines. The Forward Ammunition Point would hold 4656 rounds of 25-pounder ammunition on the ground.

Finally, there was relief and/or disappointment for those selected by the COs to be the 'Left out of Battle' (LOB or Australian Reserve Group) component. Each divisional unit 'sidelined' a specific number of officers and other ranks, usually those with desert and/or operational 'savvy' to go the rear and assist in a way which would not expose them directly to danger, either at the gun line or close to the infantry. Each regiment nominated six officers and 51 other ranks from its active strength in case heavy casualties necessitated replacements from the rear to keep the unit functioning efficiently in battle.[315]

Morale in the units was high. The night was clear with a brilliant moon. The gunners had a hot meal at dark and every lorry in the corps rolled along its designated track ready to perform its specific task. A cool south wind blew across the troops as they bent to their preparations. At the guns each gun position officer had a gun program showing the line, range, angle of sight, and fuse setting (if applicable) for those targets allocated to his Troop of guns. The gun data had been checked by the battery command post, itself an enormous clerical task for the 126 serials. In battery and Troop command posts the staffs made the latest Correction of the Moment (for meteorological conditions) adjustments to the gun programs. In the battery command post of the 2/8th and 2/12th, there was extra work involved in having the three attached gun Troops from the 146th Field Regiment, RA, working as 'Linked Troops' to bolster counter-bombardment and defensive fire.

Within each gun pit, ammunition was readied and checked, sights tested and guns given the final 'once over' before inevitably being brutalised by incessant firing. Each of the gun crews could perform in any position. The gun sergeant and layers used shielded torches to check ammunition and set sights. Guns were usually laid using a night gun aiming post (GAP). This consisted of a kerosene lamp in a small drum in which a slot 18 inches long and one inch wide had been cut. The gun layer laid his dial sight for 'line' on the slot. The GAP was usually to a flank and visible to all four guns. However, on this occasion, aiming towers were erected as GAPs and laid on during the day. Just prior to the commencement of the artillery barrage, lights on the towers (shielded from enemy view) were switched on. Camouflage netting, so essential to concealment by day, was folded back to give the ammunition numbers a little more space to work.

As the sun reached its low afternoon azimuth, the first of a huge number of lorries and other vehicles in a convoy some 30 miles long began to move along parallel tracks towards the west. The noise was concealed by the DAF which flew sorties over the enemy FDLs. As the sun was setting, a hot meal was served and the infantry began to move to its designated assembly areas, then forming-up places and finally to the Start Line. Carrying their weapons, extra ammunition, rations, sandbags and defence stores, the infantry soldiers would have a three-mile trek to their objective, each shouldering up to 40 pounds in weight. The artillery FOOs, Liaison Officers, battery and regimental commanders and their parties 'married up' with their infantry colleagues. Around them, the searchlight crews, Bofors gunners of LAA regiments, oriented their lights and weapons on compass bearings to give direction to the infantry by shining a beam and/or firing coloured tracer rounds of specific colour to signify the sequential events of the plan. Maughan (who was there) records that 'a strange, deceptive quietness reigned between opposing armies. In the deceptive peace the illusion was created that time stood still; but one by one, unarrested, the irretrievable minutes moved on.'[316] The DAF bombers supporting the attack flew over, their destination enemy gun positions and command centres. On the 9th Division's gun line, every man recognised the enormity of the enterprise taking place around him and the vital role he would play. The men waited expectantly, ready to give their all. At 9.35 pm, the 2/7th's regimental historian recorded one gunner's impressions:

> Across the Tel el Eisa flats the voices of the Gun Position Officers at other troop positions come clearly to us, calling to their guns. We stand with shaded torches, intent on the seconds of the last minute ticking around the dial of our lighted watches. Ten seconds to go! There is a soundless flickering along the skyline south of us ... Fire! Synchronisation is an artilleryman's dream.[317]

At 9.40 pm, less 'time of flight', the shells whined their way towards the enemy, the rounds landing at precisely 9.40. In the Italian lines, an observer wrote, 'without a single preliminary range-finding salvo, the entire enemy line burst into life. Soon it was one sheet of incandescence from end to end. The red flashes of exploding shells lit up our own positions ... the clear, starry night echoed with the thunder of the barrage and was filled with the blinding light of flares. At the same time bombs rained down on strong points and defensive positions ...'[318] Wellingtons from 211 Group, RAF, dropped 110 tons of HE bombs on the enemy gun lines in 15 minutes to synchronise with the artillery plan. It was the equivalent tonnage of Corps artillery (25-pounders) firing at normal rate (3 rpg) for 10 minutes. *Generalmajor* Karl-Hans Lungerhausen, commander of the *164th Light Division*, said of the fire plan's opening salvo, 'It was though a giant had banged his fist on the table.' This 'shock effect' was an important aspect of Kirkman's plans. For the remaining hours of darkness, DAF night fighters 'shot up' enemy gun flashes whenever they appeared, further assisting the infantry to advance towards its objective.

Brigadier Ramsay wrote in his diary, 'Battle started 2200 hrs—CB 2140 hrs. Moved to Tac HQ 1900 hrs.'

Chapter 16
Lightfoot and Supercharge: The Anatomy of an Artillery Battle

I have never allowed the Germans to interfere with my plans and I do not intend to start now.

Field Marshal Montgomery
(following Exercise Bumper)

It is worth recording General Morshead's thoughts on the eve of his greatest battle. He wrote to his wife:

> It is now 8.40 pm and in exactly two hours' time by far the greatest battle ever fought in the Middle East will be launched. I have settled down in my hole in the ground ... a little more than 2,000 yards from our Start Line. ... A hard fight is expected, and it will no doubt last a long time. We have no delusions about that. ... It is a supreme effort to finish the war in North Africa, and if successful, as I think it should be, it should have a very material influence on the war.
>
> The men are full of determination and confidence. ... These grand fellows have never failed to respond fully. ... At 9.40 every gun we have opens on the enemy guns. ... My CPH [Corps Photographic Liaison Officer] has just called to say that today's air photographs reveal that he [the enemy] has moved three batteries to positions previously empty. That is a usual procedure. Had he anticipated tonight's the night he would have moved nearly all his guns. So that information is encouraging.[319]

The first rounds fell on enemy hostile batteries at precisely Z-20 and continued until Z-5. At exactly Zero Hour all divisional guns switched to infantry targets. The artillery Task Table extended by 123 more serials to Z+300.[320] Each serial was a

Lieutenant General Leslie Morshead confers with General Sir Harold Alexander, GOC Eighth Army, as Brigadier Alan Ramsay, CRA of 9th Division looks on.
(AWM 024943)

pinpoint (single grid reference or coordinate) target, linear (two) or an oblong/square area target (of four coordinates). Some of these were given code words for the ease of passage of fire orders from the observers with the forward infantry companies. Code words were common to all units so that in an emergency any observer or commander could quickly call for defensive fire provided he was authorised to do so. They took the name of Australian towns, Irish towns and birds. The tracks/routes used also had code words (e.g. Boomerang, Diamond). Zero Hour was marked by bright moonlight which allowed soldiers to see for some 300 to 400 yards until shellfire and bombs kicked up dust which drifted north and reduced visibility. However it was still possible for a FOO to report where shells were falling and/or the whereabouts of enemy troops or vehicles.

For ease of understanding, the attacks which comprised the artillery battle—spanning some 13 days—have been divided into four phases:

Brigadier Sidney Kirkman, BRA Eighth Army, artillery adviser to General Bernard Montgomery, sits on the steps of his caravan at El Alamein.
(IWM E 20968)

Phase A **Initial Assault:** 23–24 October:
offensive action west to weaken enemy armoured and infantry units.

Phase B **Trig 29 and First Thrust Northward:** 25–26 October.

Phase C **Second Northward Thrust:** 27–30 October:
offensive action north to draw enemy reserves into the area.

Phase D **Supercharge:** 31 October–2 November:
offensive action north-west along the coast road and support for Operation Supercharge.

These phases should not be confused with Phase I and Phase II of Operation Lightfoot. Supercharge began on 2 November.

The descriptions that follow are extracted and synthesised from Headquarters 9th Division and HQRAA 9th Division reports, regimental war diaries, XXX Corps CBO War Diary and the *Official History*.[321] To convey some impression of the extent to

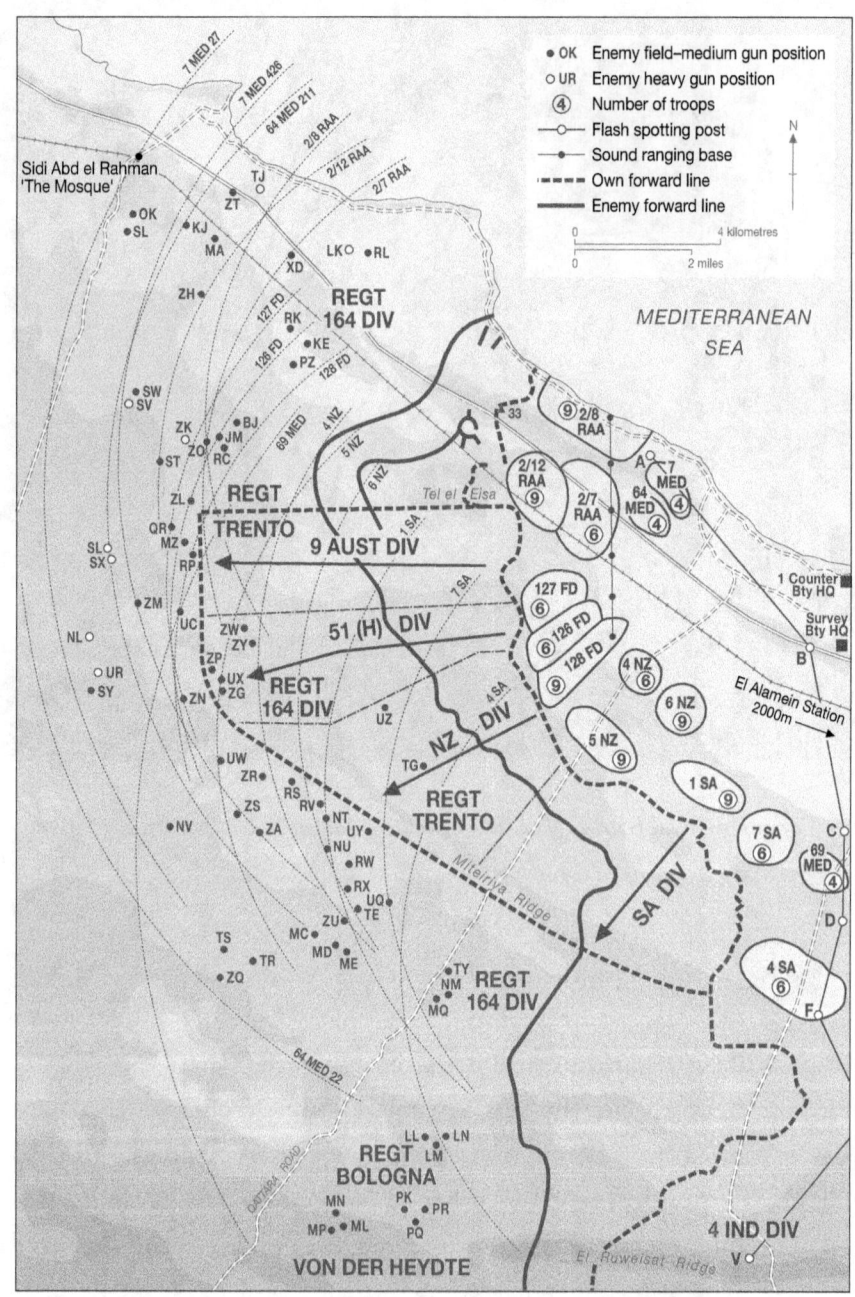

30 CORPS FRONT LINE & HOSTILE BATTERY POSITIONS

This map shows the enemy artillery positions (by code letters, at left) are concentrated against the Australian, Highland and New Zealand Divisions. The arcs of fire of 30 Corps artillery are shown by dotted lines). Enemy and own troops front lines are 5—6 kilometres apart.

(Source, Army History Unit)

Lightfoot and Supercharge: The Anatomy of an Artillery Battle 255

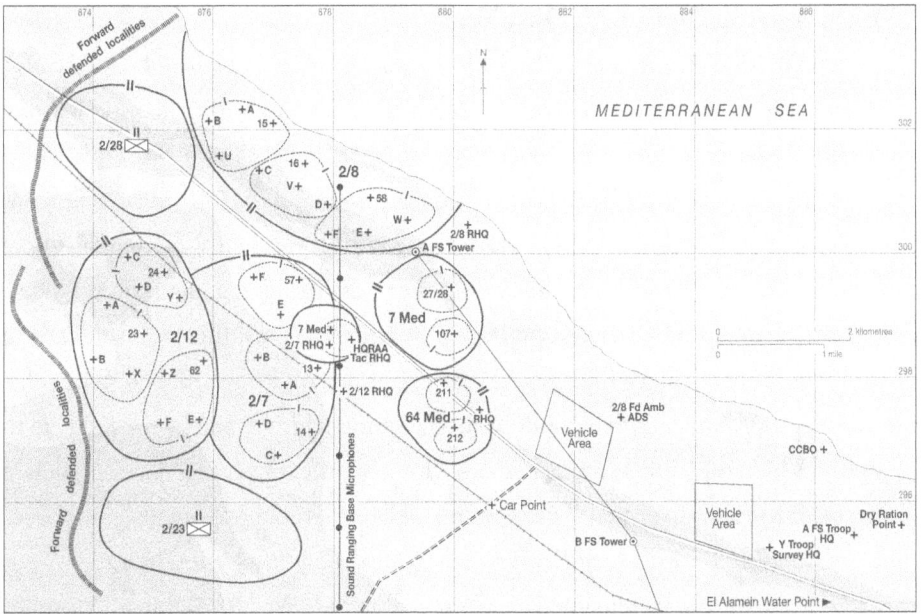

9th Division Gun Areas, showing 2/8th, 2/12th and 2/7th Field Regiment areas (by batteries), and those of the division's medium artillery, 7th and 64th Medium Regiments, RA. A sound ranging (microphone) base ran north-south from the coast to the division's southern boundary through the area.

(Source, Army History Unit)

which artillery influenced the 9th Division's success, the following pages deal with the tactical disposition of the guns and events in the divisional area, the infantry and artillery affiliations in the front line (OP parties of OPOs and FOOs, their 'Acks', signallers and drivers) and supporting armoured units over the four phases of the 13-day battle. Each phase begins with a description of the plan and the objectives sought by the commanders or the day's key events. On each day of the battle a synthesis of some key events (a log) consolidated from all headquarters involved is then provided. This is followed by a description of some facet of the action involving the gunners (sourced from regimental histories and war diaries). Defensive fire tasks are written in capitals and the daily lists that amended the grid references of their position to conform with infantry FDLs (if required) are also included. Targets from the initial fire plan were often called upon to be engaged. The numbers given are from the appropriate task table issued daily from HQRAA and are not necessarily those from the original fire plan. Finally, Brigadier Ramsay's comments written in the aftermath of the battle are provided as a form of summary and highlight the difference between objectives set and outcomes.

Phase A—23, 24 and 25 October

The 9th Division attacked with the 20th Brigade (2/13th, 2/15th and 2/17th infantry battalions) on the left and the 26th Brigade (2/23rd, 2/24th and 2/48th infantry battalions) on the right. The latter brigade was to face right (north) and west on the objective. The plan was for the brigade, with the 2/17th Battalion on the right and the 2/15th Battalion on the left, to advance 3300 yards along SQUARE route until it reached RED line just beyond the enemy FDLs. The 2/13th Battalion would then pass through and advance 3700 yards to BLUE line, around 800 yards east of Trig 33. As there was insufficient artillery to fire a creeping barrage, the artillery fired timed concentrations on known targets. The infantry was to advance 100 yards in three minutes through minefields gapped by the field engineer companies. When the infantry reached its objective the men would adopt defensive positions prior to resuming offensive operations. Their German opponents comprised the *1st* and *2nd battalions* of the *382nd Regiment* and *62nd Italian Regiment*. The *115th Regiment* of the *15th Panzer Division* remained in reserve near Sidi Abd El Rahman Mosque. By 3.10 am on 24 October, the Australian infantry reported that it had reached its objective. Captain Nelligan of the 2/12th Regiment, a FOO with the 2/24th Battalion, reported progress to RHQ and subsequent changes were made to the program as the battle developed. The 20th Brigade had to cross seven minefields (26th Brigade had to cross five). The Italian infantry force was completely destroyed.

24 October

0600	Heavy enemy fire on (all) brigade positions.
0729	15 tanks and lorried infantry counter-attack 1,000 yards west of 2/24 Bn near Trig 33. 2/7 and 2/12 engage WATERFORD, 2/8 and 2/12 WEXFORD. CCRA at 7 Medium Regiment for WOODCOCK. Lt. Menzies 2/7 joins 2/48 Bn as FOO.
1045—1335	2/7 LO (Maj. Johnston) to recce gun areas south of Tel el Eisa for night attack.
1335	Battle at 865 296. 7 Medium Regiment joins Div. Artillery.
1800	2/7 concentration on enemy FUP.
1815	Enemy counter-attack knocked out.
2040	2/7 and 2/12 on Serials 32A and 32B. Work started on new gun pits for 2/7.

Gun sergeants stood next to the gun layers calling changes in line and range. Crews brought up still more ammunition and had to wade through piles of cartridge cases to dump it in the

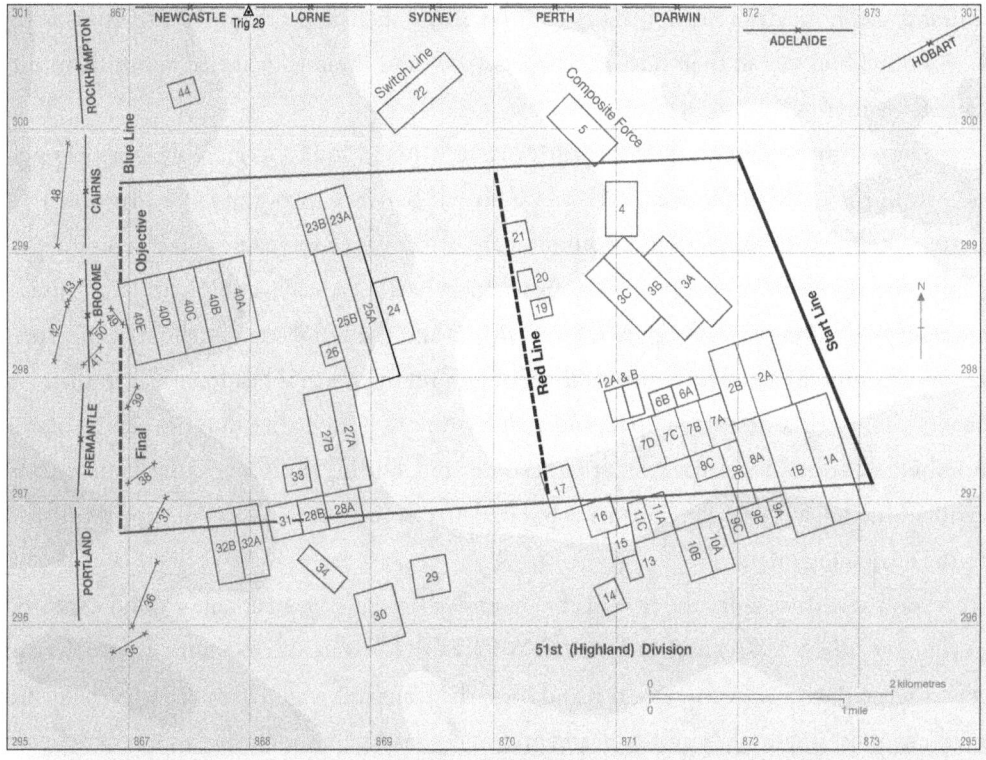

OPERATION LIGHTFOOT—ARTILLERY TASKS for 9th DIVISION'S FIRST ATTACK
This map shows the divisional attack lane from Start Line (SL) to the objective some 6 kilometres to the west. Note that 20 target areas are in the 51st Highland Divisions attack lane. Note also the four linear targets just beyond the division's objective, and eight more on the northern boundary.
(Source, Army History Unit)

gun pit. At 2.00 am on 24 October, the 2/12th Regiment's guns paused (while the infantry advanced without the need for artillery fire), by which time each gun had fired approximately 562 rounds on its set tasks. One gun of D Troop (2/7th Regiment) was out of action with a cracked recuperator (it was soon replaced from the divisional 'pool' of two spares).

Signallers were busy laying line from OPs to the guns. Two from the 2/12th Regiment, Gunners Dooley and Beard, left the Wagon Lines at 2.00 am after a cup of tea and biscuits and followed the OP party Bren Gun Carrier. They were held up by engineer mine clearance parties so they watched the RAF bombing and flare-dropping being contested by the DAK anti-aircraft guns. Once clear they laid line back to their Troop listening post (an intermediate point in the line). They then dug in and waited. They watched with alarm as a Jeep pulled the line and telephone from their pit. They could see from shell bursts that their OP was getting a 'work out' but, safe in their doover, anti-tank shot and shrapnel whistled about them but caused no injuries.

The 2/7th Regiment's first casualties came when a lone German aircraft dropped bombs in the coastal gun area, one of which fell between two guns of C Troop of 14 Battery, killing four gunners and wounding an officer, three sergeants and seven gunners. Only the day before, the Troop's previous position had been heavily bombed. The gunners were resting, having just occupied their position, and the bomb caught them in the open enjoying a mug of tea. The concentration of gun regiments and headquarters in the area around Tel el Eisa, the coast road and railway line was such that random day or night bombing was bound to cause casualties of some kind. During the hours of darkness, enemy artillery fire fell in the regimental areas but had no effect on the delivery of supporting fire to the advancing infantry.[322]

Soon after first light, the first of the defensive fire tasks to maintain ground captured during the night (WATERFORD and WEXFORD) were fired. German and Italian forces reacted with customary vigour and soon the front was engulfed in fire from field and medium guns, 'sniping' 88s, anti-tank and machine-guns. Spent shot of all calibres ricocheted across the stony landscape, careering wildly until it came to rest. Mortars contributed to the battle, their telltale 'cough' barely heard over the din of the fray. At first light the OPOs began observed shooting at enemy targets. Captain Bill Ligertwood, 2/7th Regiment, A Troop Commander, was FOO with the 2/13th Battalion. He was an experienced officer, having done sterling work in the previous battles. His Bren Gun Carrier appeared to German observers as panzerkamffwagen, a generic tracked vehicle—which it was. It was virtually unarmoured and vulnerable to armour piercing shot. Just as he was taking up his position facing the forward slopes of Trig 33 at 6.00 am, an armour-piercing shot passed through his carrier, mortally wounding him and also wounding Gunners Lewis and Parr. Lieutenant Bob Menzies was sent forward to replace Ligertwood. He established his OP in a position that offered little cover and few defences, most having been destroyed by shellfire. He and his party manned the OP continuously for 48 hours, bringing down effective concentrations and ordering defensive fire tasks when requested by the infantry. Their fire broke up enemy attacks by tanks, artillery and infantry. The citation for the award of his MC reflected (apart from his personal courage and devotion to duty) that 'he was largely instrumental in maintaining effective observation over his front and in maintaining artillery support on a vital sector of the 9th Division.'

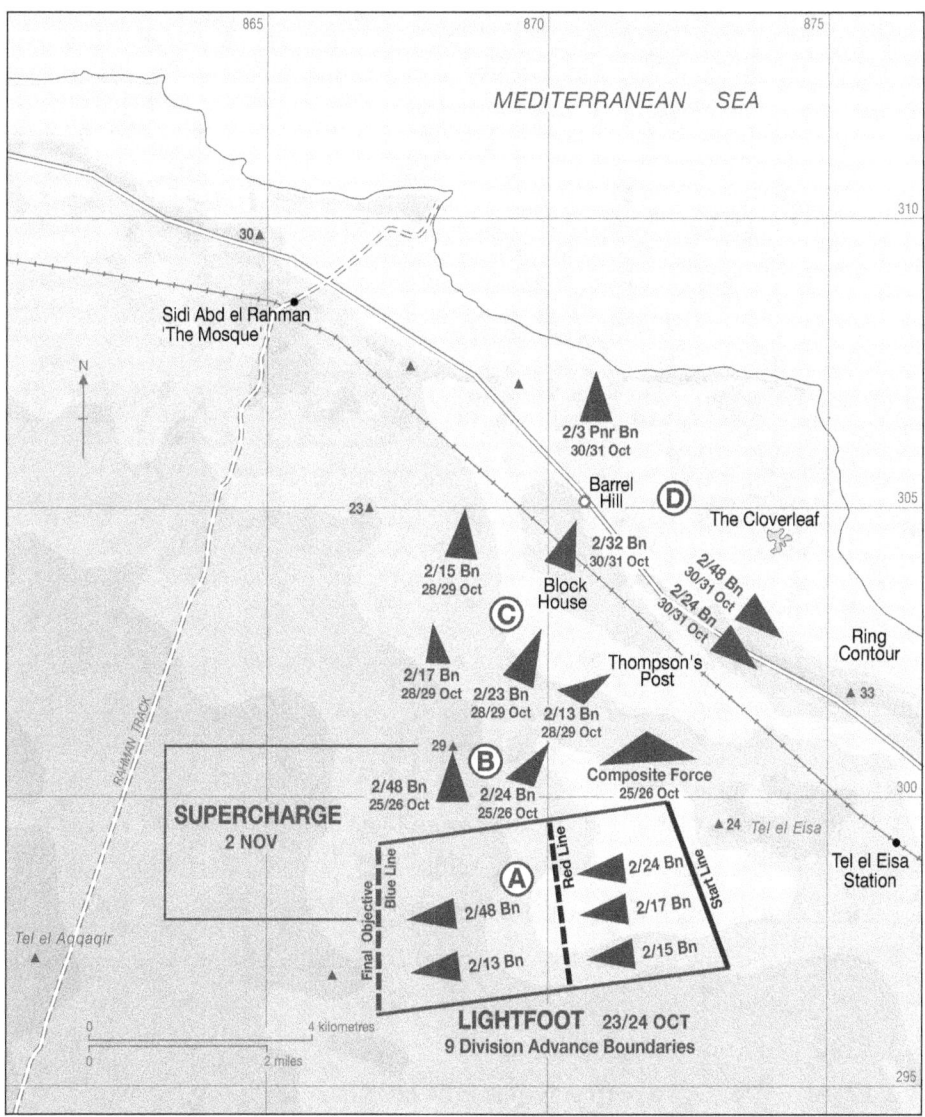

An overview of the 9th Division's battalion attacks from 23 October to 2 November, showing the relative position of them to the initial attack and the subsequent direction of Supercharge, and the armoured breakout.

(Source, Army History Unit)

25 October

0020	2/12 Regt 23 Bty have premature. 1 gunner KIA and 3 WIA.
0130	Shaded lights on FS towers guide enemy aircraft to bomb gun areas.
0200	20 Brigade attacks west and north.

Captain John Wagstaff, OPO of 2/12th Field Regiment with Gunner Murnane in their Bren Gun Carrier (Observation Post). It has been partly 'dug in'.
(AWM 024715)

0705	Enemy tanks and infantry assembling. 2/7 Regt fires WATERFORD and 2/12 Regt fires WOODCOCK.
1035	CCRA conference at NZ Tac HQ.
1045	Hostile batteries PZ XD active. 2/8 and 2/12 Regts fire bombard with 5 rounds gunfire. Strong attack on 2/48 Bn.
1230	Enemy attacks 20 Brigade.
1243	2/8 Regt reports engaging HB 6600ST.
1430	KERRY fired. 2 tanks KOd. Enemy attacks 2/48 Bn in north.
1445	4 Survey Regt reports moving forward both FS and SR bases. 30 tanks on NS line on 2/13 Bn and 2/17 Bn fronts.
1600	DF tasks fired for 20 Brigade.
1700	Lt. Rodriguez now FOO with 2/17 Bn.
1800	RAA Operation Order No.16 issued for 26 Brigade attack.

Lieutenant Rodriguez and his party reported to the 2/17th Battalion and he adopted a position from which he could observe the zone allotted to him. His battalion was taking casualties all around him, but he calmly stood and passed his fire orders to his guns.

Suddenly an airburst 88 HE shell exploded above him. A piece of shrapnel whipped under his armpit, tore open his pocket and sweater, grazed his Artillery Training Pamphlet, Vol. III, split a 50 piastre note in two inside his wallet, snipped away the eye-shields of his binoculars and finally hit the carrier's side with a dull metallic clang. Such were the margins between life, infirmity and oblivion.

The risks were undoubtedly fewer for those further to the rear, although mines laid by both sides were an ever-present hazard. An OPO party from the 2/7th Regiment was coming out of action on relief towing a Jeep that had been disabled when the OPO shouted 'Stop!' The party was in the middle of a Hawkins minefield.[323] The gunners cleared a space and a sapper came along. He told them they had already traversed an enemy minefield that had claimed 19 tanks the previous night.

The CRA's report for Phase A, 23—25 October read:

> The first objective was taken on time (24 OCT) and the right forward brigade reached its second objective on time. The left forward brigade was held up about 1,000 yards east of its second objective, its tanks having been held up on minefields. At 0440 two regiments warned to be ready to engage serials 32A, 32B, 37 and 38 (on south-western end of BLUE line), but these were not fired because of lack of information on the location of the forward battalion. It was decided to repeat concentrations 32A and 32B to assist 2/13 Battalion who reached their objectives on 24/25, but patrols reported little opposition, and the infantry advanced to BLUE line without artillery support. Further counter-attack at 0705 on 25 OCT was met by 2/7 firing WATERFORD and 2/12 firing WOODCOCK to assist armoured divisions at 0800. Further counter-attacks on northern flank were engaged by observed fire. Readjustments of infantry dispositions caused modifications to original DF tasks and regiments in support reported SOS and DF tasks arranged.

Phase B—26 and 27 October

Trig 29 was a focal point for this desperate battle between the troops of the Australian infantry and their opponents. The enemy launched no fewer than 25 separate attacks and this became the most notorious killing ground of the battle in the Australian sector. Its value lay in the fact that it allowed clear observation for some 4000 metres in all directions. The object was to draw enemy forces onto the Australian front by advancing north and cutting off a German salient and thus threatening the coastal road and supply routes of the Axis forces. Both the 20th and 26th brigades were heavily involved in this phase, in particular the 2/17th, 2/24th and 2/48th battalions. Further south of Trig 29, the 2/13th Battalion was also subjected to day-long harassing fire from 12 tanks behind Trig 33. Their opponents were from the three battalions of the German *125th Regiment* and the remnants of the Italian *Trento* and *Littorio divisions*.

26 October

0000	Barrage to support 2/48 Bn (26 Brigade) attack on Trig 29.
0700	Short shooting by 64 Medium? Calibration shoot.
0720	200 enemy infantry on 866 303. 7 Medium engage. 2/12 Regt engaging 15 Panzer Div tanks at 8660 2983.
0807	2/7 and 2/8 Regts superimposed with 7 Medium. LORNE fired.
0817	2/12 Regt and 7 Medium engage NEWCASTLE.
0830	2/7 Regt gun (57 Bty) destroyed by premature round.
0920	Infantry massing at 8640 3010. Four DF tasks to 51 (H) Div.
1200	CBO orders engagement of four 88s.
1451	2/7 Regt fires ROCKHAMPTON. 2/8 Regt on CAIRNS. 2/8 Regt calibrating on marshland.
1503	2/12 Regt engages CAIRNS.
1800	Six DF tasks (STONKS) fired.

Lieutenant Peter Turner (2/12th Regiment) was FOO with the 2/17th Battalion which had relieved the 2/48th Battalion on Trig 29. The area had good command for gunnery and also boasted some excellent trenches for cover. However enemy shellfire during the night caused casualties among the infantry and the company commander decided to withdraw his men some short distance to avoid further losses. As Turner's zone of observation was north to the coast and west to the Mosque he had to remain where he was. At first light Turner saw German gunners manning a field gun and almost immediately a shell burst near the OP. Turner then saw another gun being brought into action and infantry forming up to assault. He had close, battery and defensive fire tasks registered in front of him so he ordered them fired. The shells landed behind the Germans so Turner ordered a 'drop' correction. At this point his battery commander contacted him, anxious about the safety of 'own troops'. Turner advised that if the tasks were not fired, the infantry would be pushed off the feature. Next he burnt his map and then his signaller reported that the battery had fired.[324] When the guns stopped and the smoke and dust cleared away, German stretcher-bearers appeared. Later, Turner's OP was shelled again. A hole occupied by his two signallers took a direct hit and the roof partially collapsed on their backs. By lying on their stomachs they were able to continue passing fire orders. A shell then demolished the overhead cover used by Turner and his Ack and, to their left, a 7th Medium OP took a direct hit, wounding Captain A.F.A. Smith and killing his OP 'Ack' and signaller. For Turner and other OPs, there were many targets offering, varying from armoured cars swanning up to allied FDLs, guns in action in open ground and tanks. They engaged one target of tanks and lorried infantry with the division's guns. Turner described the fall of shot as 'devastating'.

Observers had to be prepared for anything, even their own self-defence if the situation warranted it. One OP party from the 2/7th Regiment established on a map spot found themselves sharing a trench with five dead Italians well in front of the infantry company. Their line communication was cut with their battery and the infantry. Presently the OPO saw 40 enemy troops approaching. The OP party boasted, between them, an officer's pistol and a German and Italian weapon. The former had no ammunition and the latter was jammed. The men prepared 'to sell their lives dearly'. The OPO stood up and challenged—and the Italians raised their hands first.

Back at the gun position, further bombing attacks were launched. One bomb hit F Troop command post from the 2/12th Regiment and wounded one officer and four other ranks.

The CRA's report for Phase B—the attack on Trig 29 and enemy strongpoints east of it—read:

> The infantry captured Trig 29 according to plan but reported at 0255 they were held up and asked for tasks 8F, 8G, 11 and 12. By 0300 2/8, 2/12 and 7 Medium had been warned to be ready to re-engage but this support was not required. By 0700 counter-attack by 12 tanks and 200 infantry was being engaged by observed fire from 2/7 and 2/12. At 0807 counter-attack was met by firing divisional DF tasks LORNE and NEWCASTLE with 2/7 and 2/8 on one and 2/12 and 7 Medium superimposed on the other. At 0900 four DFs were fired, 51 [Highland] Division assisting. At this time 2/7 moved a battery forward, one at a time, east of Tel el Eisa. Numerous heavy enemy attacks (enemy massing at 8640 3010) on Trig 29 were made throughout the day by both tanks and infantry and were engaged by observed fire. Division DF tasks by own artillery and 7 Medium.

27 October

The Eighth Army bridgehead originally planned was captured on 27 October. Rommel moved his *21st Armoured*, part of the *Ariete divisions*, and artillery north to counter-attack. The *90th Light Division* and *159th* and *361st battle groups* moved opposite Trig 29. A concentrated attack on the 2/48th Battalion produced the most intense shellfire seen as both artilleries supported their infantry and armour. Morshead received his orders to 'go north' and Montgomery approved his plans for the attack on 28 October. The 2/7th's RHQ and guns had moved forward the previous night and, at first light, were the farthest forward (west) of all the gun regiments. Dug in and lined with elephant iron, RHQ provided a comfortable 'office' for work around the clock. Brigadier Windeyer resumed command of the 20th Brigade.

A gun detachment of the 2/7th Field Regiment at El Alamein
(AWM 024591)

27 October

0410	Heavy German artillery fire on infantry. HBs engaged.
0425	2/12 Regt engaged FREMANTLE.
0540	15 tanks and a company of infantry attack 2/13 Bn. DF Tasks for 20 Bde—FREMANTLE fired.
0920	FOO Capt Nelligan BGC hit.
1000	88s engaged—withdrew.
1200	7 Medium OP hit during tank attack.
1410	Rommel orders neutralisation of enemy artillery.
1425	E Troop 2/7 Regt engages four 88s. *90th Light Div.* attacks 20 Brigade.
1650	HF tasks issued to 7 Medium Regt. DF (SOS) on 2/13 and 2/17 Bns. Large corrections (400 to 600 yards) to original grid refs.
1755	Rate slow. Attack held. CCMA fires 'WOODCOCK add 200 yards. Engaged.'
2100	Enemy walks into 51 (H) Div area—sick of artillery fire. 20 Brigade relieves 26 Brigade.
2130	CRA orders all troops to fire a round troop fire daily for calibration data.
2255	7 Medium fires LORNE.

The signallers, NCOs and those with OPs endured the most trying times during the battle, exposed as they were to enemy fire of all kinds and the hazards of uncleared mines. Lance Sergeant Ron Freeman (2/12 Regiment) was awarded the MM:

> ... on 26 October his battery established an OP on Trig 29, two carriers with wireless sets were knocked out, and the signallers, owing to the intensity of the shell and machine-gun fire were unable to find the OPO. He set up an OP and commenced to control fire himself until he made contact with the officer. All through the day he patrolled the telephone line under fire, mending numerous breaks and thus maintaining communications of great value.

28 October

Captain Ken Everett (2/7th Regiment) was an OPO supporting the 2/15th Battalion on Trig 29. His position was heavily shelled during the day and he and his party were cut off from their battery. Eventually his OP took a direct hit and Everett was wounded and severely concussed. His party of Gunners Edwards and Butterick were killed. He 'went missing' on the battlefield for some hours until being found and taken to hospital.

A signaller from Everett's battery, Gunner Ward, earned the MM that afternoon for continually repairing wire breaks caused by heavy shelling during an enemy counter-attack. Disregarding danger he continually crossed 400 yards of open ground swept by Spandau machine-gun fire and sniper fire to maintain the line. This enabled effective defensive fire to be brought down.

The CRA noted:

> Numerous heavy enemy attacks (enemy massing at 8640 3010) on Trig 29 were made throughout the day by both tanks and infantry and were engaged by observed fire. Division DF tasks by own artillery and 7 Medium. One of our OP team of 2/7 was killed, the FOO from 7 Med had his line and wireless knocked out and at one stage a FOO from 2/12 and his team were the only ones actually on Trig 29. Besides shooting his own regiment [i.e. directing the fire of his own regiment's guns], this FOO at 1430 engaged the enemy with 7 Medium. On 27 October further enemy attacks were engaged by division DF at 0415, 0440 and 0515. At 1400 NZ Artillery began moving into our area and were placed in support of 9 Div.
> Throughout the day enemy attacks were repeatedly engaged by observed fire. Between 1712 and 1800 divisional artillery and two medium regiments fired DFs controlled by FOOs. Further attacks on 20 Brigade were engaged by 2/7 at 2350 and later 7 Medium. Enemy attacks on 28 Oct started at 0027 and went till 0150 and were engaged by 2/12.

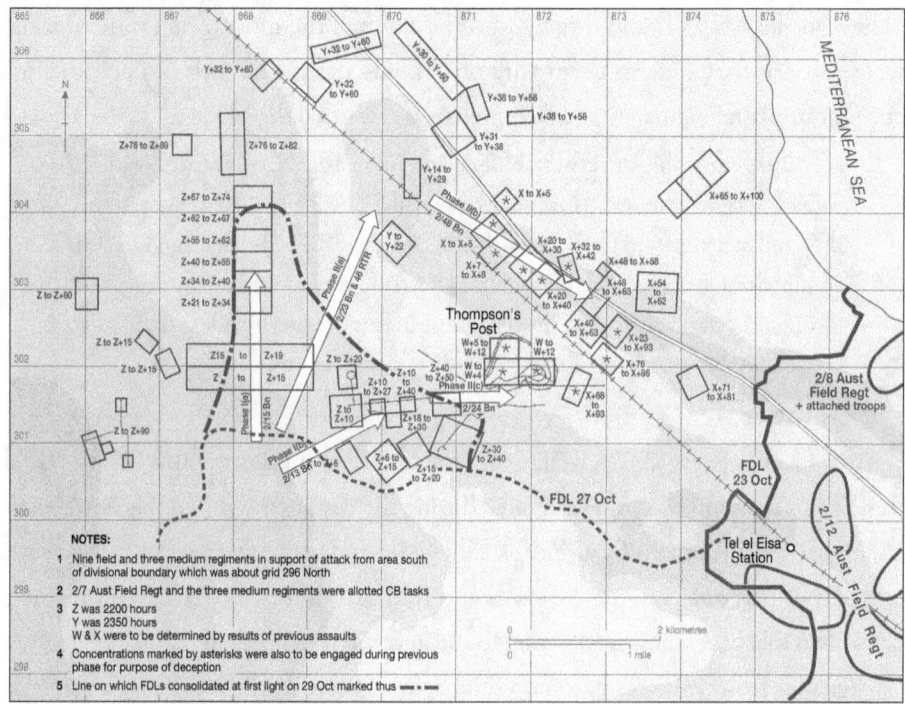

FIRE PLAN on the NORTHERN AND EASTERN FRONT LINES

This illustration shows the complex technical command post work required for this series of battalion assaults in the Thompsons Post area. It was to break the DAK front and draw their reserves north on the last four days of the battle.

(Source, Army History Unit)

Phase C—28—30 October—the second northward thrust

The northward thrust towards the coast road and the notorious area of Fig Orchard, Thompson's Post and Barrel Hill comprised an ambitious plan. Prior to any advance west up the coast road, the ground north of it and the railway would have to be 'neutralised' from enfilade fire or captured. This was not Morshead's immediate objective. The 20th Brigade relieved the 26th Brigade (with some difficulty due to enemy attacks) on 27 October. In an effort to disrupt this development, the Germans mounted a strong attack from Sidi Abd El Rahman Mosque area. The 20th Brigade (2/13th and 2/15th battalions) were attacked, sustaining heavy casualties, as did the 2/23rd Battalion and 46th RTR. New Zealand and British regiments in range supported the attack. The Australian attack of 28/29 October was launched against the *90th Light Division's 125th Regiment*, a battlegroup of the *155th Regiment* and *XI Bersaglieri Regiment*. The *15th Panzer* and *Littorio divisions* were also moved forward to counter-attack.

28 October

0001	C Troop 2/12 Regt engages tanks and fires DF (SOS).
0007	2/12 Regt fires SOS for 26 Bde. Z Troop guns out of action.
0010	2/8 Regt fires LORNE rate slow.
0135	2/12 Regt fires DFs for 26 Bde at Trig 29.
0150	Concerns re ammunition expenditure on DF Tasks (100 rpg).
0345	Z Troop reported in action.
1025	2/8 Regt smoke targets for bombers. 2/24 and 2/48 Bns counter-attacked. Hinders relief of 26 Bde by 20 Bde.
1125	Large-scale attack (infantry and tanks) on Trig 29. 2/12 Regt fires NEWCASTLE.
1430	A Troop 2/12 Regt engages enemy guns by observed fire.
2200	Concentrations fired on enemy positions to support Phase II of Lightfoot. CB tasks fired at H+49.

While the 2/12th Regiment's 23 Battery was supporting the 2/23rd Battalion, Captain Calder, a very experienced OPO, was wounded and replaced by Lieutenant Pitt. Back in the gun area, each battery mounted anti-aircraft sentries, usually two gunners manning a Bren gun mounted on a special anti-aircraft tripod. On this day there had been desultory German counter-bombardment fire over the gun area. It claimed the lives of Gunners Maine and Chuter when a shell scored a direct hit on their pit. In a Stuka attack that morning four gunners from F Troop were wounded by bomb fragments and, later that afternoon, a gunner was wounded by shellfire.

29 October

Brigadier Windeyer (26th Brigade) assumed responsibility for the northern sector and ordered the 2/23rd Battalion to advance its positions 1000 yards to link with the 2/13th and 2/15th battalions (20th Brigade). Montgomery decided to launch Operation Supercharge on the night of 31 October/1 November. He ordered Morshead to maintain pressure on the enemy. The 26th Brigade (less the 2/23rd Battalion but with the 2/3rd Pioneers plus 40 RTR) was to attack and capture Barrel Hill and the main road. Attacks on 26th Brigade positions continued.

0100	Tasks 19, 20 and 21 were repeated by 2/12 Regt from 0110 to 0140. 3,000 rounds per hour fired (till 0200).
0800	14 Tanks hull down in front of Trig 29.
1000	Enemy infantry attack. Beaten off.
1300	Enemy attack resumes.

5.5 inch medium gun of 7th Medium Regiment, RA in its loading position. It took 30 seconds to load, lay and fire a round. Note the shell on its cradle (foreground), and a British gunner wearing a slouch hat.

(AWM 024814)

1345	Enemy withdrawing. HF resumed on 2/13 and 2/15 Bns.
1700	2/7 Regt engages with NZ artillery.
1725	Stuka aircraft attack 2/7 Regt gun positions (A Troop). 2/7 Regt fires BAIRNSDALE (2/15 Bn tasks).
1850	Infantry attack.
1930—1940	2/12 Regt DF Tasks 43, 44 and 45 fired. 2/12 Regt three guns out of action—one evacuated.

A captured Allied officer had revealed under interrogation that there would be no 'left hook', but rather a northward thrust by the 9th Division. As a consequence, Rommel moved the *21st Panzer Division* north.[325]

The CRA's report described the 20th and 26th brigades' attacks of 28/29 October:

> For this attack the guns available were: 2/7, 2/8 and 2/12 Field Regiments, RAA 4, 5 and 6 Field Regiments, NZA 126, 127 and 128 Field Regiments, RA

(51 HD) 2 and 4 Field Regiments, RHA 78 and 146 Field Regiments, RA 7, 64 and 69 Medium Regiments, RA firing timed concentrations on a depth of 200 yards, lifting ahead of the infantry. Guns not well sited—concentrations in enfilade and in one series in the face of advancing infantry. For CB three field and one medium regiments on call. A deception demonstration made on enemy FDLs.

As shown, the guns were sited where they had been since 24 October. A decision was made not to move them further west where, at best, their fire would have been at right angles to the infantry axis of attack.

Ramsay declined to comment on the fact that he had 360 guns under command for the operation, the most of any Australian CRA in World War II.

30 October

Further attacks continued along the whole of the 9th Division's salient. The 24th Brigade's attack was successful, with the 2/32nd Battalion capturing the Blockhouse which was used as a medical post by both sides in turn. The 2/48th and 2/24th battalions met strong opposition and suffered many casualties in their eastward attack, finally returning to their original Start Line area. Both resisted strong enemy attacks towards The Cutting. Field and anti-tank artillery broke up German attacks on Barrel Hill and inflicted heavy casualties. Battalions reduced to less than half their original strength performed heroically. The Saucer, an area near the Blockhouse, was defended by the 2/24th, 2/32nd and 2/48th battalions. The former two were later relieved by the 2/28th and 2/43rd battalions, while the 2/32nd Battalion remained. Morshead regarded the relief of the 26th Brigade by the 24th Brigade as the highlight of the battle considering the risk involved. This was no ordinary relief in the line, as it occurred in the most hotly contested sectors of the front. Lieutenant Colonel Loughrey (2/28th Battalion) 'improvised positions adopted in the dark in a precarious situation on unreconnoitred ground.'[326]

0100	Four-phase attack begins for 24 and 26 Brigade.
0037, 0112, 0425, 0523	2/12 Regt fires SOS 8.
0158	F Troop OP with 2/15 Bn reports tank attack developing.
1010	2/7 Regt attacked by 15 Stukas. No damage. One wounded.
1830	XXX Corps advises Rommel at 8668 3050. 2/7 and 2/8 Regts advise he is out of range.
2045	RAA Operation Order No. 11 and Operational Instruction No. 61 issued at 1600.

Phase D—31 October—4 November (Breakout)

After seven days of continuous fighting involving every unit of Montgomery's and Rommel's armies, the battle was approaching a climactic period. Both sides absorbed serious casualties in men and equipment, but Montgomery's resources in all critical respects would swing the pendulum of victory his way. He was guided by Wellington's famous dictum, *'In military operations, timing is everything.'*

On 30 October the 20th Brigade with Brigadier Godfrey's 24th Brigade, the 2/28th, 2/32nd and 2/43rd battalions fought their way to be astride the road and railway line in the Blockhouse area. This threatened the enemy forces still occupying Thompson's Post and defended localities to the north-east (Tel Alam el Shaqiq and the Cloverleaf feature) opposite. Persistent enemy attacks on weakened battalions affected the timing for the launch of Operation Supercharge. Continuous artillery support for attacks and hastily mounted German counter-attacks was desperately required. The enemy withdrew during the night of 2/3 November and dawn saw both brigades push out fighting patrols to maintain contact with the enemy. The 9th Division faced the *90th Light Division* while the *21st Armoured, 125th Regiment* and *X Bersaglieri* were sited in the coast sector. The 2/7th Regiment advanced behind the Eighth Army's leading formations to El Daba and, on the night of 4 November, fired at hostile batteries, its last action in the desert war.

31 October

The 2/3rd Pioneers advanced north and cut off the *125th Regiment* and the *357th Italian Light Artillery Regiment*. The 2/32nd Battalion attacked the *361st Regiment*, taking a large number of prisoners. The first heavy enemy action began at 11.30 am with heavy artillery fire and was beaten off. Another began at 3.25 pm. Thirty Stukas attacked the positions of the 2/28th, 2/32nd and 2/43rd battalions, but were intercepted by RAF aircraft and driven off with heavy losses.

0110	FOOs report heavy shelling.
0120	CBO orders bombard forthwith. Thompson's Post tasks cancelled.
0425	2/3 Pioneers attack. No FOO (truck destroyed).
0508	2/7 Regt stops task 26. 2/12 Regt fires tasks 43, 44 and 45 three troops each. Remainder of program fired.
0720	20 Bde. Serious opposition at 86940 30512. 7 Medium engages with one battery; 2 rpgpm.
1130	15 tanks and infantry (*361st Regiment*) and artillery support attack 2/3 Pnrs.

1200—1400	2/7 and 2/12 Regts fire WAVERLEY and MORGAN.
1430	WARBURTON fired
1627	7 Medium to calibrate.
1745	2/12 Regt DF 4 fired.
1750	15 tanks attack 2/3 Pnrs.
2100	Wireless intercept of 21 Panzer Div. CRA orders HF by 2/8 Regt from 2130 to 0530.
2130	Thompson's Post held by 50 infantry, 4 infantry guns, 2 mortars and 1 Besa.
2150	2/8 Regt bombard ordered on 5 serials and an area target; 5 minutes normal at 2230, 2305 and 2345. CRA orders 2/12 Regt one battery at a time 2145 to 0530 on coast road to 872 grid.

A Stuka raid on the 2/7th Regiment's A Troop of 13 Battery destroyed No.1 gun, killing five NCOs and five gunners, the second serious calamity for the regiment's gun detachments. Having fired defensive fire tasks at sundown, the gunners rested until the following morning,

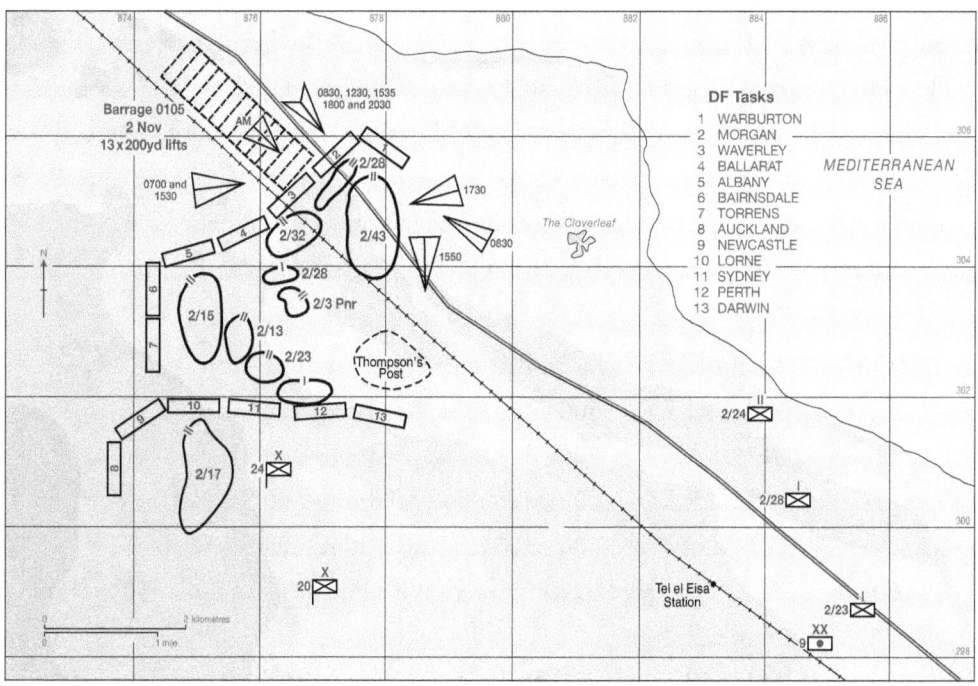

The fire plan in place for the last stages of the battle, culminating in the barrage (top left) and 13 Defensive Fire tasks to support Operation Supercharge, Montgomery's armoured thrust to the west on 2 November.

(Source, Army History Unit)

their first spell in 12 days. In the command posts, officers and gun position officer 'Acks' worked on gun programs from task tables for Operation Supercharge drawn up by the 2nd New Zealand Division Artillery Headquarters. They were long and difficult programs and very optimistic in terms of timing. The 2/7th Regiment's guns were at maximum range and the use of supercharge propellant (usually reserved for use with anti-tank shot) proved necessary. Changes to fuse length for airburst and smoke shells made life at the gun difficult for the detachment commander and his gun layer.

1 November

Strong winds and dust limited visibility over the front lines and the 2/7th Regiment was bombed again with one gunner wounded. The 2/12th Regiment rotated its OPO/FOO parties and Lieutenant Murray returned to his Troop as OPO with the 2/32nd Battalion. He established his OP (on B11) and was promptly shelled. His OP Ack, Lance Bombardier Cutler, was killed and the remainder of the party had great difficulty extracting themselves. Lieutenant Colonel Goodwin resumed command of the 2/12th Regiment and Major Houston went to 24th Brigade Headquarters as Liaison Officer to replace Major A.A. Carter who had been wounded when the headquarters was heavily shelled. His colleague, Major R.G. Trenwith, was mortally wounded.[327]

This was the most momentous day for HQRAA. As well as having to contend with the operational aspects of continued attacks on the division and Operation Supercharge, divisional and corps staffs, suspecting that the enemy's attacks could not be sustained for much longer, made frequent requests for front line tactical intelligence.

1 November

0330	Relief of 26 Bde by 24 Bde complete. DFs moved:		
	WAVERLEY	500 yds SE	Mediums 400 SE
	MORGAN	600 yds SE	Mediums 300 SE
	BALLARAT	800 yds S	Mediums 600 S
	ALBANY	600 yds SE	Mediums 600 S
1000	2/12 Regt fires DF 2 urgently (and again at 1250 and 1350) at lorried infantry.		
	Tank and infantry attack on 24 Bde with heavy artillery (8 88s, mortars).		
1245	6 tanks closing on 2/43 Bn. Driven off. 7 Medium engages.		
1435	Intercept reveals 88s on the move.		
1500	WAVERLEY fired and again at 1600.		

1525	Attack resumes. 2/28 Bn no FOO.
1700	Tanks and 100 infantry attack west from Thompson's Post. Dispersed.
1830	3 tanks and 15 lorried infantry vehicles attack.
1915	24 Bde Tac HQ heavily shelled.
2030	German artillery HF on 24 Bde.
2330	2/12 Regt fire plan to harass German relief of Grenadiers.

2 November

Operation Supercharge began at 1.05 am when XXX Corps launched its attack from Tel el Eisa on a frontage of 4000 yards towards its objective 6000 yards deep. British and New Zealand forces were to exploit westwards to Tel el Aqqaqir. A barrage was fired by Corps Artillery to relieve pressure on the 24th Brigade (right) in The Saucer and the 20th Brigade (left). The 26th Brigade remained in reserve. Later, strong fighting patrols were despatched towards the coast. The Germans attacked these with vigour, despite the fact that the German infantry had been severely mauled by artillery fire and the assaulting Allied infantry battalions. Enemy artillery was reckoned at 40 per cent strength with few tanks surviving. Harassing fire tasks by Corps guns to support the New Zealanders and on the coast road completed the 9th Division artillery support at El Alamein. The 9th Division infantry advanced west and adopted positions four miles from the start line of 12 days ago.

0105	Operation SUPERCHARGE begins. 2/8 Regt in support.
0134, 0234	24 Brigade calls for DF, followed by MORGAN.
0600	30 tanks attack 2/15 Bn. Driven off. 2/7 Regt supports 2 NZ Division.
1245	2/12 Regt fires EDEN.
1400	Tasks to support fighting patrols issued. 2/7 Regt 8 HB tasks engaged with 1200 rds.
2050	2/12 Regt fires HF tasks.

The CRA's report on the action of 1 November and Operation Supercharge noted:

> The first phase was OK and according to plan but the second was not. Tanks were caught in a minefield and the infantry lost artillery supporting concentrations. From 0050 to 0430 the division fired concentrations to help infantry forward. Two batteries given DF tasks. 2 NZ Division, 51 (H) Division and three mediums. The first counter-attack 2/12 tasked and 2/8 smoked target. Counter-attacks on 24 Bde from 1222 to 1740 from east,

north-east, north and north-west at different times. At 1800 2/7, 2/8 and 2/12 with two NZ regiments and 7 Medium supporting infantry. DFs fired by two regiments from 51 (H) Div, our own and 7 Med from 2000 to 2055. Resorted to harassing fire 2100. On 1 Nov two medium regiments moved forward to support the NZ attack, and 2/7 allotted in support. NZ attack successful and our OPs are engaging [opportunity targets]. Tanks withdrawing from our front and moving south. SUPERCHARGE Op Order No 12 to support 24 Bde (right) and 20 Bde (left) with 26 Bde in reserve. Barrage appr 1800 yards long and 500 yards wide, north-west, lifts 200 yards in 4 minutes. Rate slow. 2/8 with 2/12 superimposed. HF tasks three troops field 2300—0400 45 rpg. Support NZ with DFs. Batteries move forward one at a time. On 2 Nov 0100 attack on 2/24 Battalion 2/12 responded. Interim HF tasks issued for 2/3 Nov.

The worst of the fighting was over and the enemy was in rapid retreat.

Chapter 17
Aftermath and Analysis

The desert shall rejoice and blossom as the rose.

Micah 14:3

When the infantry brigades advanced on 4/5 November their affiliated field regiments (and mediums) moved west and took up supporting positions generally north of the coast road. There they dug in and waited for their next set of orders. The Australian field regiments were ordered to return to their previous positions and move into reserve. Their colleagues in the 7th and 64th Medium and 4th Survey regiments went west with XXX Corps, bidding them farewell forever. The regiments camped near the coast and, prior to returning to Palestine, 'sorted ourselves out a bit', as one unit historian commented, after a hectic 13 days. Details were despatched to clean up the battlefield—an X Corps task—and the opportunity was taken to locate the temporary graves of comrades for reinterment in the cemetery. Total artillery casualties were 45 killed in action or died of wounds and 60 wounded—in sum, 105 (including five 2/8th wounded in action in November):

	2/7th		2/8th		2/12th	
	KIA	WIA	KIA	WIA	KIA	WIA
Officers:	3	2			1	2
Sergeants:	5	4				
Bombardiers:	5	2				
Gunners:	23	20	1	1	7	24
	36	28	1	1	8	26

Field Engineer companies, which shared the hazards of operating with infantry and armour, lifting and sowing mines and other tasks, lost one officer killed and five wounded,

17 other ranks killed and 86 wounded for a total of 109. Between them, the three infantry brigades lost 132 officers and 1999 other ranks.[328] Total battle casualties for the division were 2694, of whom 620 were killed.

As noted for the 2/7th, Captain W. Ligertwood was awarded a posthumous MC and Lieutenants R.W.J. Menzies and T.A. Rodriguez also won the MC. Gunner S.A. Ward (a signaller) won the MM for 'gallant and meritorious service' for repairing signal cable and restoring communications for the OPO during a counter-attack. These decorations were for actions no doubt inspired by Lieutenant Colonel Eastick's exhortation to his observation parties to 'act most boldly to ensure the effective fire of the regiment'. This they had done continuously throughout the battle. The remainder of the regiment also felt the force of his leadership and personality, amply demonstrated in his 'forceful, determined, aggressive use of his regiment', as noted on the citation for his DSO.[329]

In the 2/12th, Lieutenant Colonel Shirley Goodwin and Major Geoffrey Houston were both awarded the DSO. Goodwin was described as possessing 'outstanding ability and power of handling men, wide experience in training and staff duties'. Houston was cited for 'showing initiative and resource, especially in the planning and control of observation' and for his role as Liaison Officer.[330] Major Archie Carter, wounded while serving as Liaison Officer at the 24th Brigade Headquarters, was also awarded the DSO for 'outstanding courage, coolness and judgement, notwithstanding difficulties of enemy fire, of conflicting reports and of incessant demands for fire over a wide area'. The MC was awarded to Captain Phillip Nelligan for 'three occasions when he acted as FOO'. Lieutenant Tod Cassidy was cited for his 'efforts to give utmost support to the infantry'. Lieutenant Gordon 'Black Sam' Stewart won the MC for 'his coolness and courage as FOO'. Lance Bombardier George Steele, lauded for his 'courage and perseverance during a critical period of battle', was awarded the MM, as were signallers Lance Sergeant Bill Holding, Lance Sergeant Fred Freeman and Gunner James Ryan. A common theme throughout these citations is the devotion to duty in the maintenance of communications—the lifeblood of any battle, but the essence of this one.

The 2/8th's CO, Lieutenant Colonel Walter Tinsley, was another recipient of the DSO. His citation stated that 'the quickness of response and accuracy of supporting fire provided was largely due to his insistence on high standards of efficiency, initiative and forcefulness during the battle'. His 16 Battery Commander, Major Frank Stevens, was awarded the DSO, his citation stating that 'throughout the whole operations he used the fire of his battery aggressively [and] imbued his officers with a determination to provide effective support regardless of risk or cost'.[331]

Ramsay's summaries of the actions of the 2/8th, however, contained relatively few citations and noted far fewer casualties than other units. This was primarily because the 2/8th and its attached Scots gunners comprised a dedicated counter-bombardment fire unit and generally did not deploy OP parties needing line signallers. When not engaged on counter-bombardment or high frequency tasks, their guns were laid on defensive fire tasks. When tasked for infantry support, they provided this using other regiments' OPOs. Their ammunition usage was similar to that of their sister regiments, which suggests that they were just as busy, albeit with a different target 'mix'. When it came to bombing casualties, they may simply have been more fortunate than the others.

The artillery casualty list was small compared to those of the battalions that the regiments supported with their guns; in the observation parties, however, the toll of casualties reflected the hazardous nature of their deployment. In the main, the OPOs and their parties were relatively inexperienced compared to those Troop commanders who became early casualties and whom they replaced. The latter had the confidence of their affiliated infantry, while the less experienced had to earn credibility through performance. Thus, it was not entirely unexpected that at least one battalion commander (Lieutenant Colonel Hammer, CO 2/48th Battalion) reported that the OPOs attached to his battalion 'generally lacked experience and were unsatisfactory'. It was unfortunate that these men had to 'earn their spurs' in what was one of the harshest battlefield environments of the campaign.[332] However, Brigadier Kirkman's report was full of praise for the Liaison Officers who had to reach grid references in a featureless desert, often in darkness, bringing vital orders and intelligence to company, battalion and brigade headquarters across minefields and exposed to shellfire and projectiles of all kinds.

The 2/7th and its colleagues toiled mightily night and day for 13 days. The 2/7th fired 65,594 rounds of HE and 510 smoke. Five of their 25-pounders had been put out of action and two prematures had occurred. Total rounds fired for their campaign amounted to 154,310 HE, 3249 smoke, 288 AP and 157,994 charges (12,196 super). The 2/12th averaged 1900 rpg during the final battle (approximately 45,000 rounds) and, for the four months, more than 200,000 rounds. The 2/8th fired 59,133 HE, 2318 smoke—a total of 61,451 rpg in the 13 days. Over their four-month campaign, their grand total reached 166,321 rounds.[333]

Not all the artillery units of the Eighth Army could claim similar experiences to the 2/7th, particularly those in the other two corps. However, it was the 9th Division artillery that 'took the honours' for the most continuous performance compared to its peers. Over the 13 days of battle, the Eighth Army's expenditure of 25-pounder ammunition amounted to more than one million rounds, or an average of 102 rpg per day. The highest daily

Piles of spent ammunition 'produce'—cartridges, their boxes and those of shells were a feature of Troop positions. It was difficult to conceal and a sure 'give-away' by enemy aerial reconnaissance. It was collected during the battle.
(AWM 024584)

expenditure within the Eighth Army occurred on 24 October—317 rpg per day—while the lowest—52 rpg per day—occurred on 28 October. During Operation Supercharge, expenditure amounted to 208 rpg per day.

During the October battle, the XXX Corps 25-pounders—some 354 guns — averaged 1909 rpg with an average of 159 rpg per day. Compare this with X Corps guns which averaged 1043 rpg and XIII Corps guns with an average of 518 rpg. The 2/7th's average was 228 rpg per day. Within XXX Corps, the three Australian field regiments' expenditures were half as much again as the remainder of the Corps. The medium guns performed almost as well as the 25-pounders. Their rates of fire were (overall) 157 rpg per day for 5.5-inch and 133 rpg per day for 4.5-inch. In sum, more than half a million rounds were expended by the division. These statistics confirm the different intensity of fighting in various sectors of the northern part of the XXX Corps front opposite the Australians. Little wonder Generals Alexander, Montgomery and Leese came to Morshead's headquarters on 4 November to thank him for his division's contribution, as did Horrocks of XIII Corps. Morshead alone knew why his men were withdrawn from the pursuit phase of Supercharge—the Australian Prime Minister (John Curtin) and the War Cabinet had agreed that all Australian forces would return home to help fight the Japanese.[334]

Brigadier Kirkman's report also noted that the tactical handling of divisional artillery (72 guns) had reached new heights. However, the 9th Division more often boasted 96 guns (with the 146th Field Regiment, RA, attached) and, when the 7th Medium was in 'direct support', the total was boosted to 112 guns. Those essential supports of the technical arm—signals, ordnance ammunition specialists, electrical and mechanical engineers, medical services, service corps supplies and transport—all contributed their skills and expertise during the battle. They, too, were vulnerable to random bombing attacks no matter where they were sited. They, too, shared in the well-deserved divisional plaudits.

The 2/7th RMO, Captain E.W. Sibree, noted that all ranks 'had stood up to prolonged exertion, with little sleep and general strain of combat in an excellent manner.'[335] Equipment losses (except guns) were quickly rectified and one hot meal a day helped sustain morale. It is worth noting that the War Establishment of a field regiment allowed 24-hour-a-day operation and, except during emergencies, gun detachments operated with half crews while command post staffs were divided into two 'shifts'.

While the division was resting and recuperating, Brigadier Ramsay had his regiments performing useful tasks, not the least of which was distilling the lessons learned. Lieutenant Colonel Eastick's summary report noted the following:

1. Concentrated use of artillery firepower, with guns being well forward (subject to their defence).
2. Barrages are effective if (other) defences are within the range of covering fire.
3. Fire-planned timed concentrations were the 'complete answer' (as post-battle surveys had shown) covering the front and flanks of captured ground.
4. Wireless worked well and line was generally through to the OP in good time and maintenanced.
5. Camouflage was of no value night or day (because of gun flash).
6. The value of regimental survey in their Flash Spotting role (as locators).

Providing communications were good, alteration or cancellation of one divisional artillery program could take place in 12 to 15 minutes. Calling it on again took slightly less time, up to 30 minutes if more than one division was concerned. On 19 November a conference was held to discuss times and methods of producing fire plans under stress of battle, once task tables and traces had been received. The 2/7th's solution was approved as 'standard' (for timings), and was later promulgated throughout the British Army.

On the issue of counter-bombardment policy, Brigadier Kirkman reported that:

> Since it was impossible to neutralise all the enemy guns during the attack, the policy of employing all available on CB fire immediately prior to Zero Hour was adopted. In XXX Corps during this period, 25 known Hostile Batteries within range were subjected to heavy concentrations of up to 20 to 1 and in no case less than 10 to 1, whilst during the attack more normal concentrations were again put down for short periods of time on selected batteries.
>
> Some doubt was expressed by infantry commanders as to the wisdom of CB fire immediately before Zero Hour, but they are now satisfied that it had no evil effect on surprise. Its value lay in the really heavy concentrations that were possible, and it was certainly successful as enemy artillery, though active between attacks, was particularly silent during attacks.[336]

Brigadier Ramsay paid tribute to the 2/7th, one of the five good regiments under his command: 'none was more efficient and reliable. One characteristic was its dependability. No task was too great for them ... They fully earned their share of praise passed to Lieut. General Sir Leslie Morshead ...'[337] He wrote a comprehensive post-battle report in which he noted that the main lessons distilled from the 12 days of 9th Division artillery operations (including the 146th Field Regiment, RA) focused on tactical handling and equipment. The time required to produce a divisional fire plan (three field and one medium regiments) 'was between six and six and a half hours from the infantry commander outlining his plan. Much time was saved if a list of probable targets and their coordinates was generated prior to the conference. After allotment of timings and fire units to tasks, rates etc., Liaison Officers took orders to units about two hours later. At RHQ the Adjutant allotted tasks to batteries and sometimes troops, and the battery CP split into two teams—the CPO and ACPO. The GPOs would have their Barrage Key Line forms completed in around two hours.'

Defensive fire tasks were regarded as expensive in ammunition, but effective if they were predicted fire tasks, which they most often were. Observed shooting gave equally good results, particularly when directed against tanks. Brigadier Ramsay reinforced current training doctrine in terms of the centralised control of artillery. His 'normal' command of four field and one medium regiments for neutralising enemy forays against the resolute infantry was considered sufficient to break up enemy attacks. 'Adjustments to the DF Task Lists were communicated by word rather than through the issue of new proformas.' He noted that his 6-pounders had 'no targets to shoot at'.

Accurately fixing targets on 1:25,000 maps was made easier by the good survey data available, calibrated guns and accurate meteor data. However, within RHQ, the Survey Section had an inadequate scale of equipment with which to effectively fulfil its role. On the other hand, communications within the division's artillery were good because of the extensive buried line layout, which was always preferred to wireless. OP line was difficult to maintain, and resorting to 'ladder' line layouts worked sometimes where there was less tracked vehicle traffic.[338] On wireless equipment, Ramsay commented:

101 Set	unsuitable because of its low power
11 Set	more satisfactory
18 Set	communications with infantry—useful
109 Set	Liaison Officers to HQs—worked well

Equipment deficiencies were a cause for concern. From 23 October to 4 November, eighteen 25-pounders were sent to the rear for mechanical reasons and only six replacements came forward. There were only two occasions when all 96 guns of Ramsay's command were in action. Ramsay recommended that each LAD should hold one spare gun, but this issue was not taken up by the higher levels of artillery staff. Six armoured OP vehicles (Bren Gun Carriers) were replaced during the battle and their unreliability caused him embarrassment when deploying FOOs. He recommended one more per regiment on the War Establishment.

Overall, Ramsay noted a number of underlying reasons for the superiority of the Eighth Army artillery and its tactical handling during the battle. One reason was that the artillery boasted a very comprehensive communications network, both line and wireless, between the forward observer and rearwards to Corps headquarters. The artillery's support of skilful infantry night attacks was so successful as to attract a rueful comment from Rommel. Others germane to 9th Division operations will be discussed later.

Operation Orders written by Brigadier Ramsay drew high praise from the RA hierarchy. Writing the following year, Major General Amer Maxwell, MGRA, circularised British Army establishments worldwide in May 1943 stating that:

> It is considered that these Operation Orders, which deal with fire plans for the major attack at El Alamein, are good examples and were used with great success in actual operations. From the Staff Duties point of view the following points of 'comprehensive nature of the body of the order, coupled with clearness and brevity', 'layout of Task Tables', and 'organisation of supporting data' stand out. From the tactical aspect 'the large amount of arty which was used in support for which orders were written by one CRA', and 'the combined methods used in [the fire plan] for concentrations and moving concentrations, barrages and DF tasks' are noted.[339]

Ramsay's methodical approach doubtless enabled lower level headquarters fire planning to proceed with fewer misunderstandings than might otherwise have been the case. But it was the expertise and dedication of command post staffs translating masses of data into gun programs that truly lay behind Ramsay's operational achievements.

The initial fire plan provided one 25-pounder for every 17 yards of front, with medium guns superimposed on either pinpoint or linear targets. These were primarily concentrations on prepared and well-defined defensive localities, with barrages on those areas less well defined. Barrages were fired at a slow rate to ensure that infantry could move to their Start Lines before the barrage advanced. Once on their objective, the barrage line dwelt several hundred yards ahead so that infantry could 'catch up' and knew where their objective was. While this did not apply to the Australians' front and their assigned British armoured regiments, armoured formation thrusts were not supported by heavy concentrations of fire.[340] In creating fire plans, air photos were discovered to be far more reliable than infantry patrol reports in determining whether enemy positions were occupied. Patrol reports revealing occupation at night were nonetheless very useful. Overall, air photos and counter-bombardment data were the best sources of information.

The effects of artillery fire, particularly on the Australian front, were quite remarkable. While opinions differed as to its effect on tanks, there were numerous occasions when heavy concentrations of field guns augmented by mediums had decisive effects. DAK armour and infantry attacked Trig 29 no fewer than 25 times, and a field regiment concentration once dispersed a force of 25 tanks. After the dust had cleared, five had been destroyed and several others towed away. Brigadier Kirkman noted also the substantial effect on morale of a large number of shells arriving simultaneously from different directions. However, as Pemberton comments, 'there were times when the covering fire, in spite of its intensity, failed to produce the results expected—the killing effect of artillery fire was disappointing to those without experience [of World War I].' The nub of this observation was that, against a skilled and well-trained enemy, infantry commanders had to accept some risks from friendly artillery.[341] If one assumes an intensity of ammunition expenditure by both sides during Montgomery's attack as being equivalent to the Western Front in 1917–18, then the proportion of casualties caused by three forms of destruction would approximate: artillery and mortar fire: 65 per cent; machine-gun and small arms: 30 per cent; mines, grenades, bayonets: 5 per cent.

It was also recorded that a medium gun could destroy a tank with a direct hit—on one occasion at 19,000 yards. While the 9th Division's 25-pounders never engaged tanks with direct fire, on other corps fronts they did. However, the 6-pounder anti-tank gun was far superior as a tank destroyer. The enemy's extensive entrenchments resulted in demands for time-fused airburst HE beyond the usage scale envisaged. Against troops under light cover or in the open, or against partly dug-in anti-tank guns, it was superior to ground burst. Later, a request was made for 15 per cent of all ammunition to be time fused.[342] Yet another aspect of this feature of the fighting was that the Eighth Army had deduced that the infantry's 3-inch mortars could deliver a greater weight of shell in one minute than a Troop of field guns firing at an intense rate. The Germans had made a similar deduction. Their 80-mm mortars had more killing power in their bombs per pound than 7.5-cm *kannon* using a similar 'rule of thumb'. When Rommel complained about a lack of ammunition, it was for his guns and howitzers, not his mortars.

From the outset of the campaign, Ramsay's artillery frequently used the Stonk, particularly when tank attacks threatened the infantry's hard-won gains. In almost every case, the division's 96 guns under command were augmented by the 7th Medium's sixteen 4.5 and 5.5-inch guns. This form of linear target was usually 400–500 yards long for a fire unit, and the zone of the guns gave it a depth of 100 yards or more, depending on its orientation to the line of the guns. Prior to the start of a particular phase in the battle, Ramsay and his staff would decide, based on their intelligence, where the infantry would want defensive fire and SOS tasks. They would arrive at Morshead's planning conferences meticulously prepared. Once all the details for the attacks (following the initial assault) had been finalised, Ramsay's Brigade Major would have his operation orders—typed up, complete with traces—delivered to the regiments in under an hour.[343]

Command post drills ensured that speed of delivery of fire to Stonks improved over the battle. The 14th Battery (2/7th) BHQ devised a drill that produced predicted fire in minutes. Its best time from the moment RHQ advised coordinates to 'shot' was two minutes, and invariably the battery reported first.

In comparing the tactical handling of the Eighth Army and PAA artillery, Kirkman was of the view that his own guns had achieved 'a moral superiority over the enemy artillery', even prior to the battle for which the enemy had prepared for so long. He qualified this by adding that the PAA had an unlimited supply of 25-pounder ammunition, even though it had resupply difficulties with its own ordnance ammunition types. Kirkman commented that 'the enemy rarely fired at night, whereas the Eighth Army fired "round the clock". We destroyed his flash spotting towers—he did not fire at ours, even though they were marked on his [captured] maps.'[344] The PAA knew the location of every battery on the front, yet rarely engaged any single battery for a prolonged period. This prompted the BRA Eighth

German 88 GP gun and half-track tractor destroyed by 9th Division field artillery fire
(Parsons Collection)

Army, Major General Maxwell, to comment, 'Although he [the enemy] had considerable superiority on medium and heavy artillery he made little use of it [to counter our artillery] or made use of alternate positions to avoid our CB efforts.'

Air photographs comprised a key input into the battle intelligence picture. Stereoscopic (pair) photos of a 'featureless' desert including those in the HQRAA 9th Division's War Diary, provide clear indication of the expertise of the interpreters in the Corps APIS Section in deducing tactical information. They did this, non-stop, for 13 days. Kirkman noted that oblique photos were of less use than verticals, and the DAF had to be 'educated' to produce photos of the correct scale and overlap. A Counter-Bombardment Section (an officer and two clerks) at the RAF's Army Cooperation Wing, in close consultation with an Air Liaison Officer and the Royal Engineers Photo Survey Section, produced photos for the Counter-Bombardment Section within four to five hours of the aircraft's return. Hostile Battery Lists were in the hands of the Corps CBO within 10 hours.

Counter-bombardment was a key aspect of the artillery battle. A proportion of field artillery was allotted to counter-bombardment tasks at all times and, to coordinate this aspect, the CO of the 64th Medium Regiment was nominated as the CCMA to relieve Brigadier Dennis of this role from the start of the battle. The main source of battlefield intelligence was air photos, with flash spotting towers a close second. Kirkman directed that, as soon as an enemy battery opened fire, it would be subjected to an immediate bombard. However, neutralisation shoots with predicted fire were extravagant in ammunition expenditure and not always reliable if not accurately 'fixed'. At night the flash spotter's expertise was manifest, and neutralising fire acted as a considerable deterrent. Kirkman (and Montgomery) were both critical of the DAF Army Cooperation Group's activities: 'No Arty/R was used for three days and much of the enemy's artillery had been forced to move because of our fire and many new locations were unknown.' This added considerably to the infantry's difficulties throughout the battle. However, this was only half the reason. German mortars—a feature of this battle and subsequent battles across the top of Africa—were far more deadly and almost impossible to locate and neutralise quickly. The DAF effectively engaged transport and tanks but played no useful (counter-bombardment) role in the battle. Hostile artillery fire (against friendly guns) of any magnitude was seldom experienced, and was usually limited to a few salvos from a battery before it switched to another target. As the 2/7th was the westernmost regiment most of the time, its casualty list from hostile battery fire is clear evidence of accurate German fire and bombing, albeit spasmodically delivered.

Locating (fixation) in the desert was always a problem, with formations differing by as much as 1000 yards in their estimates of their location. The most efficient method was to utilise artillery survey either by shooting onto an adjoining feature and deducing its coordinates or asking the troops concerned to put up a signal from which bearings could be resected by the flash spotters.

Nonetheless, PAA artillery and mortars did cause many casualties amongst the infantry, and *Generalleutnant* Krause's policy was clearly to give them priority. While the mortars moved frequently and caused more casualties, the artillery kept to its well-prepared battle positions (as indicated by repeated bombards on groups of batteries) and moved only when the infantry threatened its tenure.[345] Kirkman noted that 'our CB measures were most successful, and it is considered that they played a large part in the enemy's defeat. No-one with experience of this battle now doubts the value of this type of fire, or that the considerable quantity of ammunition used was well expended.'[346] The keys to this 'art and science' were experienced officers, much practice, and hard and careful work. Corps CBO Hostile Battery Lists were prepared at least every second day, and it was noted that, for each corps front, a Hostile Battery List took about six hours to prepare, such was their complexity. Survey was essential to success, not only for counter-bombardment measures but also for providing fixation in otherwise featureless terrain.

Medium artillery played a crucial role in the artillery battle and its neutralising effect was much greater than that of field artillery. The power of the 5.5-inch shell was much appreciated, particularly when airburst fuse 210 was used, although the extra range of the 4.5-inch was invaluable in reaching the enemy's heavy guns. The high ammunition expenditure of the medium regiments has already been described—300 rpg during the first 24 hours. Detachments were augmented with drivers to provide 12 to 13 men to handle the workload under these circumstances. In addition, after the first night of the battle, both calibre ordnances suffered from mechanical faults. The packings in the buffers had to be adjusted with the result that, on average, over the next 12 days, 4.5-inch batteries had only seven guns available, while the 5.5-inch guns had five. The strain this imposed on the remaining serviceable guns can only be imagined in the light of their tasks and the rates of fire assigned them. The 7th and 64th mediums also supported armour and infantry south of the 9th Division sector which involved gunners moving their split-trailed ordnance through as much as 60 degrees, and firing 22,000 rounds in 11 days. The 7th Medium's historians recorded that it 'greatly added to the labour required of the gun detachments whose response was beyond all praise.'[347]

While anti-tank artillery has not been included in this account, it is interesting to note that the use of field and medium artillery to break up tank concentrations caused anti-tank gunners much frustration. Not only did the dust raised by bursting shells obscure their target, but tanks were reluctant to advance into their 'engagement range' so as to avoid the field and medium artillery concentrations that rained down on them. Average 2-pounder ammunition expenditure was a mere 1 rpg per day in the northern sector and 2 rpg per day for 6-pounders overall. Kirkman noted that overall analysis of all the actions involving tanks revealed that 'properly used in adequate numbers, unless neutralised by artillery fire, the gun is the master of the tank, which is reluctant to advance in the face of its fire.'[348] The Armoured Corps generally expected artillery to knock out anti-tank guns, particularly the 88-mm GP gun when well concealed. This was a time-consuming task with the use of observed fire. Only when they were ill-concealed were they forced to withdraw. Yet the effect was significant and prompted the PAA's General von Thoma to attribute defeat to his high anti-tank gun losses.

Kirkman observed that the artillery of the German and Italian formations opposing the 9th Division was used very ineffectively. He noted that 'we had achieved complete superiority over his artillery'—a comment that would probably also apply to counter-bombardment fire. However, the infantry would not have agreed with his statement as Australian battalions in 'The Saucer' suffered some 500 casualties from combined German artillery. It was fortunate

DAK Self Propelled 150 mm guns captured at El Alamein.
(Parsons collection).

that there was a shortage of some types of ammunition, otherwise the Eighth Army casualty lists would have been far longer. One person who would have agreed with Kirkman's statement was *Generalleutnant* Krause, who complained that 'there was not enough petrol to transport ammunition from El Daba to the guns.'[349] However, post-battle analysis by the artillery staff refuted this claim in part.

To match the counter-bombardment assets deployed, the German artillerymen adopted the same practices as at Tobruk a year earlier. Resorting to a flashless propellant initially confused the British flash spotters, but they soon identified those German batteries using this ammunition. Another technique the Germans used was the split second firing of (for example) a battery to confuse and mislead the sound rangers in the interpretation of their recordings. This, too, was circumvented by the painstaking analysis of the times of all battlefield explosions/discharges and comparison with other data.

Having to do more with less led the German gunners to use subterfuge and deception which was inconsistent with their typically chivalrous spirit. In the fighting around Trig 29 and the Blockhouse, they sited 88s and anti-tank guns quite close together. As the Blockhouse was used as a medical post (complete with red cross on its roof), OPOs were loath to engage targets too close to it. The mosque at Sidi Abd el Rahman was the highest point for miles with commanding views over the northern sector of the battlefield. It was used as an OP, against which the only Eighth Army recourse was to screen the target (or area) or 'blind' observers with smokescreens.

Mention has been made of captured 25-pounders being used against the Australians. There was much emphasis both before and during the battle on properly calibrated guns. This was quite understandable given the high daily rates of fire. When the infantry complained about 'drop shorts'—the colloquial infantry sobriquet for a gunner—and the fact that friendly 25-pounder rounds were falling into their own FDLs, the charge was strenuously denied by the field regiments. The 2/7th historian noted that 'an infantry estimate of 80 per cent of the so-called "drop short"—bringing the Royal Artillery's good name into jeopardy—were attributed to the deliberate misuse of the same calibre captured guns by the enemy.'[350] What the Germans did was this. When an OPO was registering a defensive fire task, the Germans registered the same area and added a distance equal to where they estimated the infantry localities to be. When the Eighth Army fired the defensive fire task, the German artillery followed suit. German rounds fell on Allied infantry and, having a distinctive 25-pounder burst, it was labelled a 'drop short' by the infantry. The 25-pounder shells were later found to have been made by German ordnance manufacturer Krupp.

The policies of Rommel's artillery chief, Krause, have long been the subject of debate. In a memo he wrote at the beginning of the battle, he recognised that his intelligence had estimated his force to be inferior to the Eighth Army's in number and performance. Nonetheless, Krause was full of fight. He laid down his priorities such that, as enemy attacks were preceded by heavy artillery fire:

- immediate retaliatory action was to be taken
- batteries must keep in touch with the *33rd Armoured Survey Troop* for details of their latest reconnaissance or with the *11th Survey Troop*
- targets recorded by survey units were to be added to enemy Hostile Battery Lists pinpointed by OPs
- 'That done,' he said, 'allocation of targets can be made in a manner that counterfire can begin immediately the code word is given, thus supporting our infantry as soon as the [enemy] barrage starts.'[351]

Once the October battle commenced, Krause would appear to have changed his mind or had it changed for him by General Stumme (acting for Rommel in his absence) or by entreaty from von Thoma, PAA commander, or both. As Kirkman noted, PAA artillery was generally ineffective in a counter-bombardment role. Its artillery reacted to continuous pressure from XXX Corps, particularly in the salient occupied by Morshead's division and during its subsequent thrust towards the coast road. Instead, it was the infantry and anti-tank gunners, all positioned forward in the battle zone, who were to become the unwelcome subjects of his well-trained gunners. For example, it was estimated that artillery fire caused 500 casualties to the 24th Brigade's infantry in The Saucer.

We are indebted to the Germans for a snapshot of the state of their artillery within a month of the arrival of the 9th Australian and 2nd New Zealand divisions from Syria in July. This encompasses the arrival of Brigadier Meade Dennis as CCRA XXX Corps and, later, Brigadier Sidney Kirkman as BRA Eighth Army and, it should not be forgotten, the aggressive handling of the DAF by Air Marshal Tedder. In the three months prior to this—that is, in actions around Tobruk and Gazala—the PAA had suffered almost 14,000 casualties (2325 killed, 7498 wounded and 2746 captured). Arko 104 and *135 Flak Regiment*, now under *Generalleutnant* Krause, had lost 42 officers and 541 other ranks killed, wounded and missing from a strength of 236 officers and 6912 other ranks. It was short 107 NCOs and 859 other ranks. At this stage they mustered 51 field guns and twenty-nine 88-mm GP guns in addition to the divisional artilleries of DAK and the Italian divisions. DAK artillery units were 294 NCOs and 2011 other ranks under establishment. In terms of serviceability, *15 Panzer* had 36 guns, *21 Panzer* had 42 (establishment was 46) and *90 Light* had 19 instead of 30. Their strength of anti-tank guns was crucial to their defence during the later battles and of these guns they possessed 47, 53 and 98 respectively. The Italian divisions were in no better shape. The artillery regiments of *Brescia* had seven (instead of 17), *Ariete*, *Littorio* and *Trieste* had seven, six and eight respectively.[352]

At this time, a special artillery unit named *Project Battery Afrika* and commanded by a mere *leutnant* arrived with four each of 15-cm and 28-cm 'projectors', still in the experimental stage. This provides some indication of the mindset of DAK gunnery staff given its extended front line. The projectors developed battlefield smoke and the *leutnant* held a demonstration for Rommel and others at El Daba. Ultimately Rommel decided to hold them back as a surprise for the Eighth Army.

At this time the *Luftwaffe's IXX Flak Division* consisted of five flak regiments. Tedder's strategic and tactical bombing necessitated *Generalleutnant* Burchardt splitting his command amongst the many vulnerable points supporting the PAA at the front and their logistic tail to his rear. The latter was already under threat from the activities of numerous raiding parties and was to suck in further field artillery assets in a coastal defence role. These were primarily Italian troops and batteries whose commanders made exaggerated claims of sinking RN cruisers and destroyers etc.

Most of the post-mortems concerning Alam Halfa and El Alamein focus on armour and infantry corps encounter battles, the trials, successes and failures of both protagonists. Certainly artillery fire is mentioned as crucial—Montgomery in his report to Alanbrooke said so emphatically; Horrocks in his memoirs was less fulsome in his praise for artillery than was Erwin Rommel. Von Varst's *15th Panzers* report is important, for that formation was opposed to the northern front for most of the last battle. Insofar as the authors recognise the importance of the coordination of artillery fire against XXX Corps, their artillery planning took into account 'definition of defensive and destructive fire tasks with Army [i.e. Arko 104], 15 Panzer, 164 Light/Infantry, Trento and Littorio Division's artillery'. They knew XXX Corps had 34 battalions of infantry, 72 field guns and 16 medium guns at the start, against which they could deploy:

  4 x 100-mm guns
  24 x 100-mm howitzers
  8 x 150-mm howitzers
  5 x 25-pounders (87.6-mm) captured
  8 x 150 SP howitzers
  41 Italian guns/howitzers, incl. 76.2-mm SP gun.

As part of their defensive power they also fielded:

  70 x 50-mm anti-tank guns
  8 x 88-mm GP guns as anti-tank guns
  4 x 6-pounders (57-mm) captured.

The Germans claimed that a shortage of ammunition curtailed the effectiveness of their defence, but that would not seem to be supported by subsequent investigation. As it was, they correctly appreciated that fire to deny minefield neutralisation was important. But what they did not anticipate was the speed of minefield clearance using special techniques developed by the Australian, New Zealand and South African field engineers. This enabled the attackers to position themselves further west of the destructive fire areas and allow allied field artillery to cover deeper sited hostile batteries located by the RAF and the 4th Survey Regiment. Allied artillery observers mounted in tanks were at times able to bring down observed fire on German forward-placed anti-tank guns that had caused the RTR and the infantry much grief. During their attacks they never used fewer than 40 tanks, with two or three 'lead' tanks to entice Allied anti-tank gunners to disclose their location. They would then mark them with smoke and call down fire upon them. This is slightly at odds with Ramsay's deduction that most of the 9th Division's (and Rhodesian) losses of anti-tank gunners occurred in the consolidation phase primarily during night attacks.

Undeniably DAF tactical bombing and strafing (and communication jamming) significantly hampered the enemy artillery and, in a general summary, the staff of the *15th Panzer* attributed the root causes of their lack of effectiveness (a phrase most infantrymen would call fanciful) to:

- the late delivery of reports from the front
- a lack of planned action against [the Eighth Army] artillery owing to low stocks of ammunition
- Allied use of predicted fire which caused heavy losses to men and guns
- the uselessness of the 76.2 SP gun due to its low mobility and inadequate range. In their first action, eight were knocked out before they could fire.
- the inadequately engineered mounting of the 150-mm SP heavy howitzer
- the superiority of British guns in numbers and range which caused heavy losses [in combination with superior tanks].

The report singled out four units for special mention:

- *III Battalion, 382 Infantry Regiment*
- *I* and *III Battalions, 115 Panzer Grenadier Regiment*
- *617th Anti-Aircraft Battalion* (88-mm GP gun)
- *II/33rd Anti-Tank Battery*

By 28 October only eight 88-mm guns survived and there were presumably fewer when the 9th Division broke contact on 2 November. Losses of anti-tank guns were commented on but not quantified. Of the infantry, *I Battalion, 115 Panzer Grenadier Regiment* had 120 men standing on 28 October, and its sister battalion (*III*) 70 men. Presumably there were even fewer a week later.[353]

Battles on the scale of El Alamein with tanks, infantry and artillery were fought later in the war, but Rommel noted of Trig 29 (the Germans' Hill 28) that both sides 'were shedding rivers of blood for a miserable stretch of sand'. This is reminiscent of a verse from Sir Henry Newbolt's poem *Vitae Lampada*:

> The sand of the desert is sodden red,
> Red with the wreck of a square that broke,
> The Gatling's jammed and the Colonel dead
> And the regiment blind with dust and smoke
> The river of death has brimmed its banks.

This might equally apply to the 9th, 51st Highland and 2nd New Zealand divisions.

CHAPTER 18
THE WESTERN DESERT 'BALANCE SHEET', 1941–42

Though the mills of God grind slowly,
yet they grind exceeding small;
Though with patience He stands waiting,
with exactness grinds He all.

Friedrich von Logau
Sinngedichte
(trans. H.W. Longfellow)

At the time the 6th Division artillery returned to the Delta to embark for Greece, Major General Mackay and Brigadier Herring were unaware of Churchill's thoughts on the use of artillery. However, they had the satisfaction of knowing that they had employed their artillery and that of XIII Corps to its best effect in the campaign that had included the capture of Bardia and Tobruk before taking it on to Benghazi. Any post-mortems at that stage focused on systemic matters rather than doctrine review. The use of the 7th Armoured Division's assets had made the task of the divisional infantry far easier and, as already noted, false lessons were drawn from this on the efficacy of armour/artillery/infantry cooperation. That division was to learn salutary lessons at the hands of the DAK in the months that followed. Six months later, another Australian division (the 9th) under Major General Leslie Morshead was to sustain an aggressive and successful siege for five months with the aid of British artillery and a handful of tanks by using artillery effectively against a numerically superior and more heavily gunned force of German and Italian formations.

The 9th Division, refreshed and well trained, returned to the fray in July and contributed mightily to the defence of the Delta through its strength of arms at First Alamein, securing vital ground to enable an army to regroup under a new commander. The division's organic artillery was now grouped in operations for the first time under its gifted commander, Brigadier Alan Ramsay. However, given the summary withdrawal to Australia of units comprising Corps troops (CCRA Headquarters, medium artillery and survey regiments), the 9th Division supplemented other artillery assets from Britain, New Zealand and South Africa to enable the artillery to fulfill its role. A common doctrine and a belief in that doctrine's effectiveness were key factors in this success. Despite this, the actions that preceded the insertion of the 9th (and 2nd New Zealand) divisions into the desert demonstrated clearly that the shortcomings in cooperation between the three arms had yet to be satisfactorily resolved.

In July the 9th Division's unhappy episode with attached British armour was just one case in point and its effectiveness compared to that of its opponents was the subject of continuing debate between the three arms involved. It was widely recognised that any pursuit of or battle with PAA forces would involve a large number of tanks and other AFVs working with infantry divisions which still made up the bulk of the Eighth Army. That a hiatus still existed at this time given the scope of Allied experience to date was lamentable to say the least. This chapter explores the reasons for this hiatus and analyses the use of field, medium and (some) anti-tank artillery in desert operations. Part of this analysis involves scrutiny of the key issues of command, composition of forces, and weapons and doctrine development.

As the British Army in Egypt reacted to provocative Italian moves from late 1940 until well into 1942, the various operational commands proved themselves completely unable to develop any form of synergy. There were a number of underlying causes for this, namely the political direction of the war by the British War Cabinet; the wide scope of the supreme command HQME in geopolitical terms; the composition of Allied/British/Commonwealth forces, their standard of training, scale of equipment and individuality; and the nature of war in the desert—land, air, sea and logistic—its current doctrine and execution.

As many historians have noted, overall command responsibility in the Mediterranean was of secondary importance to Hitler and his generals who were preoccupied with the Russian campaign. The reverse was true of this theatre for the Allies for three main reasons: the importance of Iranian oil to the war effort; the Suez Canal 'lifeline'; and the geopolitical importance of and leverage of Egypt, Greece and Turkey in the long term. Thus the GOCME had a huge brief; at its peak more than a million souls were under his aegis. His staff were almost all regular soldiers of mixed British and Indian Army (and Civil Service) backgrounds, mostly professionals with World War I service. For many of them, this was the first time that they had to think conceptually and operate in a huge organisation rather than at regimental level. In addition, Churchill's intrusive 'recommendations' and prodding were not easily ignored. On the other hand, the Italian and German generals were given a free hand in a simpler geopolitical environment.

The pressure of day-to-day events at HQME tended to place doctrinal issues at a lower priority for the General Staff. Operations analysis was therefore understandably neglected when the loss of whole regiments and brigades cried out for investigation and analysis. Thus, at lower levels of command, doctrine 'modification' occurred with little supervision, often at the behest of regimental commanders based on their recent experiences. Seasoned Commonwealth commanders and their staffs were capable of completing their own analysis, but they lacked leverage with higher headquarters, primarily because they were citizen soldiers and not Staff College graduates or regular soldiers. As a result they became less enamoured of their British brethren's approach to staunching casualty lists, but were caught in a bind because, in combined operations, an agreed doctrine had to form the basis of operations. As noted, it took Morshead's rebuff of Ramsden's request to break up his division to crystallise matters.

When Britain called on Commonwealth countries to provide forces to bolster the Mediterranean theatre, Australia, New Zealand and South Africa offered divisions and some Corps troops of ultimately three, one and one respectively. These were volunteer formations with a sparse sprinkling of regular officers and senior cadre NCOs. At senior levels of command they were veterans of World War I and the formation's doctrine and systems of administration were generally harmonised with those of the British Army. The British Army (including the Indian Army) was far less homogenous than Commonwealth armies. There were regulars, recent volunteers and conscripts at various standards of training, itself a function of leadership, equipment availability and opportunity. Those from the Indian Army could claim far more experience 'on active service' than those from Britain. Such experience was gained against dissident rebels in the hinterland of the subcontinent where cavalry, artillery and infantry were in offensive mode as super-policemen against a wily but weaker foe, and thus not required to hold ground. India was the place to see action and enjoy regimental life, and so the boundary of experience of officers promoted as the Army expanded was at squadron/company/battery level. There were no divisional exercises to practise command and staff.

The Germans, on the other hand, mounted significantly larger exercises after 1935, many up to corps level—something the British never did until late 1940. Panzer formation commanders gained an appreciation of the time and space required for manoeuvre, innovated and refined organisation and doctrine, cooperated with the *Luftwaffe* in tactical air operations and, most importantly, trained personnel to operate at two levels above the rank they held. The British Army was more wedded to the infantry division in war and was unable to develop a working armoured model to match that of the Germans. With the onset of desert operations, the GOCME took an indirect interest in armoured formations and their commanders' performance and results. In their search for a solution to armour and infantry cooperation, the staff fell back on the concept of Jock columns with which all were familiar given the effectiveness of (all-arms) columns in India. Griffith argues that there was one level of command too many in Cairo 'which tended to make [British] generalship more bureaucratic and clumsy than that of the Germans ... Whereas Rommel succeeded brilliantly in moulding his army in his own image.' Montgomery, he noted, 'managed to find a style that was itself suitably bureaucratic and slow moving to fit the requirement.'[354]

Yet another factor was the stability in command and the absence of churn and change in the DAK/PAA. Rommel's *15th* and *21st Panzer* and *90th Light divisions* remained together through thick and thin alongside their Italian formations. The same could not be said of the Eighth Army until November 1942. Between Sidi Barrani and Second Alamein (two years apart) 20 different divisions appeared on its order of battle and only four of these fought in two campaigns. The divisions lacked experience of working together and there was a regular turnover of general officers. Many factors contributed to this situation.

A final comment on control and command as a feature of desert warfare concerns dispersion against air and land attack. In the early days, the 7th Armoured used dispersion to counter Axis air superiority and (its own) weakness on the ground. However, being dispersed made the grouping of units difficult to command, even when wireless communications were good—it became almost hopeless if they were bad. This encouraged a commander to remain at the rear. The provision of effective artillery concentrations was correspondingly difficult to arrange in the necessary time-frame for artillery to have any influence on the encounter. As Griffith observes, 'The Germans thought more in terms of leadership from the front to achieve concentration of armour and artillery (both direct and indirect fire) at a single *schwerpunkt*.'[355]

While the weaponry used by both sides was comparable in many ways, those weapons fielded by the Italians were inferior to both German and British. Artillery in this context extends beyond field (to include medium and heavy) guns/howitzers to that mounted on tanks for use against troops and tanks. The German Army was clearly ahead of its adversary in terms of date of introduction and utility. While German ordnance from 75-mm to 150-mm calibre was comparable to that of the British, the inclusion of the German 88-mm GP saw Rommel outgun his Allied adversary. Add to the British arsenal the 25-pounder with its high rate of fire and robust construction, the 4.5-inch and 5.5-inch guns/howitzers and the balance goes the other way. However, the strength of Arko 104 in medium and heavy guns made an enormous difference to the way Rommel could influence a battle. British artillery could not be concentrated as quickly as that of the Germans, who recognised that firepower was a battle winner. As Bailey comments, 'The Germans could concentrate greater firepower at a critical point ... This was accomplished not by moving equipment to within range of the target, or by creating fire mobility from relatively static positions, but by enticing a mobile enemy onto that source of firepower.' Once medium guns began to make their presence felt, 25-pounders could be used in indirect fire mode rather than as anti-tank weapons. Bailey lists the tactical priorities of desert warfare as: 'first, the destruction of tanks usually by anti-tank artillery; second, the destruction of anti-tank guns, usually by field artillery; thirdly, the destruction of field artillery, usually by its counterpart.'[356] Numerous panzer and lorried infantry attacks were launched at the Australians on Trig 29 which they repulsed, eliciting Brigadier Kirkman's comment that 'the gun is the master of the tank', a notion that Pemberton is at pains to qualify. This was the beginning of an accepted wisdom that it was possible to master tanks with guns, and which was later successfully demonstrated in north-western Europe by both British and American artillery.

Much has been made of the efficient way in which artillery was used by infantry divisions and the fact that putting OPO/FOOs in tanks or Jeeps gave them a (protective) mobility supporting infantry and armour. However, when it came to breaching in-depth minefields in the attack phase of war, no artillery could hasten the lifting of mines for the other two arms by the field engineer gapping parties. It was painstaking, hazardous work. The introduction of Hobart's flail tanks in the desert was a step in the right direction, but engineers on foot were still targeted by machine-guns firing on fixed lines, usually from concealed positions in enfilade and by enemy artillery. This extended the time taken to create a breach sufficiently wide to allow armour to deploy in the direction of its objectives. Both armour and infantry were vulnerable at this point and German artillerymen were quick to respond. This were probably the most crucial time for all set-piece attacks, and field artillery using observed fire and mediums engaging known hostile batteries were the only counter available to the assailants. It was pure and simple attrition until one adversary gained the upper hand.

In terms of anti-tank artillery, the 2-pounder (40-mm) and US 37-mm were each superior to the German 37-mm and short 50-mm (and Italian 47-mm) guns. German weapons could fire both AP shot and HE, whereas the 2-pounder fired only AP. The short 77-mm howitzer of the Mk IV Panzer was a close support weapon as was the 75-mm of the US Grant. For tank-mounted artillery, reliability of the prime mover was a significant factor and in this the Germans led. Griffith suggests that this had less to do with design than with maintenance. The Germans would laager at night, turn on the lights (if the DAF was not around) and service their vehicles. Griffith asserts that the myth of German tank superiority was just that—a myth. Low morale amongst British tank crews was attributed to an inferiority complex over the quality of their tanks' armour. The real villain was not German tank-mounted anti-tank guns, but their highly effective PAK 38 (50-mm) anti-tank guns when fired from close proximity. In North Africa, DAK tanks and anti-tank artillery accounted for an equal percentage of British tanks (some 40 per cent), and mines around half as much (some 20 per cent). More importantly, indirect fire and aircraft accounted for about 20 per cent of the total.[357]

There was a close interrelationship between organisational effectiveness and doctrine for the employment of offensive power. Field artillery doctrine as applied in 1940 had to be amended in terms of a field regiment's establishment—three batteries instead of two. This was an important step in the right direction and lasted some 30 years before succumbing to technical advances.[358] Initially, the essential features of training regiments were firmly focused on working with an infantry division in all phases of war. By 1942, adaptation to embrace the armoured division had yet to receive the imprimatur of the gunner hierarchy. Overlaying these matters in the desert war were the twin themes of field (and medium) artillery working with anti-tank artillery and guns on tanks, both in attack and defence, although this is not to suggest that advance and withdrawal were unimportant. Doctrinally, artillery and infantry in the attack had to learn how to help armour onto the objective. Griffith argues that, following Battleaxe, the British did not adopt the German tactic of placing all anti-tank guns well forward—as they should have—nor did they centralise/concentrate artillery or work out how infantry could help armour and vice versa. He blames lack of supervision and/or a central focus in GHQ for the absence of analysis and, for a year, the issue drifted inconclusively. The notion of an all-arms approach seems to have foundered on institutional pre-judgements. One of the more serious was to attribute to an armoured division a vision of a tank-versus-tank battle.

The Germans viewed the issue differently, recognising the limitations of tank-versus-tank in a tank battle. Instead they placed anti-tank guns well forward as a defensive screen.

More often, however, they waited for the British to charge to their destruction. The figures quoted above give the lie to the British claim that tank fire did the most damage rather than the PAK 38. This only served to strengthen the legend of panzer invincibility. Griffith cites tank battles at Gabr Saleh and at Trig Capuzzo on 21 November 1941 against Arko 104 (and later battles in Libya/Tunisia) which refuted the British tankies' belief 'that the best anti-tank weapon is another tank'. Thus, when the British tanks had expended themselves it was the associated infantry units and sometimes whole brigades that became hostages to fortune. Their only protection was the night; in daylight they were vulnerable at 1000 yards. They had to depend on the artillery either to support them onto the objective, if that were still feasible, or to withdraw. Transport could not be dug in nor could it be hazarded to stay in the tactical area; likewise, it could not bring relief stores and ammunition or withdraw laden troops to a rendezvous in their rear area. Any alternative was equally unattractive unless the supporting tank force gained the upper hand in an engagement.

An additional issue concerned artillery strength in armoured divisions, brigades and battlegroups in an army comprised predominantly of infantry divisions. These organisations fielded various strengths of infantry and cruiser tank types, supported by artillery—field, anti-tank and LAA. The 7th Armoured Division (first in the desert and last out) had, at the outset, 500 tanks and 84 field pieces (plus 72 anti-tank and 58 LAA), or a ratio of just over six tanks to one 25-pounder or seven anti-tank guns. These belonged to the Support Group and this low ratio determined that these assets would remain in the rear while the AFVs slugged it out up front—Sidi Rezegh was a case in point. At length this imbalance (being under-gunned) was rectified over a period of a year.

However, it was the battle/brigade group concept of gunner OPs in tanks that aroused most passion. Those who pointed to the loss of guns—almost as many as tanks—were strongly opposed to breaking up artillery into 'penny packets'. However, at the time, there appeared to be no realistic alternative in fluid operations and the brigade group equivalents had some successes to their credit. There were two issues: one concerned the signal network required to effect concentration, which was in place by the time of Alam Halfa. The second was the mindset of the armoured officer and his need to be trained, like his infantry counterpart, in the use of artillery. This took some time to achieve following the October battle. After Sidi Rezegh, Auchinleck proposed to restructure the armoured division to achieve a better 'balance' of armour, artillery and infantry (and engineers). Out went two (of the three) armoured brigades, in came an infantry brigade instead of the Support Group, and each brigade received more supporting arms so that it was capable of operating independently like the DAK. In the period before Second Alamein, Major General Freyberg's New Zealanders had exercised with an armoured brigade and constructed battle drills so that there was never a repetition of the dreadful hiding the division had suffered at Miteiriya when combined with British armour. These drills endured and were refined as the Eighth Army moved west so that by the time the Tunisian border was reached, one tank brigade supported an infantry division.

Most historians agree that the Germans had superior battle drills and a knack of combining all arms effectively, although in the coming Alamein battle heavy casualties in tanks, OP and infantry officer ranks reduced the effectiveness of renewed counter-attacks against the Australians and Scots. Their attacks were less well coordinated, albeit still of sufficient concern for Montgomery to delay Supercharge for a day. Griffith concludes that, over the period of the desert war, the Germans overran the British and vice versa, with the former boasting a slightly better record. German artillery was always very well handled and was a credit to its officers and men, its technical creators, innovators and doctrine developers who combined to produce a force that delivered. Up to July 1942, this is not what could have been said about the British.[359]

Finally, the accepted wisdom in British doctrine on land-air warfare dictated that RAF light and medium bombers would support the field force rather than the dive-bombers favoured by the Germans. However, any reading of the desert war would surely highlight the relationship for both protagonists that existed between effective operations and tactical air support in a parenthetical construct of logistics. The Germans had the edge in the first 18 months having refined doctrine and procedures between the PAA and the *Luftwaffe*, the key to which was good communications. General O'Connor owed his success to fighter aircraft and artillery reconnaissance (Lysanders) before the arrival of Rommel and a *Luftwaffe* air division put paid to further successes. Until First Alamein it was all Axis, much to the discomfort of the soldiers. However, the Allies grafted long-range interdiction onto tactical air support so that, by July 1942, although the Army didn't appreciate it, aircraft had been instrumental in reducing Axis reinforcements and supplies of all kinds in North Africa. When Montgomery had his DAF Headquarters moved next to his own, he was able to indulge himself in the observation (two days after the battle had begun) as bombers flew above, 'There goes the air force. They are winning my battle for me.' This was no idle remark, for the influence of aircraft on the battle was enormous. The air bombardment on 23 October had an extraordinary effect. Indeed the dislocation (moral effect) caused by one hour of air bombardment was greater than two minutes of heavy artillery preparation.[360]

Chapter 19
Accolades

Ubique Quo Fas et Gloria Ducunt
Wherever the Paths of Glory Lead

Regimental motto

During the battles of Bardia, Tobruk—and, by extension, the siege of that port that began three months later—and First and Second Alamein, the expertise and performance of the divisional RAA field regiments reached new heights. Much battle lore had been acquired in a short space of time during the first two battles—Bardia and Tobruk. The gunners received accolades from their commanders at all levels and responded to these with gusto. However their valuable desert experience could not be grafted onto the division that followed because of the strategic exigencies of those hectic months. The siege and the regiments' subsequent campaigns in the desert had built on that which preceded it. What is self-evident is that, under their inspired leadership—from Mackay and Herring, from Morshead and Ramsay down to their sergeants and bombardiers—other ranks reacted positively. The physical labour and emotional resilience necessary before and during these battles, the dangers from enemy artillery and mortar fire and bombing raids, prematures, moving about a battlefield which bristled with booby-traps and mines, was regarded by all ranks as simply a consequence of war to be taken in their stride. Both the 6th Division and its fellow 9th Division were in attack mode. At Second Alamein the 9th took part in a 'crumbling' operation that surpassed in total days the battles of Bardia and Tobruk in 1941. The Tobruk siege was of a different character, and primarily involved RA regiments. At Second Alamein the 9th Division did by far the most fighting, and its infantry, artillery and engineer casualties amounted to around one-fifth of the Eighth Army's total losses: 620 killed, 1944 wounded, 130 captured

—some 2694 men.[361] Morshead, veteran of Gallipoli and the Western Front with a total of four years' active service under his belt, said of the October battle, 'It was by far the hardest and most sustained battle I have ever experienced, and I have never known troops fight with greater courage, dash and persistence.'[362]

This was praise indeed. He had reason to be proud of his command, if only for the fact that he was acutely conscious of what he had asked his infantry to do. His frequent visits to the forward areas enabled him to judge the wear and tear on his brigades from their commanders down. After the battle he stood on Trig 29 for a long time, gazing to his front. He may have been reflecting on his generally unhappy experiences with the five RTR battalions with which his infantry had been tasked to work in those previous eventful three months. First at Tel el Eisa and now El Alamein he was to signal General Blamey his emphatic preference *not* to use tanks on 28/29 October. At times, tanks had been helpful and his infantry worked well with the tankies when they trained together. But doctrinal misconceptions and minefields had increased the casualties suffered by his men to the extent that he preferred not to use them. They tended to add to his tactical problems rather than solving them.

Morshead realised the crucial and precarious position of his division and the importance of its fighting hard to ensure the success of Montgomery's plans. The fact that all the PAA artillery, or what remained operational, was opposing him but also able to cover the 51st and 2nd New Zealand divisions' fronts was testament to the fighting prowess of his division. Morshead's casualty list in killed and wounded officers was heartbreaking, possibly the highest of the war for Australian forces in such a short time period. However, in comparison to that of other desert formations, battalions, batteries and squadrons which had been annihilated or taken into captivity, his losses were far fewer in terms of men and materiel. Even so, among those who fought or supported the fight, there was enormous satisfaction in watching their adversaries beaten to a standstill, to drift away into the night on 2 November and, ultimately, into captivity. After Tobruk's earlier capture, the 9th Division had a point to prove—that man for man, this was the best division on the field of battle at the time.

This is not to denigrate the performance of the German soldier who had fought with his typical tenacity and skill—Thompson's Post was never taken by assault despite all the shot and shell directed at its occupiers over a period of 13 days. But, in the end, the division had inflicted heavy casualties on two Italian corps and the PAA's *90th Light, 164th Light/Infantry, 15th* and *21st Panzer divisions*. The German staff firmly believed that the Australian soldier was the best on the battlefield, no mean tribute from a race that was aggressive, skilled, well structured and well led—and dearly loved a scrap.[363] A rueful Rommel wrote:

> The British artillery once more demonstrated its well-known excellence. A particular feature was its great mobility and tremendous speed of reaction to the needs of the assault troops. ... In addition [to] the advantage given by their abundant supplies of ammunition, the British benefited greatly from the long range of their guns, which enabled them to take the Italian artillery positions under fire at a range at which the Italian guns, most of which were limited to 6,000 yards, were completely unable to hit back.[364]

A notable absence from the history of the artillery battle was mention of the use of naval artillery to harass the Axis B Echelons and installations. It is usually presumed that these were air force targets. As previously described, raiding parties harassed coastal supply and storage depots further west. But, at the rear of the battlefield, enemy administrative areas were well within range of a cruiser's main armament, likewise those of battleships. The coast road was their main artery, and the enemy withdrew back down this road with relative ease because the advancing Eighth Army elements did not press them with sufficient vigour.

As the *Official History* records, British general officers were fulsome in their praise for the heroic achievements of the 9th Division. They visited Lieutenant General Sir Leslie Morshead to express their thanks and appreciation, many of them from those formations which had been seconded to the division earlier for training. Montgomery signalled Alan Brooke in London: 'The artillery fire has been superb and such concentrated use of artillery has not been seen before in N. Africa; we could not have done what we have done without the artillery; and that arm has been wonderful.' Thus it was to be in every battle Montgomery fought thereafter—concentrated artillery, flexibly applied. Kirkman had turned an infanteer into a disciple for gunnery.

This was indeed a classic Montgomery 'crumbling' operation in which, instead of using his armour to destroy the German-Italian armour once it had broken its way through the minefield belts, he would use it defensively to hold off Rommel's armour. This would allow his infantry to destroy Rommel's infantry by attacking north to destroy the German and Italian infantry in a methodical series of attacks. As enemy remnants fled via the coast road to the west, the Australians moved to the rear to rest, train, refit and prepare for their next move. A token return to Tobruk by veterans of the siege battalions was mooted, but ultimately did not occur. They were to return to Australia via Palestine, even though there was a strong rumour that they would rejoin the Eighth Army for operations involving amphibious landings—a rumour based on selected officers attending courses on 'amphibious operations'. However, as it transpired, these landings occurred in the South West Pacific Area of Operations, not the Mediterranean.

When the 9th Division finally left the Middle East, it was their old friends who were to miss their gunner colleagues the most. The 7th Medium Regiment historian wrote on 6 November:

> The regiment ended its association with 9th Australian Division through three crowded months since that formation moved into the Alamein line. In that time not only had the Australian troops earned the greatest respect for their fighting qualities but Divisional HQRAA had also proved most efficient and cooperative, eager to help the Regiment over any difficulties, appreciative of its endeavours to render them the most effective possible support and always keen to assist any 'private' operation against the enemy artillery.[365]

Four RA units, the 7th, 64th and 69th Medium and 4th Survey regiments went on to greater things as they headed west—the Battles of Mareth Line, Wadi Akarit and Enfidaville and the rout of the DAK in Tunisia. En route to Sicily at St Paul's Bay in Malta, General Montgomery addressed the 7th Medium again, telling the gunners, 'I know this regiment has been through the whole show and I consider 7th Medium Regiment to have been the backbone of the whole show.'[366] The 7th Medium supported the Eighth Army in the conquest of Sicily and, having had the distinction of firing in the first coordinated shoot in the Desert War against the Italians, fired the first shots across the Straits of Messina into mainland Italy. The 7th returned home after 11 years overseas and supported the 21st Army Group as part of 5 AGRA in the North-West European campaigns from D Day to VE Day 1945. Former second-in-command and XXX Corps CBO Brigadier Yates assumed command of 5 AGRA after Elton was killed and led it with distinction. The 7th Medium did not survive the artillery reorganisation of the late 1940s and transmogrified into the 32nd Medium Regiment, RA. The 25th Battery, the original 'Battleaxe Troop', lives on as the 74th Battery. The RA's 5.5-inch guns were fired for the last time in January 1995.

The 4th Survey Regiment continued to support the Eighth Army in the Mediterranean theatre and was part of the 21st Army Group CB resources until VE Day. As the Eighth Army CB organisation and AGRAs advanced from El Alamein to Tunisia and then Italy, the 4th Survey became far more expert in its arts than at El Alamein when, for all intents and purposes, the CB structure in which its elements operated was comparatively raw and untested. Lieutenant Colonel J. Whetton, DSO, OBE, MC, TD, the Regiment's commander since 1939, was accidentally wounded by a colleague in December 1944 and was succeeded by his second-in-command, Major R.H. Ogden, MC, TD. Lieutenant Frank Hamer ended the war as a battery commander and was awarded the MC. The unit was later removed from the order of battle. Some serving members transferred to the (Regular) Army's 94th Observation Regiment in 1948 (the former 1st Survey Regiment) then stationed with the British Army on the Rhine (BAOR) where it remained for 24 years. It hosted many Australian gunner officers and senior NCOs on attachment from 1945 onwards following their completion of the Long/Short Gunnery Staff Courses at the School of Artillery, Larkhill.

Of the senior British generals in this narrative, Field Marshal Earl Alexander of Tunis made no mention in his memoirs of the 9th Division, in stark contrast to his words to the troops at their divisional parade. Instead the 9th was wrapped up with 'XXX Corps' in his account. He took 'Tunis' into his earldom in 1943 after accepting the surrender of more German soldiers there than went into captivity at Stalingrad. He became Governor-General of Canada (1946–52), Minister for Defence (1952–54) and died in 1969.

Montgomery, when created a viscount and field marshal, took 'of El Alamein' into his title. His name passed into the army lexicon as 'a Monty'—a euphemism for 'a sure thing'. He returned to Australia (he had been a child in Tasmania where his father served there as a bishop) in 1949 and met with Eighth Army veterans. He was far more generous towards his Australian division than his superior, and is recorded as commenting, 20 years after the battle, 'When all did so well it would hardly seem right to single out any for special praise. But I must say this—we could not have won the battle in 12 days without that magnificent 9th Australian Division.'[367] His generalship of Operation Overlord, the cross-channel invasion of Continental Europe in June 1944 and the subsequent conquest of German land forces saw him take the German surrender in May 1945. He later became CIGS for three years (1946–48), Chairman of the Combined Chiefs of Staff Committee, Western Union (1948–51) and Deputy Supreme Commander, NATO (1951–58), during the Cold War. He demonstrated that he was better suited to field command than positions requiring political skills and finesse. In his retirement he became an outspoken critic of American Army generalship during the war. Montgomery, more than any other general, became a focus for psychological and military historians. His every word and deed was subjected to the utmost scrutiny and analysis.

Montgomery had hoped to revisit Australia in the last year of his life, but ill-health prevented him. Less than a month before his death, in March 1972, he told a confidant, 'I can't have very long to go now. I've got to meet God, and explain all those men I killed at El Alamein.'[368] He was doubtless alluding to an oft-repeated assertion that El Alamein was a political and not a military necessity and that, in conjunction with Operation Torch, the joint Anglo-American invasion of Morocco, Rommel could have been crushed and ejected from

Africa with fewer troops killed and wounded. This is not the place to argue that proposition; however, with his resounding defeat of Rommel he established two unarguable truisms—his own reputation as a field commander and restoration of the reputation of British and Commonwealth arms. His personal ascendancy in mastering combined operations for the invasion of Sicily and then Italy was, in Oliver Leese's opinion, Montgomery's finest hour—not El Alamein.[369] Be that as it may, tactical command of Overlord was a predictable sequelae to his career following the Battle of the Sangro. Operation Overlord was a triumph of which he was justly proud.

It is an understatement to assert that, through their achievements, six gun regiments—the 2/1st, 2/2nd and 2/3rd, 2/7th, 2/8th and 2/12th—had considerably embellished the already high reputation the Australian gunners enjoyed to this time. Their role in proving tactical doctrine and establishing high standards of performance with thorough training and inspirational leadership was to continue for the rest of the war, albeit in much different tactical and topographical circumstances. All except a significant specialist handful of the division were volunteers—they were not regulars or conscripts like the troops of the British Army. At the regimental level there was a melding of Staff Corps (Goodwin and Tinsley) with Militia (Ramsay, Eastick, and Houston). Their headquarters staff, battery commanders and almost all their junior officers were Militiamen. Collectively, they—the originals and reinforcements—had necessarily become fast learners in desert warfare and acquitted themselves with distinction, not only at El Alamein. Anecdotal evidence suggests that, within the regiments, it was at battery and Troop level that inspirational leadership was the strongest and most powerful of motivators. Moreover, the infantry and the gunners could not have achieved what they did without their British brothers-in-arms, particularly the three medium regiments and the 4th Survey Regiment, notwithstanding the morale-boosting efforts of the DAF. Both made the difference. In so doing they had merely confirmed the high standards the 'Gentlemen of the Artillery' accepted as 'the norm'. Their regimental motto said it all.

At the entry to the 9th Division section of the Commonwealth War Graves Commission El Alamein Cemetery, established six kilometres from the gun area near the former Dry Ration Point, a plaque carries this inscription:

'They fought themselves and their enemy to a standstill until flesh and blood could stand no more, then they went on fighting.'

The name and spirit of El Alamein live on in various visible forms—a fountain in Sydney at King's Cross, the names of streets, a Melbourne suburb, an Army training area in South Australia, and many other 'reminders' of heroic deeds performed in far-off lands. Annually in October at various venues around the nation, veterans have gathered in remembrances since their demobilisation, usually with their comrades-in-arms. This continues and, it is hoped, will be sustained by the gunner fraternity for years to come. At the Officers' Mess at Woolwich Barracks, the Royal Artillery's spiritual home, the October battle is remembered with a formal mess night.

El Alamein itself is now a beachside resort area for wealthy Egyptians. Memorials and tourists now grace that tortured terrain of 1942, not far from the British and Commonwealth War Graves Cemetery where more than 600 Australians lie buried:

> It is strikingly understated ... laid out so discretely behind a sandstone wall ... The ranks of marble headstones—name, rank, number and unit where the remains were identifiable. A few miles west of the British, the German memorial-cum-mausoleum offers an intriguingly assertive contrast. A massive rough-hewn stone octagon ... is visible for miles around. The names of the known dead are inscribed on bronze plaques above eight massive catafalques with the bones of Rommel's fallen. The Italian memorial and burial site, a few miles further west, is more conspicuous than the German—a flat-topped, white marble pyramid designed by Alamein veteran (and author) Paolo Caccia-Dominioni and shows names inscribed on wall panels.[370]

Chapter 20
Epilogue and Valette

Atque in perpetuum, frater, ave atque vale.
(And forever brother, hail and farewell.)

Gaius Catullus

As the Eighth Army began its long pursuit of the PAA forces remaining to their west, news of the battle was widely celebrated throughout the length and breadth of the (then) British Empire. Church bells rang in Britain and no-one would have been more relieved than Winston Churchill. His thrusting, conniving, censorious and fertile mind had at last produced a victory for him and confirmed him in his position at the pinnacle of power directing grand strategy. However, historians such as Barnett contend that the battle should never have been fought. Rommel was down and out, thanks to Hitler's directive to fight to the last man and round, and all that Montgomery had to do was to wait for his adversary, who was now tasting bitter defeat for the second time, to withdraw westward to safety and await his instructions from his superiors. While Rommel was beholden to the directives of his Führer, he was also ruing the fact that, had Hitler sent him another division instead of promoting him field marshal after Tobruk, he would have been in a far better situation at the end of October. Nonetheless, he had no illusions as to his precarious military state, or the state of his health. During Operation Crusader, he had lost many more soldiers, tanks, guns and materiel than he had during El Alamein's Operations Lightfoot and Supercharge. As a result, one of his senior staff officers had accorded Auchinleck the honour of being the best British general whom Rommel had faced in the desert, presumably on the basis of numbers rather than tactical wit.

One can only speculate as to how Auchinleck would have conducted Second Alamein. What is pertinent is that Major General Eric Dorman-Smith's appreciation of July had proved prophetic. This is not to say that he was the only one capable of appreciating what needed to be done by a commander in the theatre if he were to best Rommel in combat and save the Delta. Any intelligent Staff College graduate using the same logical approach would have arrived at similar general conclusions, as had General Marshal Cornwell, GOC British Troops, Egypt, in June 1940. However, it is doubtful, given the past performance of other generals who could then have been promoted to Eighth Army command, even Alexander or Alan Brooke, that any of these could have raised morale in the manner of Montgomery. Critical to this was the significance he accorded to focused, simulated training for the approaching battle.

Historians have also suggested that Hitler's High Command was already nominating formations to be sent to Tunis and North Africa. Operation Torch had precipitated the necessity of reinforcing the PAA and suggested that, strategically, Rommel could simply force Montgomery to fight a critical battle on the ground of the former's choosing. The PAA was nothing if not resilient and highly capable of surprising its enemy, even with depleted forces, as its 18 months of desert warfare had demonstrated. But, as far as the Australian soldiers were concerned, this was all in the future. Communiqués from the Middle East theatre during the battle alternatively thrilled and concerned the Australian public. No church bells were rung, but there was relief at the 9th Division's scheduled homecoming, both within its ranks and among the kinsfolk at home.

While, for the 9th Division, the tumult and shouting died after El Alamein, this was not the case for its companion forces in the Eighth Army. Montgomery's command swept westward across Cyrenaica, Libya, Tunisia, then crossed the Pantellaria Straits to Sicily and ultimately, Italy, inflicting further heavy losses on its adversaries. Many regimental histories of British regiments savoured the fact that the *90th Light, 15th* and *21st Panzer divisions* went into captivity in Tunisia in 1943. Morshead's division returned to Australia, recouped its strength, gathered new soldiers to its ranks and trained assiduously in Queensland for tropical (jungle) warfare campaigns in Papua and New Guinea and in Borneo.

The tactical artillery lessons that emerged from El Alamein and subsequent campaigns across Africa, Sicily, Italy and North-West Europe had little relevance to the RAA during operations in the South West Pacific Area (SWPA). While light, mountain, field and medium were all deployed in that theatre, ammunition expenditure never reached the heights of the Western Desert. Terrain, climate, transport and logistic problems limited opportunities for gun position and OP selection in very close country. Bringing survey forward was a tedious, tortuous affair. Headquarters and all units operated on establishments in which vehicles were conspicuously absent. Anti-tank units were converted to light (using heavy mortars, 4.2-inch calibre), light and heavy anti-aircraft batteries guarded vulnerable points and beachhead landings. They had been given a new designation—'tank attack regiments'—but were never used in that role. The Japanese (after Malaya) rarely obliged. A short version of the 25-pounder for tropical warfare was created by reducing the barrel and trail length, wheel size and eliminating the shield. The sighting system remained intact, its maximum range 10,200 yards. Its weight in action was one and a quarter tons. It was not a popular gun because of the absence of a shield and its inherently unstable platform. Ammunition was manhandled over steep hills in dreadful conditions and wireless sets proved less reliable in mountainous terrain. Campaigning on the plain in kunai grass, steep mountainsides and swollen rivers or the swampy ground near the coast was a far more physically draining experience for the gunners than the desert. Tropical diseases were more debilitating and malaria recurred after treatment. Turnover in units was high. Nonetheless, these obstacles were overcome by the same spirit as that which prevailed in the desert.

Following Japan's surrender in August 1945, divisional artillery units were disbanded and passed into RAA history. Gunners of all ranks were demobilised using a 'points' system based on service and other factors. They returned to 'Civvy Street', to their old jobs, or started their lives afresh. A few who enjoyed Army life 'soldiered on'. Each regiment (of the division) produced a history of its accomplishments over five years of war—some soon afterwards, others following the lapse of many years, without which this history would not have been possible. Each regiment formed an association, often with interstate branches, reflecting its more homogeneous composition at the war's end compared to the period of its formation in 1939–40. The regimental associations held reunions and 'smokos' and produced newsletters with contributions from all ranks. The affairs of each association became the focus of dedicated men who kept the spirit of the Regiment alive, but were inevitably saddened by the passing of valued comrades-in-arms. At the beginning of the twenty-first century, the gunners of El Alamein were in their seventies, eighties and a few nonagenarians defied the 'grim reaper'. But what of the spirit that remained abroad in the gunner community?

As the national government wrestled with the future of the Army (and other services and corps), first with the establishment of an 'Interim Army' and later with the Australian Regular Army (ARA) and Citizen Military Forces (CMF), the spirit was rekindled by the formation of some CMF units using the numbering convention of pre-war Militia batteries that pleased some veterans. Thus, in 1947, the 13th Field Regiment was raised at Keswick Barracks, Adelaide, and the 3rd Field Regiment at Karrakatta Barracks, Perth. The 13th was commanded (from 1950 to 1953) by Lieutenant Colonel A.L. Rogers, ED, an original officer of the 2/7th and later by Robert Rungie. Roger Fitzhardinge commanded the 3rd from 1958 to 1961.[371] The 3rd Anti-Tank Regiment was raised at Belmore (Sydney), became a Light Regiment in 1952 and emerged as the 23rd Field Regiment in 1955. Major A.A. Salter (2/8th) later became CO of the 6th Field Regiment in Hobart. Many officers and NCOs from divisional artillery regiments donned uniform again to help rebuild the CMF and later the National Service Training Scheme of the 1950s in all states of the Commonwealth. The results achieved by that volunteer force would not have been possible without them.

The best field gun of World War II was withdrawn from service, first in the ARA, and later in 1975 from CMF/Army Reserve units, and replaced in the first instance by NATO/ABCA calibre ordnance, an Italian-designed 105-mm pack howitzer. A widely held gunner opinion was that this was a retrograde step. Over its lifetime, the 25-pounder needed very few modifications to its basic design, the most obvious the addition of a two-port muzzle brake. This modification resulted from the use of supercharge for extreme range and with AP shot (initially) for anti-tank shooting and was introduced at about the same time as the 6-pounder came into service. It also reduced recoil distance, an advantage in the crowded working space of self-propelled equipments.

Following World War II, the Australian government of both major political persuasions committed its armed forces on four occasions under United Nations, ANZUS and SEATO obligations. The first commitment was during the Malayan Emergency in 1948 as part of Far East Land Forces (FARELF) which undertook counter-insurgency operations in Malaya against the Communist guerillas. Infantry (The Royal Australian Regiment) deep patrolling was supported by a field artillery battery and these units were rotated until 1963 when the Emergency was declared at an end. During the United Nations-backed Korean War from 1950 to 1953, the British Commonwealth formed a composite division—the 1st (and only) Commonwealth Division, which served alongside American and South Korean Army units. More than a dozen RAA officers and several senior NCOs were attached to the 16th Field Regiment, RNZA, as battery and Troop commanders, the Divisional CB Staff Office and as air OP pilots with 1903 Flight, RAF. The Royal Regiment, represented by British, Canadian, Australian and New Zealand gunners, achieved a standard of operational efficiency that was recognised as 'second to none' using 25-pounders, 4.2-inch mortars and later 5.5-inch guns.

In 1962 the Indonesian Government under President Sukarno began insurgency operations, known as Confrontation (*Konfrontasi*) against the newly formed (and short-lived) Malaysian Federation. FARELF forces reacted and RAA pack howitzers supported Australian and British infantry. In most cases the widely spread battery operated from specially designed forts that relied primarily on air resupply. The battery also saw artillery fire (about 100 rounds) used to create helicopter landing areas in the jungle. The final commitment was to South Vietnam in 1965. By this time, Australia's foreign policy had changed to reflect an almost complete Far East focus against which the Australian Army was structured (as were the RAN and RAAF). In brief, the North Vietnamese communist regime commenced offensive operations against its southern neighbour with Russian and Chinese assistance. America responded by initially sending advisers and large quantities of material aid to help South Vietnam's armed forces. Australia soon followed suit and a special unit was established for the purpose—the Australian Army Training Team—Vietnam (AATTV). Some members of the AATTV were gunners, and one former RAA member, Major Peter Badcoe, was awarded a posthumous VC for his heroism under fire.

The AATTV was followed by an infantry battalion and an RAA artillery battery (105) using 105-mm air-mobile pack howitzers operating as an independent battery with the US Army 173rd Airborne Division. The conflict escalated and, in 1966, an Australian all arms and services Task Force operated from bases in Nui Dat and its logistic head, Vung Tau. Eight field artillery batteries served independently and/or were incorporated into the 1st, 4th and 12th field regiments. A detachment from the 131st Locating Battery and/or other personnel rotated on tours of duty until 1972. Numerous counter-insurgency operations were conducted during this time and the telling power of field and locating artillery, including a US Army heavy artillery battalion brigaded with the Australians, was vital to the success of the Australian Army, air and naval forces deployed in the theatre. The Battle of Long Tan is perhaps the best known example.

The tactical handling of unit deployment by helicopter in this period was a new development for the RAA. New technology from survey to munitions meant that old ways and terms used were discarded and replaced by new. The senior officers who commanded the Task Force or who comprised the Military Board were variously veterans of World War II, Malaysia, Korea, Borneo and/or overseas appointments with the British, Canadian and American armies. The important point is that they remembered and applied a successful doctrine that produced fewer infantry casualties against a determined and crafty foe. Apart from subsequent triservice deployments to Iraq and Afghanistan, no Australian artillery sub/unit saw active service in a shooting war after Vietnam, although many officers and senior NCOs continued the practice of secondment to British, Canadian and American armies. Australian officers served in both Gulf Wars against Iraq and gained firsthand experience of modern combat operations of a sophistication undreamed of at the end of World War II. Further active service by secondments followed. By way of example, in April 2003, for Operation Telic, the British Army fielded an augmented field regiment (3 RHA) supporting the 7th Armoured Division with thirty-two AS 90 self-propelled 150-mm guns and supporting elements. They liberated the city of Basra (with a population of 1.5 million) for the loss of three British infantrymen. The 7th (Para) RHA, with its light field gun (105-mm), operated in the An Nasariyah area and west along the Tigris and the 29th (Commando) Regiment, RA, operated in the coastal sector (with naval gunships including HMAS *Anzac*). Together the three regiments fired more than 23,000 rounds in 17 days of operations.[372]

While field artillery has come a long way from Tobruk and Bardia in 1941, the underlying principles of its employment have changed very little (although the jargon/terminology has!). Australia's military planners must never forget the basic lessons learned through the bitter hardships of the Western Desert and revisited in these pages. Perhaps the most significant is that field (tube) artillery exists to help the infantry/armour onto the objective. That truism, expressed in the earliest editions of the *Field Service Regulations* is just as true today as it was in the 1890s, and will be in the future in one form or another. The high standards of the Regiment are due in no small measure to those gunners, Regulars or Reservists, holding rank who have demonstrated professionalism in every branch of the RAA. It continues because of a selflessness of service inherited from its mentors and their good example. The period following the end of World War II was a case in point.

The spirit and endurance of those few of the 9th Division who now survive transcends belief in this day and age. The yardstick for heroism and continuous active service in Vietnam (1965–72) with the Australian Task Force is reckoned to be the Battle of Long Tan in August 1966 involving the 6th Battalion, The Royal Australian Regiment. Significant though it may have been in Australian military history, it was measured in hours rather than days or months. It was noteworthy that field artillery played a vital role in helping rout a superior Viet Cong force. Vietnam had no Trig 29 and 25 ferocious panzer counter-attacks in three days or 200,000 rounds of (Australian) artillery ammunition expended in 13 days. It was a scale of warfare unimaginable today. For all that, the Battle of El Alamein must be remembered for what it was. First, it was the RAA's zenith; no other Australian divisional artillery would ever surpass the 9th Division's record. Second, following El Alamein, German and Axis forces were in retreat in all theatres of war and, for the artillerymen of the then British Commonwealth—now just a memory for many—their finest hour to that date. With their comrades-in-arms they can truly claim to have been part of a magnificent formation without peer, playing a key part in the 'Turning Tide'.

Lest We Forget

Notes to Sources

The major source of information on units' roles, both during the battle and in the weeks before it, and that of Headquarters, 9th Division, Royal Australian Artillery, AIF, and the divisional field regiments has been the various unit war diaries. The regimental records vary from being fairly complete to inconsistently compiled. In the case of the three regiments I have relied heavily on David Goodhart's *History of 2/7 Field Regiment* and *We of the Turning Tide*, both accounts of the 2/7th Field Regiment, and *Gunfire!*, Max Parson's story of the 2/12th Field Regiment. Goodhart's books provide a very detailed account of his unit for the four months of battle while *Gunfire!* includes more general tactical descriptions and personal citations. Taken together they provide a wonderful, composite view of a field regiment in a campaign. The 2/8th Field Regiment War Diary has also proven valuable. I am grateful for Charles and Mary Morton's unique account, *2/8 Australian Field Regiment Remembers WWII* which has been extremely useful, as has the 2/8th's war diary, particularly as the 2/8th was a designated counter-bombardment fire unit during the October battle. The 2/8th's unit association, particularly Mr Charles McKenzie, also supplied material that was particularly helpful.

I have drawn heavily on the regimental histories of three key RA regiments of the British Army closely associated with the Australian 6th and 9th division's successes, the 7th and 64th Medium Regiments and the 4th (Durham) Survey Regiment. These units played a crucial role in 1941 long before the actions from July to October 1942 with XXX Corps. The 7th Medium's history is a well-rounded general account. The descriptive detail of their (several) disasters is salutary reading for those who wonder how it felt to be placed in a hopeless situation, particularly at Gazala, where they had to destroy their guns for the second time.

A recent offering, *A Certain Excellence*, is written by a gunner from the 64th Medium Regiment and is focused at his own Troop level. He leavens his personal experiences in the desert with a diatribe on General Auchinleck's handling of Crusader. Only once does he mention his commanding officer by name, and references to other Troop officers are non-existent. Falvey's regiment (the 211th and 212th batteries) appear in our official histories as footnotes, as do other RA units who were at one time involved with Australian divisions. Our histories do not show their travails or successes, and it is unfortunate that three Australian authors have seized upon one example of an RA medium battery (the 212th) being five minutes late in executing a fire plan in support of the 7th Australian Division attack on Damour on 6 July 1941. On this occasion, the Commanding Officer of the 2/5 Field Regiment, Lieutenant Colonel John O'Brien, noted the apparently casual attitude of the RA battery commander. He was being hospitable to O'Brien at the time. He described the situation in his own book and, subsequently, both David Horner (*The Gunners*) and Colleen McCullough (*Cutler, VC*) have chosen to give O'Brien's paragraph another airing. As no harm was done by the battery's tardiness it does suggest that these authors sought to denigrate the reputation of the RA by raising the incident again. Their motive is all the more curious because there was, after all, widely reported complimentary commentary on the excellence of the RHA, RA field and medium regiments' support once the smoke and dust had cleared at Bardia and Tobruk in January 1941, much less the siege in April, prior to the Syrian campaign. This particular battery of 60-pounders had been in the theatre for four months and, like most artillery units at the time, was short of all sorts of stores and equipment. Prior to Syria it had to 'mother' and provide cadre for its sister battery that had just returned from the debacle that was Greece and Crete. The 212th Battery sustained heavy casualties at Gazala and went on to support the 9th Division at El Alamein.

The majority of photographs in this volume are from the Australian War Memorial collection, while those of the RA units are from the collection of the Imperial War Museum, London, and various other sources. The gridded maps of the El Alamein area and detail thereon have been plotted from the data, i.e., grid references given in operation orders, task tables, hostile battery lists, etc. in the war diaries and other sources. Given that, a keen eye may note some minor differences in coordinates in some instances. This in no way materially affects the main thrust of the history. Another accurate large-scale map available showing most of the tactical area of operations was included in the 4th Survey Regiment, RA (TA) 'Notes and Opinions Compiled by Lieutenant Colonel H. Hemming (on the regiment's service) from September 1942 to June 1943, Part II, Section 7(a)'. David Goodhart's maps of the PAA's defences prior to the battle, while not strictly adhering to contemporary staff duties correctness, come from *We of the Turning Tide*. The battle map showing PAA defences in the path of attacking infantry on 23 October was compiled by the 514th Corps Survey Company, RE, and printed by the 46th Survey Company, RE. Two diagrams owe their origin to the War Office publication of Brigadier A. L. Pemberton, '*The Development of Artillery Tactics and Equipment*'. Other maps of the areas covered are from the National Library of Australia collection and I am indebted to their Map Librarian for her help. Traces and overlays in war diaries were not permitted to be copied during the period of my research at the Australian War Memorial. There are several examples of Army forms used to communicate orders and information between headquarters at several levels and the end users. The 2/7th Field Regiment's 14 Battery Command Post log (AWM 52/427/7) is perhaps the only surviving battery command post log of the battle, as is its F Troop record, recently come to light in Melbourne. These may be helpful to future historians.

As this history is, in effect, a snapshot of how an artillery war was waged in the deserts of Cyrenaica and Egypt, a glossary of terms is included and appendices show honours and awards bestowed and biographical notes on the senior commanders and major figures who feature in this discussion of the Western Desert campaign.

Conventions

Just as the Axis forces were bi-national, so too was the 'British' Army represented by its diverse composition. As New Zealand, South Africa, India and Australia had all responded to the British Government's request for land and other forces, their formations were part of the larger corps and armies during their service in the theatre. In October 1942, the Eighth Army consisted of seven divisions, four of which were Commonwealth. Thus, to label artillery prior to this as British or Commonwealth would be misleading. The enemy's view was simple—it was British! With the exception of our divisional commanders, corps and army commanders were all British Army. I have adopted the convention that, where Commonwealth artillery assets outnumbered British, I make that distinction.

I have used the abbreviation PAA (*Panzerarmee Afrika*) for text where there were combined German and Italian formations. The DAK (*Deutsches Afrika Korps*) was a German formation of four divisions that was part of the PAA and commanded at that level by a *Generalleutnant* who reported to Rommel as a *Feldmarschal*. The Italian command structure was subordinated to Rommel for most of the period covered by this account. Similarly, '*Luftwaffe*' includes Italian Air Force (*Regia Aeronautica*) units.

Yet another convention I have used is the term 'gunners' which, as a general rule, refers to all artillerymen regardless of rank in a broad context. 'Gunners' connotes both the Royal

Regiment and, more specifically, its hierarchy. 'The Regiment', in that quaint British way, also means all personnel in various branches and specialists who, regardless of employment, wear a white lanyard on their right shoulder. It also means, by extension and in the era of this narrative, the artilleries of Australia, New Zealand and Canada. South Africa saw the relationship somewhat differently.

The noun 'Troop' (capitalised) is used to identify a fire unit of four guns, while 'troops' refers to a body of soldiers. Measurements are used as they occur in records and are not converted from Imperial to metric or vice versa.

Endnotes

Chapter 1

1. K. Coleman and J.T. Knight, *A Short History of the Military Forces in N.S.W. from 1788—1953*, Royal Australian Artillery Historical Society, Eastern Command, 1953; D. Brook, *Roundshot to Rapier: Artillery in South Australia, 1840—1984*, Investigator Press, Adelaide, 1986; R.M.C. Cubis, *A History of 'A' Battery, RAA*, Elizabethan Press, Sydney, 1978.
2. Cubis, *A History of 'A' Battery*, RAA, pp. 85, 91.
3. See S.N. Gower, *Guns of the Regiment*, AWM, Canberra, 1981, pp. 160—3 for details of early ordnance used in this campaign.
4. P. Dennis, J. Grey, E. Morris, R. Prior, *The Oxford Companion to Australian Military History*, Oxford University Press, Melbourne, 1999. For a description of the Compulsory Military Training Scheme, see pp. 147, 174.
5. D. Horner, *The Gunners*, Allen & Unwin, 1995, p. 187. For a broad overview of the growth, organisation and performance of Australian gunners, see chapters 4 to 8.
6. C.E.W. Bean, *Official History of Australia in the War of 1914—1918*, Vol. VI, The AIF in France 1918, Angus & Robertson, Sydney, 1942, pp. 248—50; 306—7; 495—96.
7. M. Farndale, *The Western Front 1914—1918, History of the Royal Regiment of Artillery*, Brassey's, London, 1996, pp. 221—27.
8. R.G.S. Bidwell, *Gunners at War*, Arms and Armour Press, London, 1970, p. 37.
9. A. Palazzo, *The British Counter Bombardment Staff Office*, Journal of Military History, Vol. 63, No.1, pp. 55—74.
10. Peter Chasseaud, *Artillery's Astrologers*, Naval & Military Press, Lewes, UK, 1999, p. 427.
11. J.B.A. Bailey, *Field Artillery and Firepower*, Naval Institute Press, Annapolis, 2004, p. 241; D. Zabecki, Steel Wind, Praeger, Connecticut, 1994, pp. 12—13. British figures also include the Dominion/Commonwealth artilleries. Bailey and Zabecki include Salonika, Palestine and Mesopotamia in their macro view of artillery strengths. Colonel C.N.F. Broad, who comprehensively analysed the artillery war on the Western Front, wrote that the British artillery expanded its total of field guns/howitzers from

72 batteries to 568 between 1914 and 1918, and increased its stock of six heavy/siege batteries to 440. See C.N.F. Broad, *The Development of Field Artillery Tactics, 1914-18*, Journal of the Royal Artillery, Vol. XLIX, No. 2, p. 62; Farndale, The Western Front, p. 341, Annex B. His arithmetic is at variance with other records. He also shows 289 anti-aircraft guns (3-inch, 20 cwt) that could be used in a field role in an emergency. Farndale's figures also extend to other fronts.

Chapter 2

12 R.G.S. Bidwell, 'The Development of British Field Artillery Tactics 1920—1939, Reform and Organisation', Journal of the Royal Artillery, Vol. XCIV, p. 16.
13 B.P. Hughes, *Between the Wars*, Royal Artillery Historical Series, Brassey's, London, 1992, pp. 166—76.
14 Gower, *Guns of the Regiment*, pp. 160—189. Post-World War I, a streamlined shell fired with the gun's trail in a pit could attain a range of 10,500 yards.
15 A.L. Pemberton, *The Development of Artillery Tactics and Equipment*, War Office, 1950, p. 12. The French 75-mm gun posted a creditable record in an anti-tank role in 1940.
16 M. Farndale, *The Years of Defeat: History of the Royal Regiment of Artillery, Europe and North Africa, 1939—1942*, Brassey's, London, 1996. The key figures were the Director, Royal Artillery, Major General Sir Campbell Clark; School of Artillery Commandant, Brigadier S. R. Wason; and the DCIGS, Lieutenant General Sir Ronald Adam.
17 Hughes, *Between the Wars*, p. 183. In Tunisia and Italy their fire was feared by the Germans and Italians. In essence, the reduction of targets by the heavy pieces in World War I was now regarded as the preserve of the air force. Given this background, there was little for the Australian Military Forces' heavy/medium (militia) brigades to look forward to. When mobilised there were insufficient guns for all defence purposes (some 60-pounders were deployed in a coastal gunnery role) and some medium brigade gunners had to train on the much scorned 'popguns' of field brigades.
18 Pemberton, *The Development of Artillery Tactics and Equipment*, p. 16.
19 David Horner has chronicled acquisitions and developments in some detail. See Horner, *The Gunners*, Chapter 9.
20 See Gower, *Guns of the Regiment*, p. 82.
21 Bidwell, *The Development of British Field Artillery Tactics 1920—1939*, pp. 43—45.
22 P. Ventham and D. Fletcher, *Moving the Guns: the Mechanisation of the Royal Artillery, 1854—1939*, HMSO, London, 1990, pp. 27—32; C.C. Armitage, 'Mechanising Field Artillery', Journal of the Royal Artillery, Vol. LI, No. 1, 1924, pp. 1—7.
23 Cletrac was a tractor designed by the Cleveland Tractor Company, Ohio, USA, and used to tow guns and limbers.
24 Ventham and Fletcher, *Moving the Guns*, pp. 29—54.
25 Ibid., pp. 62—85.
26 Hughes, *Between the Wars*, Chapter 6; Gavin Long, *Australia in the War of 1939—1945*, Vol. I, *To Benghazi*, AWM, Canberra, 1952, p. 82.

27 Ventham and Fletcher, *Moving the Guns*, pp. 86–95. These vehicles used the same motor as London's famous buses.
28 See Cubis, *A History of 'A' Battery, RAA*, p. 183.

CHAPTER 3

29 J. Price, 'The Great Wall of Chinagraph', *Journal of the Royal Artillery*, Vol. LXXX, No. 3, p. 210.
30 Bidwell, *Gunners at War*, pp. 80–81.
31 Ibid., p. 82. It is possible to follow the various schools of thought on mechanisation in the British Army in papers presented at the United Services Institution. Three chosen at random are: A.P. Wavell, 'Training for War' in 1933, (Vol. 78) p. 254; L.A.W.B. Lachlan, 'Mechanised Mindedness', Vol. 78; G. LeQ. Martel, 'Mechanisation', Vol. 82, May 1937.
32 J.F.C. Fuller, *Decisive Battles of the Western World*, Vol. 3, Eyre and Spottiswoode, London, 1956, pp. 278–79. Corelli Barnett (*The Desert Generals*, William Kimber, London, 1961, pp. 102–3), when analysing the 1941–42 desert battles, noted that 'the cavalry officer did display greater leadership than the tank officer, and also British soldiers preferred to be commanded by "gentlemen" than able men of their own class'.
33 Pemberton, *The Development of Artillery Tactics and Equipment*, p. 9.
34 J.P. Harris and F.H. Toase, *Armoured Warfare*, B.T. Batsford, London, 1990, p. 47.
35 R. Ovary, *The Road to War*, Macmillan, London, 1989, pp. 28, 188. For a full account, see M.R. Habeck, *Storm of Steel: The Development of Armour Doctrine in Germany and the Soviet Union, 1919–1939*, Cornell University Press, 2003.
36 See Heinz Guderian, *Panzer Leader*, Michael Joseph, London, 1977.
37 Ibid., Chapter 2, p. 46. He also developed a close relationship with the Swedish Army in the development of Panzer equipment and guns.
38 Ibid.
39 Bidwell, *Gunners at War*, p. 102.
40 A. H. Smith, 'Counter Bombardment, 1920–1941', *Cannonball*, Journal of the Royal Australian Artillery Historical Company, No. 41, November 2000, pp. 9–14. The 2nd Survey Company, a Militia unit in Melbourne, was commanded by a physicist, Major Richard Cherry, who practised his sound rangers at weekend and annual camps, innovating by using Postmaster General lines. He took his men to war as the 2/1st Survey Regiment, RAA.

CHAPTER 4

41 Bidwell, *The Development of British Artillery Tactics*, p. 3; R.G.S. Bidwell and Dominick Graham, *Fire-power: The British Army Weapons and Theories of War 1904-1945*, George Allen & Unwin, London, 1982, p. 209.
42 Pemberton, *The Development of Artillery Tactics and Equipment*, p. 28. The issue of casualties to infantry following behind a barrage is common to all artilleries, not just the British. Enthusiasm for 'leaning on the barrage', that is, following about 50 yards behind a line of shell bursts, is inversely proportional to one's rank. The French (in print, at any rate) appear the most pragmatic about accepting casualties.
43 Farndale, *The Years of Defeat*, Annex A, p. 236.

44 L.F. Ellis, *The Campaign in France and Flanders, 1939—40*, Vol. I, Official Series, HMSO, 1951.
45 Bidwell, *Gunners at War*, p. 128.
46 Ibid., pp. 121—23.
47 Ibid.
48 Farndale, *The Years of Defeat*, Annex E, p. 264; Appendix A, pp. 236—40. In sum, there were 46 field regiments (1004 guns), 18 medium regiments (288 guns) and three heavy batteries (48 pieces), a ratio of 336:1004, or 1:3.
49 J. Whetton, *Z Location and Survey in War: The History of 4th Survey Regiment, RA, TA*, unit publication, 1976, pp. 2—3.
50 Pemberton, *The Development of Artillery Tactics and Equipment*, p. 98.
51 Fuller, *Decisive Battles of the Western World*, pp. 403—5; Farndale, *The Years of Defeat*, Annex E, p. 264; Barton Maughan, *Tobruk and El Alamein*, Australian War Memorial, Canberra, 1966, p. 639.
52 Farndale, *The Years of Defeat*, Appendix E, p. 264.
53 Bidwell, *Gunners at War*, p. 87. Despite being on a more 'warlike' footing than home troops, affiliation exercises were rarely an element of training schedules.
54 See Bailey, *Field Artillery and Firepower*.
55 Ibid., pp. 273—87.
56 Ibid., pp. 138—40.
57 Ibid., p. 140.
58 Bidwell, *Gunners at War*, pp. 2—3. Parham was also heavily involved in the formation of the Air Observer Flight, RAF. With basic RAF administration to keep it flying, it was commanded by a gunner officer and pilots were all specially selected RA officers.
59 Ibid., pp.1—6.

Chapter 5

60 Craig Wilcox, *For Hearths and Homes: Citizen Soldiering in Australia, 1854—1945*, Allen & Unwin, Sydney, 1998, pp. 92—93.
61 Horner, *The Gunners*, pp. 220—22.
62 S. Sayers, *Ned Herring: A Life of Lieutenant General the Honourable Sir Edmund Herring*, Hyland House, Melbourne, 1980, Parts 1, 2 and 3. Major B. E. Klein was just one of a number of Staff Corps officers who were reposted after short periods at this stage when three divisions were being raised. Herring was to meet Klein as Counter Bombardment Officer, Lustreforce (the Greek Expedition), in April 1941 (see p. 132). Herring noted that in the Middle East he was to meet 'an increasing number of British officers in senior appointments who had known him either at Oxford or in the Salonika Army.'
63 M. Hayward, *Six Years in Support: The Official History of 2/1st Australian Field Regiment*, Angus & Robertson, Sydney, 1959, Chapter 1. The second-in-command was Major John Howard of Newcastle, a Shipping Manager. In the Middle East his expertise saw him posted to the Middle East logistics organisation to manage stevedoring operations. The Battery Commander (BC) 1st Battery was Major C. Peters; BC 2nd

Battery, Captain K. O'Connell; Adjutant, Captain J. S. Anderson. The 2/1st was the first AIF field regiment to be equipped in the Middle East with 25-pounders (the 2/3rd was the first, but was in the UK). The gunners of the 2/1st then beheld for the first time the 'Quad', the four-wheel-drive gun tractors built by Marmon-Herrington UK to a Ford America specification. Their attached troops were E Troop, Divisional Signals Regiment, 13th Light Aid Detachment (LAD), and the Electrical and Mechanical Engineers.

64   Ibid., p. 19. L.E.S. Barker had another nickname, 'Holy Joe', which belied a stern disciplinarian. Barker made Church Parades compulsory; those who opted out were sent on a long route march. He was not given to frivolity, but could take it and give it. For a gunner's view see E. J. Hewit, 'It Was Fun Sometimes', unpublished manuscript, ed. A.H. Smith, 2002, p. 14.

65   W.E. Cremor, *Action Front: The Official History of the 2/2nd Australian Field Regiment*, Regimental Association, Melbourne, 1961, Chapter 4. Cremor was a member of the six-man editorial committee that wrote this history.

66   L. Bishop, *The Thunder of the Guns: The Official History of 2/3rd Australian Field Regiment*, Regimental Association, Melbourne, 1995, Chapters 2—3. When formed, the regiment of 633 all ranks comprised men from:

Victoria:	RHQ	6 Officers 45 ORs
South Australia:	5th Battery, BHQ and C Troop	9 Officers 150 ORs
Tasmania:	5th Battery, A and B Troops	7 Officers 126 ORs
Western Australia:	6th Battery BHQ, E Troop	9 Officers 148 ORs
NSW:	6th Battery, D and F Troops	7 Officers 153 ORs

The regiment also included the 41st LAD.

67   AWM 54 422/7/8 War Establishments, A Field Regiment, RAA II/8/4. See also Artillery War Establishments, 1940-51, AWM 54 347/4/11. Wireless data from T. Barker, *A History of the Royal Australian Corps of Signals 1788-1947*, McLeod, Vic, RASigs Committee, 1987, pp. 122, 168, 173, 292—94.

68   Ibid.

69   Artillery Training Pamphlet, Vol. II, No.1, *Handling Units in the Field*; Pemberton, *The Development of Artillery Tactics and Equipment*, Chapter 1. See also B. P. Hughes, *Between the Wars*, pp. 167—82.

70   Handbook Ordnance QF 25-pounder Mark. II (War Office publication).

71   Range Tables Pt. II, 25-pounder Range Tables.

72   Artillery Training Pamphlet, Vol. III, No. 1, *Ballistics and Ammunition*. HE shell fragments were incorrectly but colloquially called 'shrapnel'.

73   Artillery Training Pamphlet Vol. III, No. 3, *Duties at RHQ and the Guns*: no. 5, *Predicted Fire*; no. 4, *Engagement of Targets by Observation*; no. 6, *Programme Shoots*.

74   Ibid.

75   Ibid.

76   Artillery Training Pamphlet, Vol. II, No.1, *Handling Units in the Field*; Pemberton, *The Development of Artillery Tactics and Equipment*, Chapter 1. See also B. P. Hughes, *Between the Wars*, pp. 167—82.

77   AWM 54 422/7/8.

78   The RA establishments were slightly different in personnel numbers, mainly in the number of batmen (one per officer). In the 7th Medium Regiment, a three-ton lorry was converted to a mobile officers' mess for a battery, complete with cooker, water trailer, tables and chairs, etc.

79   *Staff Duties in the Field*, 1950 edition (incorporating 13 previous amendments).

Chapter 6

80   J. Baynes, *The Forgotten Victor*, Brassey's, London, 1989, p. 105.

81   Richardson was posted out of 2 Battery and replaced by Kevin O'Connell who finished the war a brigadier (CRA 3rd Division). Charles Peters was captured in Greece and retired from the Regular Army as a brigadier.

82   Hayward, *Six Years in Support*, pp. 9—10, 28.

83   Elliott was a most unpopular CO. He was unsociable, uncommunicative, very self-possessed. The regiment was glad to see the back of him.

84   F.L. Johns, J.A.C. Monk and P.J.D. Langrishe, *The History of the 7th Medium Regiment, Royal Artillery, During World War II, 1939-1945*, Loxley, Sheffield, 1951, p. 4. The 60-pounders suffered from cracked trails and had to be sent to EME workshops for repair, some distance from the front. Only F Battery, 4th RHA, supported the attack. The other batteries were with the 7th Armoured Division. This differs from Horner's list (*The Gunners*, pp. 240—41). For comments on relations between the regiment and Brigadier Herring, see Sayers, *Ned Herring*, p. 138; Johns et al., *The History of the 7th Medium Regiment*, pp. 9—10.

85   Long, *To Benghazi*, pp. 154—55.

86   Ibid., pp. 144, 158.

87   J. Ellis, *World War II Data Book*, Aurun, London, 1993, pp. 136—37.

88   Long, *To Benghazi*, p. 313; HQRAA War Diary, AWM 52, Operation Order No. 6 of 1 JAN 41. Johns et al., *The History of the 7th Medium Regiment*, p. 12, gives a total of 122 guns.

89   Johns et al., *The History of the 7th Medium Regiment*, p. 11.

90   Long, *To Benghazi*, p. 177; Johns et al., *The History of the 7th Medium Regiment*, pp. 10—13. The 7th Medium's casualties were five KIA and five WIA from bombing of their wagon lines.

91   Long, *To Benghazi*, p. 143.

92   Ibid., p. 193.

93   Ibid., p. 315 and Operation Order No. 6 of 1 JAN 41, AWM WD 2/1st 4/2/1, 2/2nd 4/2/2 and 17th Brigade Operation Order No. 1. For the 17th Brigade's attack, their barrage was in 20 lifts; see Operation Order No. 2 for concentrations on posts (A to F, K to N) engaged by all fire units from Z to Z + 60 at irregular intervals for 3 rgf. A 38 serial harassing fire task table specified four bursts of 3 rgf for the preparatory period against an estimated 96 Italian field guns.

94   Cremor (ed.), *Action Front: The History of the 2/2nd Australian Field Regiment, Royal Australian Artillery, AIF*, pp. 54—58. Major Lewis Dyke was second-in-command and Major R. F. Jaboor, BC 3rd Battery; Adjutant, Lieutenant G. McL. Lee. Dyke retired as a major general after distinguished service to the Regiment. Jaboor was CO 2/2nd Field Regiment and retired with an OBE and ED.

95   Long, *To Benghazi*, p. 198.

96 Ibid.
97 2/1 Field Regiment War Diary, AWM 54 4/2/1.
98 AWM 54 521/1/7, SP. 74 Major General Iven Mackay, Report on Operations, 12 DEC 40 to 7 FEB 41; Long, *To Benghazi*, p. 200.
99 Horner, *The Gunners*, p. 200.
100 Long, *To Benghazi*, p. 205.
101 Hayward, *Six Years in Support*, p. 36.
102 Long, *To Benghazi*, pp. 204–5.
103 Sayers, *Ned Herring*, p. 149.
104 Johns et al., *The History of the 7th Medium Regiment*, p. 14.
105 Whetton, *Z Location or Survey in War*, p. 37.
106 War Diaries, 2/1st, 2/2nd and 2/3rd Field Regiments, AWM 52 4/2/1, 4/2/2 and 4/2/3. Operation Order No. 3 of 20 Jan 41. Rates of fire varied, for example, 0–15 min, 2 rpgpm; 15–25 min, 2 rpgpm and 31–38 min, 3 rpgpm. The 2/1st (for example) was allotted hostile batteries numbers 7, 9, 11 and 12. These were engaged at 3 rpgpm. Phase 2 barrage for the 19th Brigade had similar features.
107 Long, *To Benghazi*, pp. 221–34.
108 Cremor, *Action Front*, pp. 67–69. His training scheme ('bastardry') was immortalised in the regimental news-sheet 'Action Front' of November 1940. During a divisional exercise he arranged for trucks to be 'damaged', telephone lines to be cut, observation post parties captured, wireless jamming, 'prisoners' unsearched etc. to focus the regiment on how the Germans, rather than the Italians, might approach matters. See also Sayers, *Ned Herring*, p. 139.
109 2/2nd Field Regiment War Diary.
110 Interview, Captain P.J.D. Langrishe, RA, 7th Medium Regiment.
111 Long, *To Benghazi*, p. 238; C. Wilmot, *Tobruk 1941*, Penguin, Sydney, 1993, Chapters I to IV; Bidwell, *Gunners at War*, p. 156. Interestingly, Wilmot's casualty figures differ from Long's.
112 Hayward, *Six Years in Support*, p. 56. It may well be asked why he was not with Savige or his brigade commander instead of operating as an OPO. Lieutenant Colonel Barker was slightly wounded in the head at Tobruk as a result of which he sported two eyes 'as black as the Libyan night' by the time he arrived at Derna.
113 Ibid., pp. 45–56.
114 2/2nd Field Regiment War Diary.
115 Bidwell, *Gunners at War*, p. 156.
116 M. Carver, *Dilemmas of the Desert War; The Libyan Campaign, 1940-42*, Spellmount, Staplehurst, 2002. This penetrating analysis focuses primarily on armour/infantry tactics, doctrine and command and control. References to artillery are few. The book is an apologia for the reputation of General Sir Neil Ritchie, the Eighth Army Commander sacked by General Auchinlech in June 1942.
117 Ibid., p. 157.
118 Barnett, *The Desert Generals*, p. 57.
119 Baynes, *The Forgotten Victor*, p. 105.

Chapter 7

120 Maughan, *Tobruk and El Alamein*, Chapters 1 and 2 passim; M. Parsons, *Gunfire! A History of 2/12th Australian Field Regiment*, Globe Press, Cheltenham, 1991, pp. 43—45.
121 J. Moore, *Morshead*, Haldane, Sydney, 1976, pp. 3—9.
122 Maughan, *Tobruk and El Alamein*, pp.11—12.
123 Moore, *Morshead*, pp. 46—50, 60—63.
124 Maughan, *Tobruk and El Alamein*, p. 12.
125 Cremor, *Action Front*, pp. 2—3. Williams, Ramsay's BMRA for El Alamein, had a fine military pedigree. His father, Robert, was a major general (GOC 3rd Base Area, Victoria, during World War I) and Williams was an artillery veteran of that war.
126 See Appendix 7 for a more detailed biographical sketch of Ramsay.
127 D. Goodhart, *The History of the 2/7th Australian Field Regiment, RAA*, Advertiser Printing, Adelaide, 1952, p. 3.
128 Ibid., pp. 17—18.
129 Ibid., p. 115. Development of the 25-pounder between the wars is described in B.P. Hughes, *History of the Royal Regiment of Artillery Between the Wars, 1919—1939*, pp. 178—83.
130 C. and M. Morton, *2/8 Australian Field Regiment Remembers World War II, 1939-1945*, Regimental Association, Melbourne, 1991, passim. Tinsley was a widely known and respected officer who broke his leg in an aircraft accident in 1935 while on liaison duties with the RAAF. This did not stop him performing his famous mess trick (drinking a glass of beer while suspended by his toes from a door) for which he was more widely known. The Adjutant was Captain Ken Mackay, later to rise to Quartermaster General in the ARA.
131 Parsons, *Gunfire!*, pp. 42, 45 and 72.

Chapter 8

132 Maughan, *Tobruk and El Alamein*, Chapters 1, 2 and 3, passim. General Halder, Chief of Operations at OKW, thought Rommel had gone mad!
133 Ibid. See pp. 28—30 re equipment shortages. Morshead had 15 anti-tank and 19 LAA guns and British armour used captured Italian tanks.
134 L.J. Morshead, *Report on Operations in Tobruk, Parts 1-5*, AWM 54 522/7/29, Chapter 3, sections V and VI. See also HQ 9 Div Reports, AWM 54 522/6, File 2 G 269/159/1. The Support Group of an armoured division comprised a field regiment RHA (25-pounders); an anti-tank regiment RHA (2-pounders); a LAA regiment (40-mm Bofors); two lorried (motor) infantry battalions; and engineers and signals squadrons.
135 Maughan, *Tobruk and El Alamein*, p. 49. The 339th Battery took less than three minutes to swing into action.
136 F.L. Johns, J.A.C. Monk and P.J.D. Langrishe, *The History of the 7th Medium Regiment, Royal Artillery, During World War II, 1939-1945*, Severn Press, UK, 1951, pp. 16—17.
137 Moore, *Morshead*, pp. 77, 80—87; Maughan, *Tobruk and El Alamein*, p. 44.
138 Maughan, *Tobruk and El Alamein*, pp. 105—7 provides details of the 2/3rd at Mechili.
139 Ibid., p. 120. Todhunter was later captured near Gambier-Parry's headquarters group.

140 Ibid., pp. 68—69.
141 J. Herington, *The Air War Against Germany and Italy, 1939-43*, Australian War Memorial, Canberra, 1954, pp. 68—75.
142 Maughan, *Tobruk and El Alamein*, p. 142, provides a graphic account of the attack on the 2/17th Battalion.
143 Bidwell, *Gunners at War*, p. 182; D. Graham, *Fire-Power, The British Army and Theories of War 1904-1945*, Pen and Sword, Barnsley, 2004, pp. 235—43.
144 Maughan, *Tobruk and El Alamein*, p. 159.
145 Bidwell, *Gunners at War*, p. 242; Maughan, *Tobruk and El Alamein*, pp. 159—61.
146 Maughan, *Tobruk and El Alamein*, pp. 112—13.
147 Brigadier L.F. Thompson, *Report on Operations at Tobruk*, AWM 54 /522/6/2, Chapter 2, Artillery. Some of these points have been incorporated in Pemberton, *The Development of Artillery Tactics and Equipment*, Chapter 5.
148 Maughan, *Tobruk and El Alamein*, pp. 133—34.
149 Ellis, *World War 2 Data Book*, pp. 136—37, 205.
150 Parsons, *Gunfire! A History of the 2/12 Australian Field Regiment 1940-46*, pp. 43—45.
151 Ellis, *World War 2 Data Book*, p. 205. For more detail on the organisation of the panzer division in 1941, see Heinz Guderian, *Panzer Leader*, Michael Joseph, London, 1977, pp. 518—19, Appendices XXIV, XXV. The structure was altered after the Polish campaign to include more field, anti-aircraft and anti-tank artillery, engineers and motorcycle companies and the organisation remained intact when the division went to Tunis after Flanders.
152 Maughan, *Tobruk and El Alamein*, p. 151.
153 Pemberton, *The Development of Artillery Tactics and Equipment*, pp.12—15.
154 Bean, *Official History of Australia in the War of 1914—1918*, Vol. VI, *The AIF in France*, pp.1072—96; Chester Wilmot, *Tobruk 1941*, Penguin, Sydney, 1993, pp. 89—90.
155 Maughan, *Tobruk and El Alamein*, p. 141. Loder-Symonds, an outstanding individual, had a 'nose' for battle, arriving in a crisis and transforming the situation. He was a CRA at the age of 32 and was killed in Java, of all places, in 1945.
156 Bidwell, *Gunners at War*, p. 151.
157 AWS 7/29; see Wilmot, *Tobruk*, p. 96.
158 Parsons, *Gunfire!*, p. 46. There were those who, irrespective of rank, never became used to it and took shelter in a way that saw them labelled 'cave happy'.
159 Ibid., p. 48.
160 K. Glasson, *Locating the Enemy! Australian Artillery Surveyors at War, The History of 2/1st Australian Survey Regiment, RAA*, AWM monograph, 1993, pp. 50—52.
161 Maughan, *Tobruk and El Alamein*, p. 269. By September the ratio was 209:118.
162 Whetton, *Z Location*, pp. 66—68.
163 Maughan, *Tobruk and El Alamein*, pp. 294—95.
164 Whetton, *Z Location*, p. 71.

165 Parsons, *Gunfire!*, p. 81.
166 Ibid., p. 67; Glasson, *Locating the Enemy!*, p. 51.
167 Whetton, *Z Location*, p. 72; Glasson, *Locating the Enemy!*, p. 57. The comparator is now in the Artillery Museum at Puckapunyal in Victoria.
168 Whetton, *Z Location*, pp. 69–73.
169 G. Morely-Mower, 'Artillery Reconnaissance at Tobruk, 1941', *Aeroplane*, December 2004, pp. 32–36.
170 AWM 54 Item 75/6/3, 'Report on CB Operations, Lieutenant Colonel B. E. Klein', May to August 1941.
171 Glasson, *Locating the Enemy!*, p. 51.
172 Maughan, *Tobruk and El Alamein*, pp. 174–78.
173 Wilmot, *Tobruk 1941*, pp. 242–43.
174 B Pitt, *The Crucible of War, The Year of Alamein*, Cape, London, 1982, p. 334. Jon Latimer, *Alamein*, Murray Bros, London, 2002, p. 321, itemises the units involved as: the *1st* and *2nd Afrika Korps Artillery Regiments*, *221st Artillery Regiment (Motorised)*, *408th Heavy Artillery Regiment (Motorised)*, *364th, 408th, 528th, 533rd* and *902nd Artillery Battalions*. Michael Carver, *Dilemmas of the Desert War-The Libyan Campaign of 1940-1942*, Spellmount, Staplehurst, 2002, gives the strength of the Arko as: 9 x 210-mm guns, 38 x 150-mm and 12 x 105-mm ordnance. No Italian 149-mm weapons are mentioned.
175 Parsons, *Gunfire!*, pp. 66–67. Wilmot, *Tobruk 1941*, p. 248 records that one group of Bush Artillerymen charged '2 piastres a pop' for the doubtful privilege of firing a gun—and were annoyed when told to move on! He also mentions (p. 250) that two captured 88-mm guns were used, along with 60-pounders, to shell Rommel's bypass road and the El Adem aerodrome.
176 Ibid., e-mail to author, 29 December 2003, quoting F Troop records held by F Tutton.
177 Wilmot, *Tobruk 1941*, p. 284. The gunners were shocked to hear that the 144th RHA was overrun and destroyed at Gazala in June 1942.
178 Bidwell, *Gunners at War*, p. 158.
179 Moore, *Morshead*, p. 107.
180 Bidwell, *Gunners at War*, pp. 80–81.
181 Colonel J. N.L. Argent, *Target Tank, The History of the 2/3rd Anti-tank Regiment, RAA*, Regimental Association, Cumberland Newspapers, 1957, p. 118.
182 Maughan, *Tobruk and El Alamein*, Chapter 6 passim.
183 Parsons, *Gunfire!*, p. 78.
184 Wilmot, *Tobruk 1941*, pp. 248, 250.
185 Ibid., p. 76.
186 Carver, *Dilemmas of the Desert War*, p. 237.

Chapter 9

187 The principle of *Schwerpunkt* or 'concentration of force' gave the attacker numerical superiority at the point of the main effort, handing the attacking forces the advantage and tactical and operational superiority even if the attacker was numerically inferior along the entire front.

188 Pemberton, *Development of Artillery Tactics*, p. 88; Carver, *Desert Dilemmas*, p. 24.
189 Maughan, *Tobruk and El Alamein*, p. 354. Rommel took seriously the threat of high velocity gunfire to his tanks, concerned that the British would copy the Germans by employing their 3.7-inch heavy anti-aircraft guns in an anti-tank role. He need not have worried.
190 Bidwell, *Gunners at War*, p. 161.
191 Barnett, *The Desert Generals*, pp. 72—73.
192 Pemberton, *Development of Artillery Tactics*, p. 88.
193 Bidwell, *Gunners at War*, p. 172.
194 Morton, *2/8th Australian Field Regiment Remembers World War II*, p. 30.
195 Maughan, *Tobruk and El Alamein*, pp. 355—57.
196 Goodhart, *The History of the 2/7th Australian Field Regiment*, pp. 80—82.
197 Morton, *2/8th Australian Field Regiment Remembers World War II*, pp. 38—39.
198 Goodhart, *The History of the 2/7th Australian Field Regiment*, pp. 81—82.
199 Ibid., pp. 106—8.
200 Ibid., pp. 108—9.
201 The DAF comprised both RAF, RAAF and USAF wings and groups and there were variations in the DAF order of battle. The Australian component comprised, together with RAF units in direct support for the land battle:

Air HR Egypt; No. 2 Photo Reconnaissance Squadron (RAF), No. 208 Army Cooperation Squadron (Hurricanes). The Hurricane was the sixth aircraft type used for artillery reconnaissance, the others being the Gladiator, Tomahawk, Kittyhawk, Albacore and (later) Mustang.

239 Wing: 3 and 450 Squadrons, RAAF (Kittyhawk)

201 Group: 459 Squadron, RAAF (Detachment of Hudson)

235 Wing: 459 Squadron, RAAF (Hudson), 458 Squadron (Wellington)

245 Wing: 462 Squadron (Halifax)

202 Maughan, *Tobruk and El Alamein*, pp. 360—64.
203 Goodhart, *The History of the 2/7th Australian Field Regiment*, p. 88.
204 Ibid., p. 111.
205 Maughan, *Tobruk and El Alamein*, pp. 467—84. This covers the Crusader operation and the 2/13th Battalion's role at el Duda.
206 Ibid.

Chapter 10

207 Carver, *Dilemmas of the Desert War*, p. 31. ZBV translates as 'For Special Purposes' (Zu Besonderen Verwending).
208 AWM 54 HQRAA 7 Division, Report on Operations, Syria. File S/J3/1413, by Major J.G.N. Wilton, BMRA. For lessons see pp. 6—7.

209 Pemberton, *Development of Artillery Tactics*, pp. 110–11; Bailey, *Field Artillery and Firepower*, p. 308.
210 Maughan, *Tobruk and El Alamein*, p. 429.
211 Pemberton, *Development of Artillery Tactics*, pp. 104, 122.
212 Bidwell, *Gunners at War*, p. 172.
213 Barnett, *The Desert Generals*, p. 121.
214 Ibid., pp. 104–5, 130.
215 See Carver, *Dilemmas of the Desert War*, p. 52, for a discussion on the armour/anti-tank technical and performance characteristics of the ordnance of both protagonists.
216 P.P. Griffith, 'British Armoured Warfare in the Western Desert 1940-43', Chapter 4 in *Armoured Warfare*, Harris and Toase (eds.), pp. 82–106.
217 Johns et al., *The History of the 7th Medium Regiment*, pp. 78–83. The 25th/26th Battery had the doubtful distinction of being the only RA battery that had to destroy its guns TWICE!
218 Bidwell, *Gunners at War*, p. 173. This episode is recorded in the RA Commemoration Book.
219 Barnett, *The Desert Generals*, p. 176.
220 Ibid., pp. 171–72.
221 Carver, *Dilemmas of the Desert War*, pp. 131–32.
222 Barnett, *The Desert Generals*, Part V, Chapter 2. Carver (*Dilemmas of the Desert War*, p. 137) maintains that it is misleading of Barnett to emphasise the importance of Dorman-Smith's appreciation because General Montgomery, when he assumed command, immediately ordered the 44th Infantry Division into the line from Egypt to form a continuous front.
223 Bidwell, *Gunners at War*, pp. 188–89. The term 'Peter Principle' had yet to be coined by the Canadian psychologist who gave his name to it after analysing managerial behaviour and competence in business organisations during the 1950s. Simply stated, he postulated that 'managers [i.e., officers] rise to their level of incompetence'. Within the AIF there was a trite truism that 'a major is a captain promoted for inefficiency'.
224 Pemberton, *Development of Artillery Tactics*, pp. 99, 105.

Chapter 11

225 Fuller, *Decisive Battles of the Western World*, Vol. 3, pp. 484–503. See also Maughan, *Tobruk and El Alamein*, chapters 12–15 passim.
226 Barnett, *The Desert Generals*, pp. 21–36, Appendix p. 298, pp. 304–5.
227 Ibid., pp. 214–36.
228 Lieutenant General Sir Henry Wells became Chief of the General Staff of the Australian Army (1954–8).
229 Morton, *2/8 Australian Field Regiment Remembers World War II, 1939-1945*, p. 20.
230 Maughan, *Tobruk and El Alamein*, p. 533. The record of defensive 'box' employment in the desert was generally poor. Rommel's astute commanders exploited their many weaknesses with relative ease.
231 Parsons, *Gunfire!*, p. 95.

232 C. McKenzie, *Gimme the Guns*, unpublished ms, 1999, p. 14.
233 Morton, *2/8 Australian Field Regiment Remembers World War II, 1939-1945*, p. 243.
234 McKenzie, *Gimme the Guns*, p. 15.
235 Rommel was promoted field marshal by Hitler after the capture of Tobruk in June 1942.
236 Goodhart, *The History of the 2/7 Australian Field Regiment*, p. 151. Morshead, who was trying valiantly to get through the advancing division, called from his staff car, 'That is how battles are lost!' Barnett cites Auchinleck as commenting that the troops he met were in good spirits and looking forward to getting to grips with Rommel again.
237 Maughan, *Tobruk and El Alamein*, pp. 555—62. Lieutenant General Norrie did not have Auchinleck's or Sir Alan Brooke's confidence. He returned to London and joined the planning staff of Operation Overlord, the cross-channel invasion. He fell out with Montgomery and was appointed Governor of South Australia in 1944.
238 P. Caccia Dominioni, *Alamein 1933 — 1962, An Italian Story* (trans. Dennis Chamberlin), Allen & Unwin, London, 1962, pp. 71, 81. There are numerous references to 'enemy 88s' and their effects in the text. These refer to 25-pounders, which boasted a 3.45-inch calibre (87.6 mm). See M. Johnston and P. Stanley, *Alamein—The Australian Story*, Oxford University Press, South Melbourne, 2002, p. 69.
239 Morton, *2/8 Australian Field Regiment Remembers World War II, 1939-1945*, p. 240; McKenzie, *Gimme the Guns*, pp. 19—21.
240 See Maughan, *Tobruk and El Alamein*, p. 563; and Goodhart, *The History of the 2/7 Australian Field Regiment*, pp. 159 —61 for a more detailed, if more light-hearted account. Major Day's nickname was 'Appla' and the area in which the skirmishes were fought passed into regimental history as 'Appla's Acres'. The plan originated in Headquarters XXX Corps—a round trip of 30 miles to Miteiriya (Ruin) Ridge and beyond through hotly contested enemy territory.
241 Pemberton, *The Development of Artillery Tactics and Equipment*, pp. 109—10.
242 Morton, *2/8 Australian Field Regiment Remembers World War II, 1939-1945*, p. 20.
243 Goodhart, *The History of the 2/7 Australian Field Regiment*, pp. 123, 167. Colonel E.C. Mansergh, MC, RA (later GOC 4th Indian Division and Master Gunner at St James Park) was CO and Chief Instructor of the School. Lieutenant Colonel Eastick was CO/CI for a short time followed by Major Millege (then second-in-command of the 2/7th) Chief Instructor in Gunnery. Other complimentary messages were received by the regiment for their technical proficiency and gunnery.
244 Parsons, *Gunfire!*, p. 106.
245 AWM 52/4/1/16, WD HQRAA 9 Div., OO No. 8 and Op. Instruction, and AWM 54/526/4/13 Brigadier Ramsay's Report 3 Aug 42, Appendix 1, Ref. 302. Ammunition expenditure and casualty records cited in the text are from regimental histories unless otherwise noted.
246 Goodhart, *The History of the 2/7 Australian Field Regiment*, pp. 177—80 and Maughan, *Tobruk and El Alamein*, pp. 594—95.
247 Ibid., pp. 593—96. Both regimental histories cover this engagement in detail.
248 Parsons, *Gunfire!*, p. 112. Simulated flashes were relatively easy to distinguish from the genuine article,

although they could cause confusion when used effectively. Reports of enemy shelling (Shelreps) require a measure in seconds of 'flash to bang'.

249 Morshead's biographer, David Coombes, regards this nicety amongst generals as a shallow gesture. Yet Morshead would have been aware of the battering the PAA had received in this period at Auchinleck's hands. Besides, Morshead's stated philosophy was to give credit where due.

250 Maughan, *Tobruk and El Alamein*, p. 610. The other reason offered was that the panzers ran out of fuel before they reached their point to go north-east. Cited by Major E. Baillieu, *Both Sides of the Hill: The Capture of Company 621, A German Intercept Wireless Unit By the Sea near Tel el Eisa, 10 July, 1942*, 2/24th Battalion Association, Victoria, 1987, p. 20. Barton Maughan wrote an introduction to Baillieu's title in 1985.

251 Bernard Montgomery (Field Marshal the Viscount, of El Alamein), *El Alamein to the River Sangro*, PSS, BAOR, 1946, p. 9. Macksey gives the relative artillery strengths as Allied: 892 and PAA/DAK: 552. See Kenneth Macksey, *The Afrika Korps*, Moewig, 1980.

252 Goodhart, *The History of the 2/7 Australian Field Regiment*, p. 168; H. Hemming, *Notes and Opinions of North African Campaign with 4th Survey Regt, RA—Report on Operations*, undated, pp. 1—4; Letter to author, A. de G Benson (7th Medium Regiment, RA); Johns et al., *The History of the 7th Medium Regiment, Royal Artillery During World War II, 1939-1945*, p. 104.

253 Montgomery, *El Alamein to the River Sangro*, p. 196; Maughan, *Tobruk and El Alamein*, p. 595. The main nub of the doctrinal argument was armour- infantry cooperation. One important aspect of this was the control of engineer companies lifting mines on the axis of attack. The engineers were organic to the infantry division/brigade and armoured brigades had none. See comments by MGRA Report (para 3) at WO 201/2877/770/8—RA Notes on the Offensive by Eighth Army from 23 Oct—4 Nov on the El Alamein Position, by Major General A. Maxwell, MGRA, HQ Middle East: Brigadier S. W. Kirkman, BRA Eighth Army (cited as S. Kirkman), CRME/37965/2/RA, dated 14 Dec 42.

254 Maughan, *Tobruk and El Alamein*, p. 579; Barnett, *The Desert Generals*, p. 212; Bidwell, *Gunners at War*, p. 159.

255 Maughan, *Tobruk and El Alamein*, pp. 600—1.

Chapter 12

256 B.H. Liddell Hart (ed.), *The Rommel Papers*, Collins, London, 1953, pp. 199—200.

257 Hugh Skillen, *Spies of the Air Waves*, self-published, Pinner, Middlesex, 1989, p. 147; Army Intelligence Report No. 245 of 13 July 42. For a fascinating account of the 'cat and mouse' signals intelligence war at this time, see chapters 11 and 12.

258 Ibid., pp. 286—87.

259 Ibid., pp. 199—200. For further insights into the 'Int war' in Cairo, see J. Bierman and C. Smith, *Alamein, War Without Hate*, Viking, London, 2002, pp. 195—205. One of the espionage 'players' was Lieutenant Anwar Sadat of the Egyptian Army, later Egyptian President.

260 Baillieu, *Both Sides of the Hill*, pp. 20—21.

261 Skillen, *Spies of the Air Waves*, p. 148. Relations between Rommel and his *Fliegerkorps* commanders were acrimonious, and *Generalmajor* Ramcke delighted in the fact that his paratroops were better fighters than the *Wehrmacht*.
262 Ibid., p. 207.
263 Liddell Hart, *The Rommel Papers*, p. 202.
264 Latimer, *Alamein*, pp. 320–38. Each division had an anti-tank battalion. In his detailed appendixes, Latimer notes that DAK bakery companies appeared on the DAK order of battle!
265 In 1926 Fuller wrote that 'the most important thing on the battlefield was the gun and the gunner' and that siege (i.e. heavy artillery) would be the only artillery survivor in a future war that involved armour.
266 Bruce Gudmundsson, *On Artillery*, Praeger, Westport CT, 1993, p. 118.
267 Ibid., p. 119.
268 Edward. B. Westermann, 'Fighting For the Heavens From the Ground', *Journal of US Military History Society*, Vol. 65, No. 3, p. 653. The ordnance was designed by Krupp at its Essen factory.
269 Ellis, *Data Book of World War II*, pp. 205–209; Ian Hogg, *German Artillery of World War II*, Hippocrene, New York, 1975, pp. 17–87.
270 Guderian, *Panzer Leader*, p. 118. This ordnance was similar in design to the very efficient PAK 38.
271 Liddell Hart, *The Rommel Papers*, p. 209. At Alam Halfa, Rommel complained that the British artillery fired 10 shells for every one of his. There are numerous references to his ammunition supply problems in chapters X to XV.
272 Gudmundsson, *On Artillery*, p. 133.

Chapter 13

273 HQRAA 9 Division War Diary, AWM Series 52, 4/1/16—July to November 1942.
274 Maughan, *Tobruk and El Alamein*, p. 622. The division's 2/3rd Anti-Tank Regiment deployed sixty-four 6-pounder guns. During the battle, other batteries/companies of 16 guns were placed in support. They came under command of the infantry brigade they supported, not HQRAA.
275 See Appendix 2 for the 9th Division order of battle.
276 Because of the high officer turnover, 'Learner' (L) staff officer positions were created for non-Staff Corps officers. These were highly regarded lawyers, salesmen and accountants, men who knew 'the system' and could adapt well and quickly. The system appeared to work well and I never heard of any of these men being 'shunted out' because their performance was inferior to a member of the ASC.
277 Army List and CMA Dept. Record of Service. Williams' father was Major General Robert E. Williams, CMG, VD (journalist and company director), GOC Victoria (3rd Base Area) 1915–1919 (he died in 1943). He forbade Hylton's enlistment in World War I until his 18th birthday.
278 XXX Corps Counter Bombardment Section War Diary and Intelligence Summary, WO 169/4039.770/8. Entry for 27 Oct 42.
279 Goodhart, *The History of the 2/7 Australian Field Regiment*, pp. 200–202.
280 Fuller, *Decisive Battles of the Western World*, Vol. 3, p. 492; Maughan, *Tobruk and El Alamein*, p. 639. By

September-October 2453 tanks and 8700 vehicles had reached the theatre. In the period January to August, the services were reinforced: Army—149,800; Air Force—32,400; Navy—9800 men. From August the equivalent of six field and three medium regiments reinforced the Eighth Army, primarily to make good losses suffered in June.

281 Whetton, *A History of 4th Survey Regiment*, p. 101. Montgomery's tank also had 'Monty' painted in large white letters on the side.

282 Montgomery, *El Alamein to the River Sangro*, p. 15. See also Bierman & Smith, *Alamein—War Without Hate*, pp. 288—89; Latimer, *Alamein*, pp. 154—56.

283 Maughan, *Tobruk and El Alamein*, p. 655; Caccia, *Alamein, 1933-1962*, pp. 211, 233.

284 Parsons, *Gunfire!*, p. 120.

285 Montgomery, *El Alamein to the River Sangro*, p. 17. See also Latimer, *Alamein*, p. 158. One feature of camouflaging guns was the invention of a 'cannibal'. A 25-pounder, its limber and tractor (Quad) would be hidden under covers designed to resemble a lorry. Some 360 guns were hidden in this way.

286 Fuller, *Decisive Battles of the Western World*, Vol. 3, p. 492.

287 Parsons, *Gunfire!*, p. 117.

288 Goodhart, *The History of the 2/7 Australian Field Regiment*, pp. 202—3.

289 Parsons, *Gunfire!*, p. 119.

Chapter 14

290 Farndale, *The Western Front, 1914-1918, History of the Royal Regiment of Artillery*, p. 264, Annex E.

291 R. Walker, *Alam Halfa and El Alamein*, NZ *Official History*, Wellington, 1967, p. 296; N. Orpen, *South African Forces in WW II*, Vol. III, *War in the Desert*, Purnell, 1971, pp. 33, 459. Fred Theron was killed (with Major General Pinnaer) in an aircraft accident in South Africa in 1943.

292 Whetton, *Z Location or Survey in War; A History of the 4th Survey Regiment, RA (TA)*, pp. 68—75. See also Chapter 13.

293 Johns et al., *The History of the 7th Medium Regiment, Royal Artillery During World War II, 1939-1945*, pp. 93—95, Chapter II (Greece and Crete). On 31 May, 25/26 Battery was overrun during the defence of the Ualeb Box (Gazala) as was the 107th Regiment RHA (South Notts Hussars), TA. Elton (ex-RHA) had acted as XXX Corps CCMA on a number of occasions in June and July. He was appointed Commander 5 AGRA when its headquarters was formed later in 1942. He was killed in an aircraft accident at Sollum in December 1942. Brigadier Meade Dennis should have been on the aircraft with him, but missed the flight.

294 Morton, *2/8th Australian Field Regiment Remembers World War II, 1939-1945*, p. 80, for the thoughts of an RMO during a campaign. Douglas had been sent to the Middle East from Milne Bay (Papua New Guinea). He practised in Queensland post-war.

295 Johns et al., *The History of the 7th Medium Regiment, Royal Artillery During World War II, 1939-1945*, p.109 and A. deG. Benson, letter to author.

296 D. Falvey, *A Certain Excellence; The History of the 64th Medium Regiment, RA (TA)*, Brassey's, 2002, pp. 123-29. The guns had an effective full charge (a measure of barrel wear) of 2000 rounds before they

were withdrawn and reconditioned. The 64th's guns exceeded this limit for the second time. Guns were then replaced. The guns also suffered breech and recuperator problems. On 23 October they would fire 45 serials from their program, 43 of which were on counter-bombardment tasks.

297 Latimer, *Alamein*, pp.103, 131.
298 Whetton, *Z Location or Survey in War*, pp. 68—75.
299 PRO WO 201/2877/ 770/8. *RA Notes on the Offensive by Eighth Army from 23 Oct—4 Nov 1942 on the El Alamein Position*, Major General A. Maxwell, MGRA HQME, and attachment by Brigadier S.C. Kirkman, BRA Eighth Army (cited as S. Kirkman), CRME/37965/2/RA, 14 DEC 42, para.5(ii). See also WD 9th Division, loc. cit., Operation Order No.18 for 2/8 para 5 (b) 'under control of CCMA for CB tasks from Z minus 20 to Z minus 5, Z plus 115 to Z plus 175 and after Z plus 185. Its 'Intention was to support 24 Bde and Composite force'. (see para 4) Bungay covers this question in *Alamein* (John Murray, London, 2002) and cites sources such as Montgomery, who describes the employment of 'over 1,000 field and medium guns'. Hamilton (Montgomery's biographer) writes that 'over 800 guns' were used, while Kirkman lists X and XXX Corps as employing '408 field, 48 medium' with XIII Corps utilising some '744 in total' with some in reserve. Playfair, the Official Historian, writes of '456 in the north, 136 in the south', while Bierman and Smith describe '1,000 gns, 474 behind XXX Corps' (p. 276). Latimer cites the Eighth Army order of battle, ref. J (p. 325 et seq), stating that XXX Corps employed 348 field, 48 medium and 320 anti-tank guns, while XIII Corps utilised some 144 field, 16 medium and 172 anti-tank guns. X Corps used 720 field, 64 medium and 684 anti-tank guns. These figures do not include any 'Army reserve'. See Appendix 3 for a comparison of British and German guns.
300 Artillery Training (WO) Vol. I, Pam. No. 2, Counter Bombardment.
301 L. Chrime, *History of 5 AGRA*, privately published, pp. 20—24.
302 XXX Corps CB Log, PRO, Summary.
303 Fuller, *Decisive Battles of the Western World*, Vol. 3, p. 493; G. Richards and H. S. Saunders, *The Fight Avails*, Vol. III, pp. 10—12, Appendix XI. For the RAAF contribution see Herington, *The Air War Against Germany and Italy, 1939—1943*, pp. 368—74. Early on in the battle, Montgomery was sitting in a chair at his headquarters as flights of aircraft headed west to harass the PAA. He remarked, 'There goes the air force. They are winning my battle for me.'
304 XXX Corps CB Log, 22 Oct 42. See Zabecki, *Steel Wind*, p. 111.
305 *4th Survey, Notes and Opinions*, pp. 3—4, Appendix, A Troop Record.
306 Ibid., p. 6.
307 Ibid., Appendix (2 pp). Accuracy given by alphas (accuracy in yards): Z: 25, A: 50, B: 100, D: 200 or 'unconfirmed'.

Chapter 15

308 I am indebted to Colonel T.A. Rodriguez for this 'thumbnail' sketch of Eastick. Rodriguez was Regimental Survey Officer on RHQ for two years and was later on Eastick's staff when he was 9th Division CRA in Borneo and during the island campaigns.

309 'February' is one of the two terms used by soldiers to refer to a disciplinary sentence of '28 Days CB' (confined to barracks, i.e. no leave). February has 28 days in a non-leap year.

310 See Army Lists—various dated editions.

311 Goodhart, *The History of the 2/7 Australian Field Regiment*, p. 215. Captain Nelligan was later killed in New Guinea.

312 See Appendix 4 for the full text of Operation Order No. 19 of 21 October 1942.

313 AWM 54, Operation Order No. 19 of 21 Oct 42. Meteor Telegrams were generated by the Meteorological Section of the 4th Survey Regiment commanded by a RAF meteorologist. The telegrams detailed wind speed and direction, barometric pressure and temperature for 10, 20 and 30-second 'time of flight' i.e., horizontal layers of the air above the battle zone rising to 10,000 feet or more.

314 Dominioni, *Alamein 1933 —1962, An Italian Story*, p. 209. On 23 October 'The aerial recce revealed a slight slackening in the enemy's transport movements. On the previous day the figure had been more than 12,000 counting transport and armour together.'

315 AWM 52 4/1/16 HQRAA War Diary, Sep-Oct 1942.

316 Maughan, *Tobruk and El Alamein*, p. 665.

317 Goodhart, *The History of the 2/7 Australian Field Regiment*, p. 212.

318 Dominioni, *Alamein 1933 —1962, An Italian Story*, pp. 210—11.

Chapter 16

319 Moore, *Morshead*, p.143. Morshead's watch must have been one hour fast.

320 Falvey, *A Certain Excellence*, p. 129. Both 7 and 64 medium regiments were tasked for 45 serials, 43 of which were for counter-bombardment tasks. This amounted to approx. 500 rpg. This figure had been established (at the School of Artillery, Larkhill) as the maximum a full gun detachment could prepare for firing in one day.

321 See also Johns et al., *The History of the 7th Medium Regiment, Royal Artillery During World War II, 1939-1945*; Whetton, *Z Location or Artillery Survey in War, A History of 4th Survey Regiment, RA, TA*; War Diary 4th Survey Regiment, RA (TA), WO 169/4039.770/8 XXX Corps and Intelligence Summary and RAA regimental histories.

322 Goodhart, *We of the Turning Tide*, p. 214.

323 Ibid., p. 212.

324 There are numerous references to this 'map burning' in personal and regimental accounts. The maps were marked with all the locations of the artillery—a 'gift' to the enemy if the map fell into their hands. Similarly, the German airmen used the lights on the towers to work out how far the gun line would be from them, then bombed it.

325 Skillen, *Spies of the Air Waves*, p. 208.

326 Maughan, *Tobruk and El Alamein*, pp. 725—28. For Operation Supercharge, creeping barrage timings were predicated on trials conducted in Syria by Brigadier 'Steve' Weir's New Zealand field regiments. As a result, the early lifts were timed at 75 yards per minute through minefields followed by 100 yards in three minutes beyond it.

327 Ibid., pp. 694—96.

Chapter 17

328 Maughan, *Tobruk and El Alamein*, p. 742.
329 Goodhart, *We of the Turning Tide*, p. 227.
330 Houston reverted to his rank as major when Goodwin returned.
331 Morton, *2/8th Australian Field Regiment Remembers World War II, 1939-1945*, p. 148. A full list of decorations and awards for July/November is included in Appendix 5.
332 Johnston and Stanley, *Alamein—The Australian Story*, p. 289, citing footnote 39, War Diary of the 2/48th Battalion, Notes, AWM 54 8/3/6. These related to Hammer's battalion on Trig 29.
333 Parsons, e-mail to author 29 December 2003 detailing F Troop ammunition expenditure for the campaign. Parsons would later claim that the 2/12th's quantum would surpass 200,000 rounds.
334 Maughan, *Tobruk and El Alamein*, pp. 745—48.
335 Goodhart, *We of the Turning Tide*, p. 227.
336 WO 201/2877/770/8—RA Notes on the Offensive by Eighth Army from 23 Oct—4 Nov on the El Alamein Position, by Brigadier S.W. Kirkman, BRA Eighth Army, CRME/37965/2/RA, dated 14 Dec 42, para 5 ii.
337 Goodhart, *We of the Turning Tide*, Foreword.
338 AWM HQRAA War Diary, Ramsay's Report on Operations. 'Ladder line' was a method of laying line communications by having parallel lines of cable 100—200 yards apart joined about every 400 yards by a connecting line—from above, it looked like a 'ladder'. If shellfire or vehicle tracks cut a line, there was always an alternative circuit for the message to get through.
339 PRO GHQ ME 9209/22/ MT1 0f 30 May 1943. Skillen, *Spies of the Air Waves*, p. 209.
340 Kirkman, WO 201/2877/770/8, CRME/37965/2/RA, dated 14 Dec 42, para 1(v).
341 Pemberton, *The Development of Artillery Tactics and Equipment*, p. 148.
342 Kirkman, WO 201/2877/770/8, CRME/37965/2/RA, dated 14 Dec 42, para 5(iv).
343 Johns et al., *The History of the 7th Medium Regiment, Royal Artillery During World War II, 1939-1945*, p. 112.
344 There are no two reports available that agree on the strength of Axis artillery by numbers, calibre or location. In the case of the 88-mm GP gun, a distinction is not made in its role at the time of the count. The Eighth Army October assessment of PAA artillery amounted to 'about 400 guns'.
345 War Diary of *Panzerarmee Afrika*, 28-7, AWM 52 418/4/1 SP 1265, Enemy Artillery, translated by Air Ministry, April 1952; *15 Panzer Division* Report on El Alamein and Retreat to Marsa el Brega, 23 October to 20 November 1942, Air Ministry Translation, May 1952. Bombing probably accounted for as many guns destroyed or disabled as artillery fire.
346 Kirkman, WO 201/2877/770/8, CRME/37965/2/RA, dated 14 Dec 42, para 5(iv).
347 Johns et al., *The History of the 7th Medium Regiment, Royal Artillery During World War II, 1939-1945*, p. 112.
348 Kirkman, WO 201/2877/770/8, CRME/37965/2/RA, dated 14 Dec 42, para 5(iv).
349 War Diary of *Panzerarmee Afrika*, 28-7, AWM 52 418/4/1 SP 1265.
350 Goodhart, *We of the Turning Tide*, p. 277.
351 War Diary of *Panzerarmee Afrika*, 28-7, AWM 52 418/4/1 SP 1265.
352 Ibid.
353 Ibid.

Chapter 18

354   Griffith, 'British Armoured Warfare in the Western Desert, 1940-1943', pp. 84—86.
355   Ibid.
356   Bailey, *Field Artillery and Firepower*, pp. 308—9.
357   Ibid., p. 318; Griffith, 'British Armoured Warfare in the Western Desert', p. 74. In an attempt to increase the penetrating power of anti-tank shot, a tapered bore weapon made its appearance in the weaponry of the German Army. A simple physics formula demonstrates that taper increases muzzle velocity and hence penetrating power.
358   Indeed, Artillery Training Vol. III, Pamphlet No. 3 authored by Major Kennedy in 1926 had undergone some amendments by 1942, whereas other pamphlets required more continuous amendment due to changes in equipment and ammunition development.
359   Ibid., pp. 81—82. A Panzer division in 1941 comprised a reconnaissance unit, one panzer regiment of two battalions of tanks, a rifle regiment of three battalions of lorried infantry, an artillery regiment of three battalions, and combat support services. The 7th Armoured Division in October 1942 comprised a light armoured brigade of three regiments of light tanks/armoured cars and a field regiment; an armoured brigade of three battalions of tanks, a battalion of lorried infantry and two field regiments; an infantry brigade of three battalions of lorried infantry; a field engineer squadron, two field regiments, and divisional troops including extra anti-aircraft, anti-tank, armoured reconnaissance, signals and all services. Note the much higher ratio of field guns to tanks. Later, the self-propelled field regiment would replace a tractor-drawn field regiment in each of the brigades.
360   Bailey, *Field Artillery and Firepower*, p. 319. For excellent coverage of the desert campaigns, see pp. 292—315.

Chapter 19

361   Maughan, *Tobruk and El Alamein*, p.742. See also Appendix 6 for a full list of gunner casualties.
362   Moore, *Morshead*, p. 149. See Morshead's signal to General Sir Thomas Blamey, GOC Army, after the battle. He was no stranger to the rigours of soldiering.
363   Liddell Hart, *The Rommel Papers*, pp. 330—31. Ironically, Rommel once commented that the best infantry in the world would comprise German officers and New Zealand other ranks.
364   Attributed to Hans-Otto Behrendt, a DAK staff officer, in David Kahn's *The Codebreakers: The Story of Secret Writing*, Scribner, New York, 1967.
365   Johns et al., *The History of the 7th Medium Regiment, Royal Artillery During World War II, 1939-1945*, p. 114. The 7th Medium's Intelligence Officer noted that there was a high correlation between the number of Australians visiting them and the arrival of their ration (beer) truck from Alexandria.
366   Ibid., pp.145—46. Brigadier Yates noted that, as at June 1945, the 7th Medium had fired more ammunition than any other RA unit during the war.
367   Derek Jewell (ed), *Alamein and the Desert War*, London, 1967, p. 93; cited in Nigel Hamilton, *Montgomery, The Making of a General*, Hamish Hamilton, London, 1981.

368 Bungay, *Alamein*. p. 237.
369 Oliver Leese, Memoirs, unpublished manuscript, p. 834. Oliver Leese, later knighted and promoted full general, who took over the Eighth Army following the Battle of the Sangro, paid tribute to Australian troops and Lieutenant General Sir Leslie Morshead in his (unpublished) memoirs.
370 Bierman and Smith, *Alamein—War Without Hate*, pp. 411—12.

Chapter 20

371 See Appendix 7 for details of the post-war lives of the major figures in the Western Desert campaign.
372 N.D. Ashmore, D.R.K. Francis and N.P. Sargent, *Journal of the Royal Artillery*, Vol. CXXX, No. 2, pp. 6—10, 12—15, 18—26.

Appendix 1
Establishment of a Field Regiment, RAA

(AWM 54 327/4/11, II/8/4 14 March 1942)

This table provides an outline of a field regiment in action and/or on manoeuvre. Each vehicle has an identification sign, an alpha and numeral indicating the officer, senior NCO occupying it and/or function (if the vehicle occupant did not carry any rank). This code was as follows:

RHQ Group:

Z:	Commanding Officer, second-in-command
S:	Signals Officer, NCOs and attached battery signals personnel
MC:	motorcycle—orderlies, dispatch riders
Svy:	Survey Officer and parties.
A:	administrative e.g. sergeant clerk, Medical Officer, hygiene dutyman, orderlies

At battery level, nomenclature was:

X:	Battery Commander
R (A—F):	Troop Commander (A—F Troop)
H:	Battery HQ, Command Post Officer, Assistant CPO, Wagon Lines Officer
M:	Battery signals personnel
J:	Battery/Troop Sergeant Major
G (A-F):	Gun Position Officer of (A—F) Troop
K:	Battery Captain and clerks.
TL (A—F):	Troop Leader of (A—F) Troop
A to F1—4:	Troop guns

These vehicles were part of F Echelon (i.e., Fighting) of RHQ Group and three batteries. A and B Echelon were supportive elements of RHQ.

Q 1—5: Quartermaster, RQMS, water dutyman, butcher, cooks, storemen, clerks, driver, vehicle mechanics/motor transport, gun fitter.

At battery level:

Amn 1—3: ammunition lorries

P: vehicle mechanic.

Q 1—4: BQMS, storemen, equipment repairers, cooks and water dutyman

The attached EME LAD vehicles were designated L 1—6. The LAD (14 personnel) provided vehicle mechanics, electricians, fitters, welder, technical clerk, storeman, armourer and a cook. Attached RA Sigs personnel provided radio mechanics and wet-cell (12v) battery charging sets (one per gun battery).

These groupings were flexible depending on the tactical situation, casualties to equipment and personnel, shortages, etc. Regiments seldom had their full establishment allocation of everything.

The vehicle types on establishment were:

- Scout Car/BGC (armoured); 5 cwt Car (4x4)*; 15 cwt (4x4); motorcycle; 3 ton (4x4) in following adaptations: office, load carrying lorry, signals troop GS**, binned for carrying stores; 4 seater car (4x4); 10 cwt and 1 ton trailers.
- The LAD special vehicles were two 3 ton (4x4) store lorries (binned) and 6x6 breakdown/tractor.
- * Jeep, 4 wheel drive, towed 10 cwt trailers was substituted for motorcycles and Austin 5 cwt vans.
- ** Light truck GS = General Service.
- 4 x 4 wheel drive 'Blitz Buggy' (Chevrolet) 3 ton lorry/truck.
- Marmon-Herrington tractors towed guns.
- 25-pounder limber (24 rounds HE and 8 rounds AP shot) for first line ammunition. Three ammunition lorries of each battery carried 480 rounds of HE, or 60 rounds 'first line' per gun.

Note: During the Middle East campaigns a feature of the provision of vehicles from Base Ordnance Depots was one of 'handover'—that is, as one regiment was reassigned to another formation or even theatre of operations, it 'handed in' its vehicles and these were issued to another. As an example, when the 2/12th came out from Tobruk, it was allotted vehicles, almost its complete War Establishment. Their previous owners were the 144th Field Regiment, RA, which had come up from the Abyssinian campaign. In many cases the vehicle identification was suitable, but where changes had to be made, new alphas and numerals were painted on doors. The divisional 'logo' was a white platypus on a black oblong background and was (subject to orders) painted on the mudguard of vehicles and the shield front of guns.

Weapons—Personal/close and AA defence:

Weapon	RHQ	Battery	Total
Pistols	17	10	47
Rifles 303	9	18	63
Carbines	2	2	8
Bren AA LMG	2	2	8
A/Tk Rifle	1	2	7
	31	34	133

A total of 1000 rounds of 303 ammunition was carried for each pair of Bren AA LMGs at RHQ and each battery; ie, 4000 rounds. Only one in five men was allocated a personal weapon.

Wireless Sets:

Four types of wireless sets were used:

WS101 and WS 11 for communications within regiment, battery and Troop, and WS 18 for working with infantry. WS 109 was used by Liaison Officers. These were wet-cell battery-powered superheterodyne single sideband transceivers operating in the Medium Wave band, ie, 5 to 10 Kilocycles (KiloHerz)

Command Group.

Party	Officer or NCO	Type of vehicle	Wireless Net
COs	Commanding Officer, Intelligence Officer	Scout, Armoured Car, BGC	Regimental CRA
2 i/c	Second in Command, Signal Officer, Survey Officer	Jeeps, MCs, 15 cwts	Regimental

RHQ Group

Party	Officer or NCO	Type of vehicle	Wireless Net
	Adjutant, Liaison Officer	As above	Regimental CRA
A and B Echelon, LAD	Quartermaster RSM RQMS RMO	eeps, MCs, 3 Ton Trucks, Signal, Water, Office, Stores, LAD	Regimental
	Number of Personnel: Officers: 9 Snr NCOs: 11 ORs: 54 Total: 74	Number of Vehicles: 24 (Trailers not included)	

Battery Group

Party	Officer or NCO	Type of vehicle	Wireless Net
BCs	Battery Commander	Scout Car, BGC	Bty, Regiment, Infantry/Armr
OP Parties (2)	Troop Commander	As Above	As Above
CPOs Party	CPO, WLO, BSM	MCs, 15 cwt, Jeeps	As Above
Gun Group	Battery Capt. ACPO, TL, BSM	15 cwt, MC, Tractors	As Above
A and B Echelons	BQMS	3 ton GS 4 x4 15 cwt, Water Truck.	
	Number of Personnel: Officers: 10 Snr NCOs: 14 ORs: 179 Total: 203	Number of Vehicles: 38	

EME

Party	Officer or NCO	Type of vehicle	Wireless Net
EME Light Aid Detachment	Number of Personnel: Snr NCO: 1 ORs: 13 Total: 14	Number of Vehicles: 8 and 1 Trailer	

Total strength of a field regiment was: 39 officers, 53 warrant officers and sergeants, 80 bombardiers and 504 other ranks, a total of 683. One regimental history details an additional officer at battery level, ie, Wagon Lines Officer. This gave more flexibility in manning than a Transport Officer on RHQ. On operations additional officers from a Reinforcement Draft were supernumerary to Establishment as 'Learners'. The following NCO and other ranks personnel classifications were used in the establishment:

Trade Group I: Gun Fitter

Trade Group II: CPO's assistant, cooks, mechanics motor transport, observation post assistants, signallers

Trade Group III: surveyors, butcher, clerks, driver mechanics, equipment repairers, gun position officer's assistant, gun layers, tailor, technical storeman

Non-Specialists: AA LMG numbers, ammunition numbers, batmen, drivers mechanical transport, RMO's orderly, motorcycle orderlies, sanitary dutyman, storeman, water dutyman

The six most common trades were drivers (128), signallers (120), gunners (72), gun layers (48), batmen (41) and driver mechanics (25).

Attached Personnel

The establishment provided for a Corporal Shoemaker (Ordnance Corps), Pay Sergeant (Army Pay Corps), Sergeant (Postal Services). An omission from this document is that of Chaplain (Chaplain's Department of A Branch).

Appendix 2
9th Australian Infantry Division Order of Battle, 23 October 1942

Abbreviations:

Cav:	Cavalry	AASC:	Australian Army Service Corps
Fd:	Field	Tpt:	Transport
Lt:	Light	Wkshp:	Workshop
Coy:	Company	Ord:	Ordnance (Corps)
Div:	Division	AGH:	Australian General Hospital
LAD:	Light Aid Detachment	Amb:	Ambulance
Pk:	Park	Hyg:	Hygiene
Trg:	Training	CCS:	Casualty Clearing Station
Bde:	Brigade	Empl:	Employment
Bn:	Battalion	Sigs:	Signals (Corps)
Tps:	Troops	Det:	Detachment

Headquarters: HQ 9 Div

9 Aust Div Intelligence Section
Armoured Corps: 9 Div Cavalry Regt,

82 Aust LAD (EME)

Artillery: HQRAA 9 Div

2/7 Aust Field Regiment, 63 LAD (EME)
2/8 Aust Field Regiment, 64 LAD (EME)
2/12 Aust Field Regiment, 61 LAD (EME)
3rd Aust Anti-Tank Regiment, 71 LAD (EME)
(also known as 2/3rd Anti-Tank Regt)
4th Aust LAA Regiment, incl. RA Sigs and
EME Det (also known as 2/4th LAA Regt)

Engineers: HQ RAE 9 Div

2/3 Aust Fd Coy	2/4 Aust Fd Park Coy
2/7 Aust Fd Coy	9 Aust Div Camouflage Training Unit
2/13 Aust Fd Coy	72 LAD (EME)

Signals: HQ RA Div Signals

Attached Signal Troops to Brigades, Corps HQs, Arms and Services	67 LAD (EME)

Infantry:

HQ 20 Aust Inf Bde: 2/13, 2/15 and 2/17 Bns
58 LAD (EME)
HQ 24 Aust Inf Bde: 2/28, 2/32 and 2/43 Bns
76 LAD (EME)
HQ 26 Aust Inf Bde: 2/23, 2/24 and 2/48 Bns
78 LAD (EME)
2/3 Aust Pioneer Bn

Machine-Guns:

2/2 Aust MG Bn
77 LAD (EME)

Supplies and Transport (Service Corps):

HQ AASC 9 Aust Div	11 Coy AASC
Div Tps Coys	12 Coy AASC
10 Coy AASC	101 Aust Gen Tpt Coy

Medical:

2/3 Field Ambulance	14 Dental Unit
2/8 Field Ambulance	20 Dental Unit
2/11 Field Ambulance	AIF Surgical Team (6 AGH)
2/4 Fd Hyg Sec	AIF Surgical Team (7 AGH)
2/3 CCS	

Ordnance:

2/1 Aust Adv Fd Wkshp
Corps Sec 1 Aust Ord Fd Park

G, H and J Div Sections

Provost:

9 Div Pro Coy

Miscellaneous:

A Coy 2/1 Aust HQ Guard Bn
9 Aust Div Postal Unit
9 Aust Div Empl Pl
9 Aust Div Salvage Unit

9 Aust Fd Cash Office
Mil Hist and Int Section

Appendix 3
El Alamein–Characteristics of German, British and Italian Guns

Artillery nomenclature for German field guns and howitzers was FK (kannon is the German word for gun) and FH (howitzer), K for medium and H for heavy artillery. Shell types were HE, armour piercing (AP) solid shot and hollow charge and white/coloured smoke with impact or time fuses. They all fired separate ammunition (i.e. shell and propellant were united in the chamber) and, with variable charges (propellants), were capable of reaching the gun positions of their opponents.

German Artillery

Identity Weight	Traverse	Elevation	Shell	No of Chgs	Muzzle Velocity and Range			
					Minimum		Maximum	
Calibre cms Kgs	Degrees L to R	Degrees - to +	Wgt Kgs		m/s	m	m/s	m
Infantry Artillery								
IG 18 7.5	12	-10 +75	6	5	92	800	210	3795 400
IG 42 7.5	60	-6 +32	6				4600	590
IG 33 15	11	0 +73	38	6	122	1475	240	5140 2000
Field Artillery								
FK 16 7.5	4	-9 +44	5.83	4	290	5975	662	12300 1524
FH 16 10.5	4	-10 +40	14.80	5	199	3450	395	9225 1450
FH 18 10.5	56		14.80	6	220	3575	470	10675 1985
K 41 10.5	60	-5 +45	15				665	15000 2640
Heavy Artillery								
K 18 10	64	0 +48	15.1	3	550	12785	835	19075 5640
FH 36 15	56	0 +45	43.5	7	210	n/a	485	10230 3280
All Purpose								
F15 '88'	72	-3 +85						

Note: FH16 and FH18 are Howitzers. The '88' was rated at 15rpm

British Artillery

Identity Weight Calibre	Traverse Degrees L to R	Elevation Degrees - to +		Shell Wgt Kgs	No of Chgs	Muzzle Velocity and Range			
						Minimum		Maximum	
						m/s	m	m/s	m
Field Artillery									
25 Pounder QF Gun 8.76	8	-5	+40	11.3	4	200	1350	518	12250	1778
Medium Artillery									
4.5 inch 112mm	60	-5	+45	25	4			685	18745	5842
5.5 inch 140mm	60	-10	+40	45.5	5	550	1480	762	14815	620

Italian Artillery

Identity Calibre cms	Shell Wgt Kgs	Muzzle Velocity and Range	
		m/s	m
Gun 75/27 7.5cm	6.35	502	10240
Gun 149/40 14.9cm	46	800	23700
Gun 149/19 14.9cm	46	600	14250
Gun 210 21.0cm	133	560	15400

As this comparison suggests, there was some standardisation of German calibres and, with the exception of the 88-mm dual purpose gun, only three sizes of ammunition were required for infantry and field artillery, some of which did not suit the Italian ordnance. Given that the supply position of the PAA became increasingly perilous as the campaign continued, small tonnages delivered had to cater for all calibres. The British required four types: 25-pounder, 4.5 and 5.5-inch and 155-mm (for American ordnance).

APPENDIX 4
9TH DIVISION OPERATION ORDER FOR THE BATTLE OF EL ALAMEIN

The historians of the 2/7th Australian Field Regiment regarded this document as the most important in their history. All operations generated Operation Orders from higher headquarters. The necessities of less important events were covered by Operation Instructions.

OPERATION ORDER No. 19 of 21 OCT 42

Ref. Maps 1/50,000 EL ALAMEIN   SECRET   COPY No

EL HIQEIF. GHAZAL. 1/25,000 Tel El EISA AREA Sheets 1,2,3   21 OCT 42

INFORMATION

1. ***Enemy***: See latest I summaries. Enemy defence overprint map.

2. ***Own Tps***: Eighth Army is attacking with the objective of destroying the enemy forces in their present position.

30 Corps will attack with 9 Aust. Div., 51 HD., 2 NZ Div., and 1 SA Div. To secure Point 27 867 298 —Trig 30 870 292—MITIERIYA Ridges to Point 33 877 287.

13 Corps is attacking simultaneously in the SOUTH.

On capture of this bridgehead by 30 Corps, 10 Armoured Corps is to debouch through 30 Corps.

3. ***Outline Plan 9 Aust. Div.—Intention***: Capture and hold area including enemy locations at 8712 2991—Point 23 872 296—8760 2960, Point 27 867 298.

Facilitate passage of 10 Corps through bridgehead.

Exploit.

Method—Attack with 2 Brigades forward—26 Aust. Inf. Bde. On the right, 20

Aust. Inf. Bde. Left—in two phases with one hour pause between.

Phase I—Capture enemy's forward defence to RED LINE—See Trace A.

Phase II—Capture final objective—BLUE LINE—See Trace A.

Tasks—26 Inf. Bde. (less 2/23 Bn.) with under command Composite Force comprising one Sqn. Div. Cav., two troops A/T guns, one coy. MG Bn., one coy Pioneer Bn., and one A/T platoon, is to —

Capture and hold NORTHERN 800 yards of Div. Area. Protect NORTHERN flank from BLUE LINE EAST to present FDLs.

Exploit to Trig 29 868 300.

20 Aust. Inf. Bde. Is to —

Capture and hold remainder of Div. Area. Exploit towards Trig 33 866 296.

24 Aust. Inf. Bde. is to —

Hold present FDLs in Coastal Sector, carry out diversionary op., and maintain one Bn. As Div. Reserve.

2/3 Pioneer Battalion is to mine the NORTHERN flank to include WEST POINT 24 and WEST as far as poss., prior to attack, and to extend the minefield during and after the attack to excluding the enemy FDLs at 8714 2993.

Inter Bde. and Inter Div. boundaries, Inf. Start Line, Inf. Pause Line—See Trace A.

Air: Fighter cover over NORTHERN AREA after first light on morning after attack is launched. Hy. B attacks on enemy gun areas on previous nights and from Zero and on 9 Aust. Div NORTHERN flank after first light.

4. **Intention**: 2/7 Aust. Fd. Regt. will spt. the attack of 9 Aust. Div. and spt. it against counter-attack. On capture of final objective 2/7 Aust. Fd. Regt. will spt. 20 Aust. Inf. Bde.

5. **Method:** CB: The Regt. will come under command CCRA from Z minus 20 to Z minus 5 for CB tasks, and at Zero will revert to command CRA 9 Aust. Div.

6. **Zones:**

(a) Fire: Within the limits of range—from the sea to the southern boundary of 51 HD. Area at Trig 30 870 292.

(b) Obs: WEST of 872 Easting Grid and NORTH of 296 Northing Grid.

(c) A/T: All round.

APPENDIX 4—9TH DIVISION OPERATION ORDER FOR THE BATTLE OF EL ALAMEIN

7. **OPs:** To be established by first light on 24 OCT 42.

(a) 13 Bty. Will supply 1 FOO and 1 OPO to spt. 2/13 Bn.

(b) 14 Bty. Will supply 1 FOO to spt. 2/13 Bn. and 1 OPO to spt. 2/15 Bn.

(c) 57 Bty. Will supply 1 OPO to spt. 2/17 Bn. and 1 OPO to remain at Bde. HQ with Major Johnson until ordered forward.

8. **Bty Areas:** 1st Phase as indicated on ground and 2nd Phase to be recced on 24 OCT 42.

9. **Preparatory:** Gun Pits and CPs will be dug and camouflaged and cable laid during the hours of darkness commencing on night OCT 20-21. Cables will be buried deep under arterial routes. All preliminary moves will be in accordance with the Vehicle Movement Table. Routes will be recced and all ranks impressed with the necessity for strict adherence to the timings and routes indicated. NO movement which will indicate a move of guns forward or of increase in Arty. Spt. will take place in daylight.

10. **WLs:** 1st Phase—As Movement Table already issued.

2nd Phase—To be recced on 24 OCT 42.

11. **Registration:** Nil.

12. **Tasks:** Traces to be issued when available.

2/7 Aust. Fd. Regt. will be prepared to answer calls for defensive fire on HD front. Orders to engage will be given by RHQ.

(b) Sp. will be continued by observed fire at first light.

13. **Meteor:** Meteors will be made available as follows:

Date	Period for which	Time due to Btys Predicted
OCT 23	2140—2330	1930 hrs
OCT 23—24	2330—0630	2300
OCT 24	0630—0800	0600

14. *LOs and FOOs:* Major R L Johnson will be LO at 20 Aust. Inf. Bde. HQ. Btys will arrange movement of FOOs and OPOs with Bns.

15. *Forward Moves:* Sub Units will be prepared to move from their fwd. Area to a gun area SOUTH of TEL EL EISA in vic. Sq. 873 298 as soon as poss. After first light on 24 OCT. No move until ordered by RHQ.

Survey Party will tee in Pivot Guns as soon as poss. After gun posns. are selected.

16. *A/T:* 9 Aust. A/T Bty is in spt. of 20 Aust. Inf. Bde.

17. **Time to be Ready**: 2100 hrs on night of OCT 23 –24.

18. **Time of Zero**: To be notified.

19. **Ord:** Losses and Cas. to Equipment will be reported to RHQ as soon as they occur.

20. **Medical:** Location RAP 8778 2991.

(a) Evac. Of WOUNDED will be through Aid Posts to RAP if necessary.

(b) 2/8 Aust. Fd. Amb.—ADS 8827 2981, MDS 8904 2943;

2/11 Aust. Fd. Amb.—ADS 8802 2980, MDS 8860 2948

Lt. Sec 2/3 Aust. CCS 433 9024.

Surgical teams are at both MDS.

(c) Car Post at X Roads 8805 2975 for all cas. Passing that pt. Cas. From EAST of these X rds to most accessible ADS or MDS.

(d) Med. Vehs are to patrol tracks as far as RAPs for collection of walking wounded.

(e) Units will ensure that, to enable our medical organisatons to concentrate on own tps:

Enemy medical personnel are allowed to continue work.

Enemy medical equip. is NOT touched except by medical personnel.

Enemy medical vehs. Are not used for other than medical purposes.

21. **PW:** Div. PW Cage at 8891 2963.

22 **Intercom:** Tac RHQ will open at 8779 2991 by 2100 hrs. 23 OCT 42.

23. **W/T and R/T:** Normal traffic until 1000 hrs 23 OCT 42. Except for essential op. purposes complete wireless silence from 1000 hrs 22 OCT 42 until Zero.

All frequencies will change at a time to be notified prior to 23 OCT.

24. **Call Signs:** After OCT 23 wireless call signs within Btys will NOT change until ordered by RHQ. Btys. Will continue to use call signs allotted for OCT 23.

25. **Master Frequencies:** Object of the Master Frequency is to provide a link when others have failed. Div. Master Frequency Set is located at Div. Tac. HQ. Frequency 4910 K/cs (See 9 Aust. Div. Sig. Instrn. No.3 attached).

26. **Pigeons:** Are allotted to 20 and 26 Aust. Inf. Bdes. And will be used ONLY in an emergency.

27. **Visual:** A central visual station will be established on Point 26 879 880 for use if nec.

28. **Passing of Information**: Reports from Arty. LOs, FOOs and OPOs of the progress of the battle and enemy dispositions and movements have been of great assistance to the Div. G Staff in previous ops.

The importance of passing this information back QUICKLY will again be stressed. Information to be of value must be accurate, detailed and up-to-date.

29. **Code Words for Objectives:** See Appx. E.

30. **Success Signals:** By. Fw. Bns. On RED LINES and BLUE LINE will be signal rocket yellow repeated twice at one minute intervals. This rocket which shows a large number of yellow STARS will not be used for any other purpose.

31. **SOS:** Will be called for by light signals and through Sigs. Light signal will be signal rocket three star with trailer Mk. I TP which shows three white stars with a very pronounced trail.

32. **Synchronise:** BBC time will be used. Regts. Should check with HQRAA at 2100 hrs on night OCT 23-24 and will pass on to Btys.

33. **Location of HQs:**

Tac HQ RAA 9 Aust. Div. 878 298—time of opening to be notified.

Main HQ RAA 9 Aust. Div. 8883 2943—time to be notified.

2/7 Aust. Fd. Regt. 8779—2991—time to be notified.

2/8 Aust. Fd. Regt. 8803 3003—no alteration.

2/12 Aust. Fd. Regt. 8783 2979—time to be notified.

2 Aust. A/T Regt. 8907 2934—no alteration.

4 Aust. Lt. AA. Regt. 8783 2979—time to be notified.

ACKNOWLEDGE

Time of Signature 2030 hrs A. F. Forwood, Capt.,

Method of Issue SDR Adjt. 2/7 Aust. Fd. Regt.

Appendix 5
Artillery Honours and Awards

6th Division, January—March 1941

HQRAA

Brigadier E. F. Herring	CBE	
Major G. O'Brien	DSO	

2/1 Field Regiment

Lieutenant Colonel L.E.S. Barker	DSO	CO
Lieutenant N. A. Vickery	MC	OPO

Mentioned in Despatches

Lieutenant Colonel L. E. S. Barker	CO
Captain A. G. Hanson	Battery Captain
Captain K. F. Dwyer	OPO
Lieutenant N. A. Vickery	OPO
Lieutenant G. Y. D. Scarlett	OPO
Sergeant A. E. Pearse	Gun Sergeant

Commander in Chief's Card

Bombardier H. L. Morton

2/2 Field Regiment

Lieutenant Colonel W. E. Cremor	OBE	CO

9th Division, July—November 1942

Headquarters RAA 9th Division

Brigadier A. H. Ramsay	DSO	CRA	July—October

Mentioned in Despatches

Sergeant A. C. Allen   July—October

2/7 Australian Field Regiment

Lieutenant Colonel T. C. Eastick	DSO	CO July—October	
Captain W. L. Ligertwood [KIA 24 Oct 42]	MC	OPO (24 Bde, 2/28 Bn)	October
Captain T. A. Rodriguez	MC	OPO (2/17 Bn)	October
Gunner A. Manning	MM	OP Signaller	July
Gunner T. Hill	MM	RHQ Surveyor	July
Lieutenant R. W. J. Menzies	MC	OPO (2/13 Bn)	October
Gunner S. A. Ward	MM	Line Maintenance Sig	October
L/Sergeant C. S. Minson	MM	In the role of OPO	October

Mentioned in Despatches

Lieutenant Colonel T. C. Eastick
Major J. Day
Sergeant R. W. Fowler
Lance Sergeant Bird
Bombardier G. D. C. King

2/8 Australian Field Regiment

Lieutenant Colonel W. N. Tinsley	DSO	Regiment CO	July—October
Major F. D. Stevens	DSO	Battery OC	July
Captain T. Roberts	MC	OPO (2/23 Bn)	July
Lieutenant R. E. Richardson	MC	OPO (2/48 Bn)	July—September
Sergeant I. Hay	MM	Gun/Sgt TSM	July—November
L/Bombardier C. A. Dennis	MM	OP 'Ack'	July
Gunner A. Kinghorn	DCM	OP 'Ack'	July

Mentioned in Despatches

Lieutenant Colonel W. N. Tinsley

Warrant Officer II C. J. Smart
Sergeant I. Hay
Sergeant F. J. McQueen
Bombardier J. L. Carroll

Commander in Chief's Card

Major H. C. Kaiser
Gunner S. T. George
Gunner R. W. Hoskin

2/12 Australian Field Regiment

Lieutenant Colonel S. T. W. Goodwin	DSO	Regiment CO	July–October
T/Lieutenant Colonel G. D. Houston	DSO	Temporary CO	Sept–October
Major A. A. C. Carter	DSO	Bde LO/OC Bty	July - October
Captain P. W. Nelligan	MC	OPO (2/43, 2/28, 2/15 Bn)	July–October
Captain A. B. Cassidy	MC	OPO duties	July
Captain G. Stewart	MC	OPO (2/28 Bn)	July
Lieutenant P. Turner	MC	OPO (2/17 Bn)	October
L/Bombardier G. L. Steele	MM	OP Asst	July
L/Sergeant W. E. Holding	MM	Signals NCO	October
L/Sergeant F. R. Freeman	MM	Troop Signals NCO	October
Gunner J. M. Ryan	MM	F Troop Signaller	October

Mentioned in Despatches

T/Lieutenant Colonel G. D. Houston
Major A. A. C. Carter
Major R. M. Jones
Lance/Sergeant G. H. Brien
Bombardier L. W. Maddison
L/Bombardier A. E. Trengrove
Gunner W. R. Gillespie
Gunner F. E. G. Pearce

Commander in Chief's Card

Major A. A. C. Carter
Gunner W. J. Lloy

Appendix 6
Artillery Casualties in the Western Desert

Casualty figures listed in the various sources for this period vary, doubtless calculated on different bases including those compiled in the interim (1942 to 2002). The summary of Axis losses provided by Johnston and Stanley gives a total of 30,539:

	German	Italian
Killed in Action	1149	971
Wounded in Action	3886	933
POW/Missing	8050	15,552
Total	13,085	17,456

Bierman and Smith supply a POW/missing total of 30,000 (from 100,000 in the field) with 20,000 KIA and WIA, conceding that the official number was lower. Their breakdown of XXX Corps casualties from all causes is:

9th Australian Division	2827*
51st (H) Division	2495
2nd New Zealand Division	2388
1st South African Division	922
Total	8632

Maughan gives the total casualties for the 9th Division as 2694 comprising 620 KIA and 1944 WIA with 130 POW.

---

* Maughan, *Tobruk and El Alamein*, p. 294.

6th Division Artillery

2/1st Field Regiment

Killed in Action/Died of Wounds		Wounded in Action	
NX 8138	Bdr G A Thomas	VX 12777	Lt Col. L E S Barker
NX 5547	Gnr W K Smith	NX 3354	Lieut E V Haywood
NX 3413	Gnr J H O'Sullivan	NX 8717	Sgt A E Pearse
		NX 3385	Sgt N Freeman
		NX 803	Sgt J H Greenwood
		NX 3123	Gnr C Duggan
		NX 708	Gnr F Ashton
		NX 785	Gnr C W Smith
		DX 162	Gnr D J Bullen
		NX 8118	Gnr M S Krumbeck
		NX 3348	Gnr E Mooney
		NX 7105	Gnr J T Ryan
		NX 147	Capt F Dwyer
		NX 830	Bdr B R Vidal

2/2nd Field Regiment

Killed in Action/Died of Wounds		Wounded in Action	
VX 94	Lieut C R Nethercote	VX 83	Major A E Arthur
		NX 3315	Lieut J M Crawford
		VX 4884	Bdr P Russell

2/3rd Field Regiment

Killed in Action/Died of Wounds		Wounded in Action	
		SX 386	Gnr A E Newell
		WX 66	Gnr W A Vinden

9th Division Artillery

2/7th Field Regiment

Killed in Action/Died of Wounds		Wounded in Action	
SX 3265	Gnr E J Cook	SX 3131	Sgt K C Batty
WX 4347	Gnr N Drummond	WX 2697	Gnr J N McCashney

2/7th Field Regiment (continued)

Killed in Action/Died of Wounds

WX 3537	L/Sgt G W Phillips
WX 3046	Gnr A Oldfield
WX 3130	Gnr W H Cunnold
SX 3373	Gnr R F Pilgrim
VX 51925	Lt N F Smith
WX 12094	Gnr S R Wall
SX 4846	Gnr W Souter
SX 3377	Sgt M R Brookman
SX 3819	L/Bdr C C Scull
SX 3614	Gnr K W Mahoney
SX 5271	Capt A W Fielding
SX 7743	Gnr C H Gentle
WX 3675	Gnr S C Newham
WX 11825	Gnr T C Newham
SX 10323	Gnr J L O'Shea
WX 1583	Capt W L Ligertwood
SX 9787	Gnr E Edwards
WX 6459	Gnr F A Butterick
WX 2863	L/Sgt E J Skeels
SX 3597	Sgt J A Gilbert
SX 3555	Gnr G H Howard
SX 2996	L/Bdr J S Poyntz
SX 9638	Gr D H J Watson
SX 4454	Gnr T F Wilson
SX 12820	Gnr J J Dadds
Sx3383	A/Bdr R H Pope

Wounded in Action

WX3117	L/Bdr H C Parker
WX 3517	Sgt V Burghall
WX 4382	Gnr F Collins
X 4687	Gnr H M Baker
WX 12101	Gnr R B Hannar
SX 8296	Gnr H D Dawson
SX 3558	Bdr W D Monkhouse
SX 12383	Gnr H H Sneath
SX 11597	Gnr W M Hoskin
WX 5734	Gnr G Weller#
WX 2701	Sgt R H Hazell
SX 4849	Gnr W S Crawford
WX 3654	Gnr P J Bradley
WX 4700	Gnr J Gregory
WX 12041	Gnr D J Ingram
WX 4631	L/Sgt P StJ Kennedy
WX 2823	Gnr H Farmaner
VX 51688	Lt S W Mummery
QX 2803	Gnr G L G Hooper
SX 11356	Gnr H Delbridge
SX 8404	Gnr G Huntingdon
SX 8270	Gnr A R Murray
WX 12873	Gnr J E Alexander
WX 4659	Sgt C D C King
WX 3949	Sgt A D Holder
WX 4755	Gnr A P Fisher
WX 3935	Sgt A L Trott
WX 3930	Gnr R A Gordon
SX 9754	Gnr T A Lewis
SX 4850	Gnr K W Parr
SX 9093	Gnr A B Feuerheerdt

2/7th Field Regiment (continued)

Killed in Action/Died of Wounds		Wounded in Action	
		WX	Capt K W Everett
		WX 3062	Gnr W J O'Brien
		WX 2972	Gnr L Mitchell
		NX 13970	Bdr R W M Baker
		NX 42537	Gnr E M Miller
		SX 3381	A/Sgt T E Chapman
		SX 11423	L/Bdr W R Horsell
		WX 3688	Gnr D E Luck
		SX 3393	Gnr S P R Hollitt
		VX 17477	Gnr W H Lindsay
		WX 6380	Gnr S W Woods
		WX 11533	Gnr H J Schnaars
		WX 3285	Gnr W R Millen

2/8th Field Regiment

Killed in Action/Died of Wounds		Wounded in Action	
VX 17245	Gnr R P Davey	SX 12685	Sgt A R Hay
VX 14716	Gnr J M Ryan	VX 14732	Gnr J R Knight
VX 37041	Gnr C P Browne	VX 14729	Gnr M R Blair
VX 53797	Gnr S Major	TX 1449	Gnr M A A Lord
VX 14872	Bdr D L Griffiths	TX 5511	Gnr W D Cohen
TX 3540	L/Sgt A T R Wilson	VX 18196	Sgt C Hill
TX 4469	Gnr L G Caville	VX 14382	Sgt R W Hunter
VX 18576	Gnr R Mountjoy	TX 525	Lt J L Steer
VX 16936	L/Sgt W K Sheenan	VX 17212	Gnr J C O'Halloran
TX 2588	L/Bdr P K Rogers	TX 1670	Lt J G Cuff
VX 37089	Gnr W B Fiddler	TX 1650	Gnr A D Johnson
TX 2753	Gnr T J Crisp	TX 1523	Gnr Johnson H D
TX 1960	Gnr R A Norris	TX 1527	Gnr D W Smith
VX 30276	Gnr W A L Gunton	TX 295	Sgt W Lingerwood
TX 5396	Gnr A C Jackson*	TX 2511	Gnr J R Jones
VX 37089	Gnr W B Fidden*	VX 17413	Gnr J V Robinson

2/8th Field Regiment (continued)

Killed in Action/Died of Wounds

TX 2588	L/Bdr P K Rogers*
VX 16936	L/Sgt W E Sheehan*

* Died on Active Service

Wounded in Action

TX 3129	Gnr H D Fysh
SX 12316	Gnr L A Coffey
VX 15969	Bdr L J Currell
VX 14787	Sgt E J Hussey
VX 63222	Gnr G W McKay
VX 40445	Gnr I Milner
DX 4	Lt T D Smith
VX 14128	Capt J S Elder
VX 14371	Lt C R Morton
VX 14433	Bdr F F Fairthorne
VX 17241	Gnr A R Hills
TX 1461	Sgt R W McKercher
VX 30051	Gnr T A Forster
VX 62335	Gnr K Hutchinson
VX 17239	Gnr A C Stephen
VX 14388	Sgt T H Wallworth
VX 15048	Gnr L W Hanna
VX 15629	Gnr A B Turner
TX 1765	Gnr N R Knight
TX 1640	Gnr J N W Nicholson
TX 4523	Gnr G T Buckney
TX 1940	Gnr W H Fulton
TX 2254	Gnr G T Gabriel
TX 1443	Gnr H G Glatte
TX 3131	Gnr H J Lees
TX 2372	Gnr J A Prouse
VX 15972	Sgt J M Hayward
VX 15963	Gnr A L Sharp
VX 51805	Gnr S M Wragge
VX 62311	Gnr J J Reynolds
TX 3007	Gnr L V C Dixon
VX 56944	Gnr R E Talbot

2/8th Field Regiment (continued)

Killed in Action/Died of Wounds	Wounded in Action
	VX 14429 Lt F W Rowlands
	TX 1949 Gnr V T Wilcox
	VX 17278 Gnr A T Saunders
	TX 2205 Bdr L J Harrison
	TX 1934 Gnr P R Upcher
	TX 1523 Gnr H I Johnson
	VX 50491 Gnr J L Nankervis
	TX 3157 Bdr L H Barnard
	TX 1539 Gnr R T Ware
	TX 2063 Capt T L Roberts
	VX 15964 Gnr E C Allen
	VX 31426 Gnr R R Stammers
	TX 2636 Gnr C E Drysdale
	TX 1438 Lt F E L Adlard
	VX 31970 Gnr F J G Gleeson
	VX 36956 Gnr W J Barry
	TX 5219 Gnr A H Byrne
	TX 4704 Gnr F J Conway
	VX 16270 Gnr S G Moore
	TX 5266 Gnr J R Bishop
	TX 1684 Sgt C S Craw
	TX 5304 Gnr K G Horton
	VX 15629 Gnr A B Turner
	TX 2196 Lt R E Richardson
	TX 3040 Gnr H C Garwood
	TX 315 Gnr M E Heathcote
	TX 5396 Gnr A C Jackson
	TX 3282 Gnr R A Little
	TX 5428 WO 2 C H Clarke
	VX 15470 Gnr D J McCann
	VX 50347 Lt J R Dench
	VX 51378 Bdr G H Palmer

2/8th Field Regiment (continued)

Killed in Action/Died of Wounds	Wounded in Action	
	TX 1725	Gnr R A Roberts
	VX 41704	Sgt L A Mason
	NX 35832	Gnr L G Helyer
	TX 2797	Gnr L G Bond
	TX 1455	Sgt B B Richardson

2/12th Field Regiment
(Tobruk campaign)

Killed in Action/Died of Wounds		Wounded in Action	
VX 33108	L/Bdr B McD Butler	VX 19992	A Main
VX 17492	L/Bdr J R Kenna	VX 15678	A E Stigant
VX 22499	L/Bdr RA Lewin	VX 30901	D A Ambrose
VX 15696	Gnr A E Cowell	VX 22256	L R Lillingstone
VX 18085	Bdr R A Shmith	VX 17925	D J Ryan
VX 17601	Gnr S Ogilvie	VX 23516	K V Woodstock
VX 17784	Gnr E Harvey	VX 18779	C Smith
VX 27367	Gnr C J Wright	VX 32979	R J Shovelton
VX 17734	L/Sgt J E Downie	VX 19434	W A Morrow
VX 19398	Gnr H P Doyle	VX 18445	R J Jones
VX 18539	Gnr A Gilchrist	VX 18772	R T Smith
VX 20984	Sgt B V Nash	VX 25123	P D Price
VX 25885	Gnr W S Wicks	VX 20012	A J Barker
VX17603	Gnr H T Cooke	VX 22194	R Lewis
VX 22700	Gnr W Rigby	VX 15654	L A Matthews
VX 14930	Gnr K P Connolly	VX 17743	W H Parker
VX 22181	Gnr H S Gregory	VX 28061	H F Kerry
VX 18020	Gnr J Mackay	VX 19535	G G Harders
VX 15684	Bdr C J Lackmann	VX 18145	A G McRobert
VX 20768	Gnr L B Olney	VX 17390	E G F Francis
VX 19455	Gnr H C Orr	VX 21203	W Basset
VX 23488	Gnr H G Peacock	VX 15688	E J Curran

2/12th Field Regiment
Tobruk campaign

Killed in Action/Died of Wounds

VX 38253	Gnr B C Pilgrim
VX 22133	Gnr R J Saunders
VX 15791	WO2 R D Williams

Wounded in Action

VX 46821	L V Frank
VX 14446	J F Melville
VX 19480	H W Nugent
VX 24029	C M Cheeseman
VX 13910	H S A Kennedy
VX 21858	E A Starkey
VX 18886	R C Ward
VX 17508	H J Billing
VX 39720	C D Mowbray
VX 34065	A E Searle
VX 22910	H B Quinn
VX 23718	E L Matthews
VX 22198	W P Greville
VX 8227	L N Field
VX 14860	A G Clarkson
VX 19474	J M Stephen
VX 19427	R Lyle
VX 13914	W Clark

First and Second Alamein

Killed in Action/Died of Wounds

VX 21741	R E Ware
VX 22820	F E G Pearce
VX 16845	G H Brien
VX 37089	W J Fiddler
VX 50558	B P Kelly
VX 18166	C R Rumberg
TX 4317	A G Hayes
VX 18660	A P Chuter
VX 19927	E K Main
VX 21994	E W Cutler
VX 24346	V Monro

Wounded in Action

VX 19723	L Damm
VX 35217	M G Delecca
VX 52748	H J Adams
VX 18782	A M Payne
VX 13685	J K Tutton
VX 50942	W A Jordan
VX 21749	A E Warren
VX 51356	K D Fleming
VX 4858	D Reid
VX 37078	F D Wood
VX 20042	E Challenor

First and Second Alamein (continued)

Killed in Action/Died of Wounds

VX 13679	R G Trenwith
VX 27106	R T Crozier
VX 13049	C A L McKenzie
VX 28288	W J Teirney *

Wounded in Action

VX 19699	A D Connley
VX 14225	A B Cassidy
VX 18359	E C Norton
VX 14451	J C Phillps
VX 30493	D G E Eyre
VX 31042	J G Bailey
VX 19868	E O Billing
VX 20295	G R Burkhill
VX 20812	J P Kimpton
VX 20002	H J Maltman
VX 21918	T A Taylor
VX 16144	F B Morgan-Taylor
VX 52931	D F Daly
VX 47824	L R Hale
VX 59772	K Fisher
VX 13094	V C Harrison
VX 51105	W J Meloury
VX 33873	A J Seater
VX 15887	J Trounson
VX 46616	W C B McKinnon
VX 32528	E W McVeigh
VX 17750	R C McG Fisher
VX 22259	J F Muller
VX 21533	W J Martin
VX 15978	F C Gibbs
VX 22332	R J Harris
VX 19731	H G Laverty
VX 17659	R T Luke
VX 23426	V E A Nairn
VX 14637	W L Ross
VX 19474	J M Stephen
VX 14859	F R Whinfield

First and Second Alamein (continued)

Killed in Action/Died of Wounds	Wounded in Action	
	VX 54153	B F Caelli
	VX 55135	R K Sewell
	VX 2899	J W Calder
	VX 55343	H W Lloyd
	VX 6011	J P Moroney
	NX 44319	C Olsen
	VX 21736	R A E Sewell
	VX 27266	G G Thurley
	VX 33993	H Fawcett
	VX 57549	W J Lloyd
	VX 51027	J H North
	VX 13696	A A C Carter
	VX 23221	A M Cummings
	VX27369	C C Osborne
	VX 13913	D Johnson
	VX 58142	K F Lonne
	VX50392	T S Jenkins

Appendix 7
The Post–War Lives of the Major Figures in the Western Desert

Long after the end of World War II, the major players directing events in these pages continued to contribute their expertise and energy to worthy causes and the national interest. General Sir Thomas Blamey was retired by an ungrateful government in 1945. The subject of two biographies and countless articles, he continued to arouse strong passions of both polarities until his death. He chose his generals with a deft touch and, through his recognition of military skills and leadership in subordinates, he never chose a person who did not win a battle given an even chance. He was appointed field marshal in June 1950 and died in Melbourne on 27 May 1951.

Lieutenant General Sir Iven Mackay returned to Australia following the campaigns in Greece and Crete and was appointed Commander-in-Chief Home Forces, then New Guinea Force. He became Australia's first High Commissioner to India from 1944 to 1948.

Lieutenant General Sir Edmund Herring, KCMG, KBE, DSO, MC, ED, succeeded Mackay as GOC 6th Division and returned to Australia to become GOC Northern Territory Force before assuming command of New Guinea Force and I Corps in September 1942. He retained Blamey's confidence despite having intrigued to have him removed as Commander-in-Chief. Herring was credited with several successful campaigns in Papua New Guinea. In 1944 he was appointed Chief Justice of Victoria and died on 5 January 1982.

Brigadier Lewis Barker, CBE, DSO, MC, became Director of Artillery, CCRA I Corps, BRA First Army, New Guinea Force and LHQ. He died in 1981. Brigadier Bill Cremor, CBE, ED, was CRA 3rd and 6th divisions, and CCRA I Corps and later II Corps and died in 1962. Colonel Athol Hobbs, ED, was appointed Liaison Officer in South Africa before retiring. He died in 1979. Brigadier Horace Strutt, DSO, ED, a businessman from Hobart, was appointed acting CRA, NZ Division in Crete, CRA 6th Division, BRA Northern

Territory Force, and later became Speaker in the Tasmanian House of Assembly.

Lieutenant General Sir Leslie Morshead, KCB, CMG, KBE, ED, was knighted in November 1942. He was appointed GOC II Australian Corps, GOC New Guinea Force and Second Army, and GOC I Australian Corps for the complex amphibious OBOE operations to recapture Japanese-occupied territories in Borneo. Following the war, as a senior executive of the P&O Shipping Company in Sydney, he headed the Morshead Report, a 1957 government review of the grouping of Defence-related departments which the government proved ultimately loath to progress. Morshead died on 26 September 1959 and was accorded the largest military funeral ever seen in Sydney. Gavin Long, Editor-in-Chief of the World War II *Official Histories*, wrote that he was 'the most trusted of the senior [Australian] commanders'. He was his division, and his division was him.

Major General Charles Lloyd, known to the artillery fraternity by the sobriquet 'Gaffer', went from Tobruk to Wavell's ABDA staff before returning to Australia in 1942. Appointed Director of Staff Duties the following year, he then became Adjutant General, the youngest ever major general, for the next two years. He resigned from the Army in 1946 and went into business (newspapers) before joining the United Nations Reconstruction Agency as Chief of Mission in Korea.

Lieutenant General Sir Henry Wells, KBE, CB, DSO, was Brigadier General Staff HQ 2, then I Australian Corps and HQ Australian Military Forces from 1943 to 1946. After service as a senior officer on AHQ, he became Commandant of the Royal Military College of Australia, GOC Southern Command and Commander-in-Chief British Commonwealth Force, Korea, from 1953 to 1954. He was knighted during his appointment as Chief of the General Staff from 1954 to 1958. Wells presided over the inception of the National Service Trainee Scheme and a vastly increased CMF and ARA Training Cadre in that time.

Major General Sir Alan Ramsay, CB, CBE, DSO, MSM, ED, returned to Australia and was appointed CCRA II Australian Corps. He gained General Sir Thomas Blamey's confidence and was appointed GOC 5th and later 11th Infantry Division from 1944 to 1945, directing operations on the north coast of New Guinea. He returned to his profession as Principal of Melbourne High School in February 1946 and, two years later, became Director General of Education for Victoria until 1960. He was a Trustee of the Shrine of Remembrance, active in the RSL and his community. He died in September 1973.

Brigadier Sir Thomas Eastick, CMG, DSO, ED, was appointed CRA 7th Division in 1943, CRA 9th Division in 1944 for its final campaigns, and commander of Kuching Force at the war's end. He soldiered on in the CMF and, from 1950 to 1953, was Commander HQ

Group Central Command. He was State President of the RSL for 15 years, a member of the Australia Day Council for 25 years and a leader of a movement 'Call to the Nation'. The citation for his knighthood simply stated 'For Service'. He died in 1985 aged 85.

Brigadier Walter Tinsley, DSO, was appointed CRA 9th Division in 1943, Commander Fremantle Fortress from 1944 to 1945, and CRA 5th and 11th divisions. He died in Melbourne in 1969 aged 71. Brigadier Shirley Goodwin, DSO, returned to the appointment of CRA 9th Division and was killed in action at Scarlet Beach, Finschhafen, on 25 October 1943 when Japanese aircraft bombed HQRAA.

Of the division's regimental historians, David Goodhart (2/7th) wrote *We of the Turning Tide* in 1947 and was Editor and a member of the Editorial Board (with Brigadier Eastick and Major R. H. Rungie) for the 2/7th's official history, published in 1952. A journalist who married but had no family, Goodhart appears to have been a somewhat enigmatic, if humorous and laconic figure. He died many years ago. Subsequently, James A. Quilliam, who edited the regimental newsheet 'Dial Sight' for many years, produced 'Dial Sight Revisited—1967 to 1997' in 1998.

Charles Morton, a former long-serving officer of the 2/8th, compiled an anecdotal historical record of the 2/8th for the 50th anniversary of El Alamein. It is in a form quite unique for regimental histories, but nonetheless useful for future historians. In 1991, Max Parsons (2/12th) wrote *Gunfire!*, a very comprehensive and meticulously compiled record. He also produced a pictorial record, *Take Post—2/12th Field Regiment 1940-46*. Max served his association for many years after a successful career in retail and has been involved in the editing and/or production of a number of military histories since his retirement. There is no doubt that, had Alan Ramsay been alive when the last two regimental histories were completed, he would have been fulsome in his praise for their achievements.

Other survivors of this extraordinary battle are few and far between. Colonel 'Tim' Rodriguez, MBE, MVO, MC, became BM (L) 7th Division for the Nadzab/Ramu Valley campaign, BMRA 9th Division under Eastick and was awarded the MBE. He soldiered on in the ARA and commanded 'A' Battery in Japan (BCOF) and later the 1st Field Regiment. He trained in the UK in ammunition proofing and finished his career as Comptroller General at Yarralumla to four Governors-General, of whom Field Marshal the Viscount Slim of Burma was the first. He retired to Sydney and, at 94, marched the distance on Anzac Day 2003. Lieutenant Colonel Roger Fitzhardinge, a contemporary of Sir Roden Cutler VC from Manly, became Senior Instructor in Gunnery at Northam after the Western Desert and, later, SORA 2 Fremantle Fortress. He rejoined the 3rd Field Brigade/Regiment (CMF) for the third time in 1952 and rose to its command in 1958. Captain Ken Everett transferred to the Amphibious Bombardment Branch and returned to a career with the Commonwealth Bank. Captain Bill Stevens remained with HQRAA 9 Division and was standing next to Brigadier Goodwin when he was killed. During a tour of Europe in the 1970s, he met a German DAK Artillery officer who was at El Alamein and who was playing bridge in his OP on 23 October. The German was about to bid 'Two No Trumps' when the battle opened!

Glossary

A

AA	anti-aircraft; assembly area
AATTV	Australian Army Training Team Vietnam
'Ack'	assistant
ACPO	Assistant Command Post Officer
ADC	aide-de-camp
Adjt	Adjutant — Headquarters Staff Officer, a captain
AFA	Australian Field Artillery (Militia)
AFC	Australian Flying Corps
AFV	armoured fighting vehicle
AGRA	Army Group, Royal Artillery
AHQ	Army Headquarters (Australian)
AIC	Australian Instructional Corps —Permanent Army, NCO cadre staff
AIF	Australian Imperial Force
AMF	Australian Military Forces
AP	armour piercing
APIS	Air Photo Interpretation Section
ARA	Australian Regular Army
Arty/R	artillery (airborne) reconnaissance
AWM	Australian War Memorial (Research Centre files)
A/T	anti-tank (artillery)

B

'balls of fire'	a method of counter-bombardment artillery fire
BC	Battery Commander, usually a major
BEF	British Expeditionary Force
BGC	Bren Gun Carrier, a lightly armoured tracked vehicle (used by Observation Post Officers)
BHQ	battery headquarters
BK	battery captain
BM	Brigade Major, Grade 2 Staff Officer on a brigade (infantry) or divisional (artillery) headquarters

BRA	Brigadier, Royal Artillery—adviser to the Army Commander		
Bren	a light machine-gun for local and close air defence		
BSM	Battery Sergeant Major, senior non-commissioned officer in a battery		

C

CAGRA	Commander Army Group, Royal Artillery
CB	counter-bombardment — of enemy guns and mortars
CBO	Counter-bombardment Officer
CBSO	Counter-bombardment Staff Office
CCBO	Corps Counter-bombardment Office(r)
CCMA	Corps Commander Medium Artillery
CCRA	Corps Commander Royal Artillery
CIGS	Chief of the Imperial General Staff (British)
CMF	Citizen Military Forces (Australian)
CO	Commanding Officer
CP	command post
CPO	command post officer
CRA	Commander Royal Artillery— divisional artillery commander
cwt	hundredweight

D

DAF	Desert Air Force—RAF, RAAF and USAF squadrons, wings and groups
DAK	*Deutsches Afrika Korps*— German and Italian formations under command of General Erwin Rommel
DAS	died on active service
DCM	Distinguished Conduct Medal
DF	defensive fire
DF (SOS)	defensive fire tasks on the most likely enemy approaches to forward infantry localities
direct fire	fire aimed directly by the gun at a visible target
'doover'	temporary earthwork for concealment or protection — a personal slit trench
DOW	died of wounds
DRA	Director Regimental Artillery
DSO	Distinguished Service Order

E

EME	Electrical and Mechanical Engineers, Corps of
FARELF	Far East Land Forces
FDL	forward defended locality, usually a cluster of infantry platoon/company/battalion areas closest to the enemy lines.

F

FFE	fire for effect: after ranging, applying fire to the target to achieve the desired result
fixation	fixing by grid reference a point on the ground
FOO	forward observation officer
FS	flash spotting — locating enemy guns by taking cross bearings from two or more observers from accurately surveyed posts
FSR	*Field Service Regulations* — the gunners' 'bible'
FUP	forming-up position — an area where infantry deployed in their battle formation prior to moving to the start line

G

GAP	gun aiming post
GHQ	General Headquarters
GOC	General Officer Commanding
GOCME	General Officer Commanding the Middle East
GP	general purpose
GPO	Gun Position Officer
GR	grid reference, a point on a map identified by an easting and northing measurement from a point of origin
GSO	General Staff Officer, either Grade 1, lieutenant colonel (GSO1); 2, major (GSO2); or 3, captain (GSO3)

H

Hawkins Mine	an anti-tank mine weighing 15 pounds laid on the ground to form a temporary minefield
HB	hostile battery — enemy guns and mortars
HE	high explosive
HQME	Headquarters Middle East
HQRA	Headquarters Royal Artillery
HQRAA	Headquarters Royal Australian Artillery

I

indirect fire	guns firing at a target which is not visible to the gunners. Fire at the target is directed by an observer using a standard procedure

K

KIA	killed in action
KIAA	killed in aircraft accident

L

L	'Learner' — a position 'twinned' with a staff officer to understudy his duties and to provide relief during extensive operations
LAA	light anti-aircraft
LAD	Light Aid Detachment (EME)
LO	Liaison Officer

LOB	left out of battle — a reserve group of experienced officers and other ranks to replace serious casualties and/or help reform a unit		corporal, sergeant (including lance ranks), Warrant Officer Class 1 or 2
		NZA	New Zealand Artillery

M

MBE	Member (of the) British Empire, a decoration
MC	Military Cross: a decoration for gallantry awarded to officers
Meteor Telegram	message giving current meteorological conditions of the air through which the shell passes for a specific time period
MG	machine-gun, H: heavy; L: light
MGO	Master General of the Ordnance
MGRA	Major General Royal Artillery
MID	Mentioned in Despatches
Militia	volunteer (part-time) army personnel who staffed units between World War I and II
MM	Military Medal: a decoration for gallantry awarded to other ranks
MO	Medical Officer
mph	miles per hour
MSM	Meritorious Service Medal
'Murder'	a method of counter-bombardment fire

N

NCO	non-commissioned officer, i.e. bombardier, bombardier/

O

OC	Officer Commanding
OP	observation post
OPO	observation post officer
Orientation	aligning the direction of the easting grid line to grid north

P

PAA	*Panzerarmee Afrika*
PMF	Permanent Military Forces — full-time professional officers and NCOs who staffed headquarters and Militia units between World War I and II
POL	petrol, oil and lubricant
POW	prisoner of war
predicted fire	fire applied without preliminary ranging. The grid references of both gun and target are known, and the sum of variables that affect the shell's trajectory are calculated and applied to the gun's sights.

Q

QM	Quartermaster

R

RAAF	Royal Australian Air Force

RA	Royal Artillery (British)	Stonk	standard concentration (of fire units) on a linear target
RAA	Royal Australian Artillery		
RAAMC	Royal Australian Army Medical Corps	Svy	survey — fixation and orientation of the battle zone using triangulation etc.
RAAOC	Royal Australian Army Ordnance Corps		

T

TA	Territorial Army (British 'Militia')
TEWT	tactical exercise without troops
TOF	time of flight
Tp	Troop — two to a battery of field guns
TSM	Troop Sergeant Major — senior NCO of a Troop

RAC	Royal Armoured Corps (British)
RAE	Royal Australian Engineers
RAF	Royal Air Force (British)
RAN	Royal Australian Navy
RAP	Regimental Aid Post (medical)
RA Sigs	Royal Australian Corps of Signals
RE	Royal Engineers (British)
RFA	Royal Field Artillery (British)
RFC	Royal Flying Corps (British)
RGA	Royal Garrison Artillery (British)
RHA	Royal Horse Artillery (British)
RHQ	Regimental Headquarters
RMO	Regimental Medical Officer
RN	Royal Navy (British)
RNZA	Royal New Zealand Artillery
rpm	rounds per minute
RSM	Regimental Sergeant Major
RTR	Royal Tank Regiment (British)
RTC	Royal Tank Corps (British)

U

UK	United Kingdom

W

WIA	wounded in action
WLO	Wagon Lines Officer
WO1	Warrant Officer Class 1
WO2	Warrant Officer Class 2

Z

Z	zero hour — start time for a major event

S

SAA	South African Artillery
Sec	section/sub-section — a small detachment
Spandau	a heavy German machine-gun

Index

A

Aberdeen, Operation, lessons, 168–169
Adams, H.J., 378
Adlard, Lieutenant, F.E.L., 376
Adye, Brigadier J.F., 165
Air Photograph Interpretation Section (APIS), 127, 224, 231, 234, 239, 286
air photography, 76, 86, 131, 139, 167, 191, 206, 208, 214, 219, 234, 251, 284, 286–287
Alam el Halfa, attack on, 190–193
Alam Halfa, see El Alamein, First Battle of
Alamein Annie naval gun, 184
Alexander, General Harold, 177, 201–202, 252, 280, 309, 314
Alexander, Gunner J.E., 373
Allen, Brigadier, 78, 93, 99
Allen, Gunner E.C, 376
Allen, Sergeant A.C., 368
Ambrose, D.A., 377
Amiens, Battle of, 7–8
ammunition, types
    airburst, 73, 172, 182, 184, 187, 192, 206, 272, 285, 288
        for Correction of the Moment calculations, 228
        effectiveness of, 231
    German lethality, 207
    Italian, 73–74
    smoke, 6–7, 17, 28, 59, 61, 81, 188, 207, 272, 279, 292, 357
    solid shot AP, 17, 121, 226, 229, 288, 358

Andersen, Captain J. 'Hans', 72
anti-tank artillery
    Allied, 30–33, 284–285
    DAK, 121, 166, 206
Argentino, General, 81
Arko 104, see Artillerie Kommandeur 104
armour/artillery/infantry cooperation, 295–298
armoured assault tactics, 205–206
    British, 166
    DAK, 161–162, 198–200, 205–206
Arthur, Major A.E., 79, 372
Artillerie Kommandeur 104, 135, 156, 162, 169, 184, 203, 205, 291–292, 299, 301
    1st African Artillery Regiment, 203
    2nd African Artillery Regiment, 203
    221st Artillery Regiment (Motorised), 203
    408th Heavy Artillery Regiment (Motorised), 203
    606th Anti-Aircraft Battalion, 203
    612th Anti-Aircraft Battalion, 203
    617th Anti-Aircraft Battalion, 203, 293
artillery development, interwar period, 13–22
    anti-aircraft artillery, 33
    anti-tank artillery, 30–33
    artillery-aircraft cooperation, 33
    artillery-tank cooperation, 28–30
    command and control, 26
    communications, 35
    equipment development, 15–16
    German, 31–32
    mobility, 18–22
    organisational structures, 17

sound ranging, 34–35
artillery development, WWI and earlier
    ammunition development, 6
    Anglo-Boer War, 2
    calibration, 9
    communications, 10
    First AIF, 3–11
    infantry-tank cooperation, 5–7
    interwar period, 49–55
    Kitchener Review, 2
    predicted fire, 9
    pre-Federation, 1–2
    Sudan, 2
    support to infantry, 9
    survey, 9
artillery weapons, relative effectiveness, 298–300
Ashton, Gunner F., 372
Atkinson, Signaller E., 182
Auchinleck, General Sir Claude, 148–149, 165–166, 168–171, 175–176, 181, 187, 189–190, 301, 313–314, 322, 331, 337–338
Australian Field Artillery Brigades (AFA)
    3rd Field Artillery Brigade, 101
    13th Field Artillery Brigade, 101, 241–242
    13th Field Brigade, 50–51
Australian formations
    9th Australian Division
    artillery order of battle, 100–104
    Headquarters Royal Australian Artillery, xiii, 100, 184, 212, 214, 219, 226–227, 239, 242–244, 246, 253, 255, 272, 286, 308, 353, 365, 386
    leadership, 96–100
    Operation Order for Battle of El Alamein, 353–355, 361–365
    order of battle, 96, 353–355
    20th Infantry Brigade, 93, 96, 107, 109–111, 114, 140, 178, 181, 188, 215, 217, 244–245, 256, 259–261, 263–268, 270, 273–274, 354, 362–364
    24th Infantry Brigade, 95–96, 110, 113, 118, 178–179, 181, 185, 217, 220, 228, 269–270, 272–274, 278, 291, 341, 354, 362, 368
    26th Infantry Brigade, 96, 110–111, 114, 179–181, 217, 244, 256, 261, 266–269, 272–274, 354, 361, 364
Australian infantry battalions
    2/3rd Infantry Battalion, 79–80, 85
    2/4th Infantry Battalion, 86, 89
    2/7th Infantry Battalion, 79, 86
    2/8th Infantry Battalion, 86
    2/12th Infantry Battalion, 96, 136
    2/13th Infantry Battalion, 96, 109, 111, 113, 149, 155, 165, 178, 244, 256, 258–261, 264, 266–268, 271, 335, 354, 363, 368
    2/15th Infantry Battalion, 96, 109, 178, 188, 218, 244, 256, 259, 265–269, 271, 273, 354, 363, 369
    2/17th Infantry Battalion, 96, 109, 122, 134, 178, 218, 244, 256, 259–262, 264, 271, 333, 354, 363, 368–369
    2/23rd Infantry Battalion, 96, 134, 182–183, 186, 255–256, 259, 266–267, 271, 354, 362, 368
    2/24th Infantry Battalion, 96, 110, 180–182, 200–201, 218, 244, 256, 259, 261, 266–267, 269, 271, 274, 338, 354
    2/28th Infantry Battalion, 96, 117, 136, 185, 187, 255, 269–271, 273, 354, 368–369
    2/32nd Infantry Battalion, 96, 259, 269–272, 354
    2/43rd Infantry Battalion, 96, 181, 269–272, 354, 369
    2/48th Infantry Battalion, 96, 110, 134, 141, 244, 256, 259–263, 267, 269, 279, 343, 354, 368
    2/1st Pioneer Battalion, 96
    2/3rd Pioneer Battalion, 267, 270, 354, 362

B

Bailey, C.C., xiii
Bailey, J.B.A., xxv, 43, 299, 325
Bailey, J.G., 379
Baker, Bombardier R.W.M., 374
Baker, Gunner H.M., 373
Bale, Major R., 84
Balfe, Captain J.W., 122

Index

Bardia, attack on, 74–82, 108
    air support, 78
    ammunition resupply, 76
    artillery deception, 76
    artillery plan, 75–79
    artillery weapons captured, 80
    casualties, 79
    Italian losses, 108
    lessons, 81–82
    naval gunfire support, 78
Bardia Bill gun, 76, 130–131, 141
Barker, A.J., 377
Barker, Brigadier Lewis, 51, 53–54, 72, 73p, 81–82, 88, 92, 329, 331, 367, 372, 383
Barnard, Bombardier L.H., 376
barrages, xix, 5–6, 38, 61, 171–172, 187, 208, 248–249, 284
    box, 77, 86
    creeping, 5, 78, 84–86, 88, 256, 342
    divisional, 26
    effectiveness of, 281, 327
Barry, Gunner W.J., 376
Basset, W., 377
Battleaxe, Operation, 140, 145–148, 161, 163, 165, 172, 225, 300, 308
    lessons, 148
Batty, Sergeant K.C., 372
Bean, C.E.W., 97
Beard, Gunner, 257
Benghazi, advance to, 88–92
    artillery plan, 88–89
    lessons, 90–91
    results, 92
Benson, Captain Alan de G, xiii, 191
Beresford-Peirse, Lieutenant General Noel Monson de la Poer, 116, 146–148
Bergonzoli, General 'Electric Beard', 92
Berryman, Colonel Frank, 80, 160
Bertram, Operation, 219
Bidwell, Shelford, xxv, 8, 25, 28, 33, 88, 91, 122, 138, 147, 163, 168, 172, 193
Billing, E.O., 379
Billing, H.J., 378
Birch, Lieutenant General Sir Noel, 8, 19, 29, 47

Bird, Captain J., 83
Bird, Lance Sergeant, 368
Birdwood, Lord, 55
Bishop, Gunner J.R., 376
Black, Wing Commander Eric 'Digger', 132
Blair, Gunner M.R., 374
Blamey, General Sir Thomas, 52, 71, 95, 97, 99, 112, 141, 156–157, 193, 306, 344, 383–384
Boettcher, Generalmajor Karl, 135, 156, 162, 286
bombards, 131, 134, 192, 215, 227, 232, 287
Bond, Gunner L.G., 377
Bradley, Gunner P.J., 373
Bradshaw, Flight Lieutenant, 129, 177
Bragg, Major Lawrence, 34
Braid, Sergeant I.L., 127
Braund, Lieutenant Colonel, 97
bren gun carriers for observation post officers, 21, 62, 183, 185, 257–258, 260, 283, 389, see also OP tanks
Brien, Lance Sergeant G.H, 369, 378
British armoured formations and units
    4th Armoured Brigade, 146
    6th Royal Tank Regiment, 110, 183
    7th Royal Tank Regiment, 134
    40th Royal Tank Regiment, 188, 267
    46th Royal Tank Regiment, 266
    50th Royal Tank Regiment, 176
    Armoured Support Group, 83, 147
British artillery weapon characteristics, 358
British Expeditionary Force (BEF)
    artillery lessons, 43–44
    artillery losses in France, 42
    artillery order of battle, 38
    doctrinal implications, 43–44
British forces, other
    51st Highlanders Division, 217, 229–230, 244, 257, 293
    70th Division, 163
    22nd Guards Brigade, 96, 103, 145, 149, 154–156, 160
Broad, Colonel C.N.F., 325
Broad, General, 29
Bromley, Lieutenant, 125

Brooke, General Alan, 8, 44, 139, 177, 307, 314, 337
Brookman, Sergeant M.R., 373
Browne, Gunner C.P., 374
Bruchmuller, Colonel Georg, 8
Buckney, Gunner G.T., 375
Bulimba, Operation, 183, 188, 191, 193
Bullen, Gunner D.J., 372
Burchardt, Generalleutnant, 291
Burgess, William Sinclair, 8
Burghall, Sergeant V., 373
Burkhill, G.R., 379
Burrows, Lieutenant Colonel F.A., 155–156
'Bush Artillery', 115–117, 120, 134–135, 334
Butler, Lance Bombardier B.McD., 377
Butterick, Gunner F.A., 265, 373
Byrne, Gunner A.H., 376

C

Caelli, B.F., 380
Calder, Captain J.W., 244, 267, 380
calibration, 9–10, 59, 126, 132, 228, 236, 262, 264
Cambrai, Battle of, 7
Campbell, Bombardier, 151
Campbell, Brigadier 'Jock', 148–149
Carey, Major P., 223
Carroll, Bombardier J.L., 369
Carter, Major A.A.C., 104, 124, 214, 272, 278, 337, 369, 380
Cassidy, Captain A.B., 369, 379
Cassidy, Lieutenant Tod, 278
casualties
    Australian artillery, 371–380
    Bardia, attack on, 79
    Crusader, Operation, 166
    El Alamein, First Battle of, 186, 191
    El Alamein, Second Battle of, 277–279, 305–306
    Tobruk, defence of, 137
Caunter, Brigadier J.A.L., 91
Caville, Gunner L.G., 374
Challenor, E., 378
Chapman, Acting Sergeant T.E., 374
Cheeseman, C.M., 378
Cherry, Lieutenant Colonel R.O., 228, 327

Churchill, Prime Minister Winston, 176–177, 295, 297, 313
Chuter, Gunner A.P., 267, 378
Clark, Major General Sir Campbell, 326
Clark, W., 378
Clarke, Warrant Officer II C.H., 376
Clarkson, A.G., 378
Cleland, Lieutenant P.F., 151
Clements, Lieutenant Colonel H.T.M., 138
Clowes, Lieutenant Colonel Cyril, 51
code words, use of, 201, 245, 252, 290, 365
Coffey, Gunner L.A., 375
Cohen, Gunner W.D., 374
Cole, Captain W.J.L., 244
Collins, Gunner F., 373
Collins, Padre Wilfred 'PK', 137
command, control and communications, 26, 296–298
Commander Royal Artillery, role, 167
communications equipment, 13, 349
    Allied, 283
    continuous wave Morse Code, 35
    dispatch riders, 38
    line, 40
    single sideband super heterodyne, 35
    telephone lines, 35, 38
    teletype machines, 208
communications practices
    Allied, 283
    German, 115
    jamming, 135, 293, 331
    ladder line, 283, 343
    wireless silence, 201–202, 245, 364
Connley, A.D., 379
Connolly, Gunner K.P., 377
Conway, Gunner F.J., 376
Cook, Gunner E.J., 372
Cook, Lieutenant Colonel T.P., 116
Cooke, Gunner H.T., 377
Cornwell, General Marshal, 314
Counter Bombardment Staff Office, 8–10, 86, 102, 127, 130, 134, 167, 223, 286, 328
counter-bombardment, 7, 40
counter-preparation fire, 6

Cowell, Gunner A.E., 377
Cox, Lieutenant Desmond, 53, 86
Coxen, Walter, 8
Crawford, Gunner W.S., 373
Crawford, Lieutenant Colonel, 122
Crawford, Lieutenant J.M., 79, 372
Creagh, Major General, 146
Cremor, Brigadier W.E., 54, 73, 76, 79, 84, 86, 88, 92, 367, 383
Crisp, Gunner T.J., 374
Crisp, Lieutenant Colonel Alan, 101–102
Crozier, R.T., 379
Crusader, Operation, 156, 160, 162–167, 313, 322, 335
    artillery, method of employment, 163, 165–166
    casualties, 166
    lessons, 166–167
    survey, 164
Crute, Lieutenant, 86
Crüwell, Generalleutnant, 121, 160
Cuff, Lieutenant, J.G., 374
Cummings, A.M., 380
Cunningham, General Alan, 159, 162–163, 170
Cunnold, Gunner W.H., 373
Curran, E.J., 377
Currell, Bombardier L.J., 375
Curtin, Prime Minister John, 281
Cutler, Lance Bombardier E.W., 272, 378
Cutler, Sir Roden, 385

D

Dadds, Gunner J.J., 373
Daly, D.F., 379
Damm, L., 378
datum point shoots, 172, 228
Davey, Gunner R.P., 374
Dawes, Lieutenant, 1
Dawson, Gunner H.D., 373
Day, Major John, 183, 242, 337, 368
de Graz, Colonel, 331
deception, 219
    dummy gun flashes, 76, 151, 187
    dummy guns, 76, 117, 231
    German, 202–203
    intelligence, 190
defensive fire, 6, 184, 188, 245, 247, 255, 258, 262, 265, 271, 279, 290, 314, 363, 390, see also stonk
    code words for, 245, 252, 290
    for counter-bombardment, 138
    effectiveness of, 39, 282
    SOS, 6, 246, 261, 264, 267, 269, 285, 365, 390
Delbridge, Gunner H., 373
Delecca, M.G., 378
Dench, Lieutenant J.R., 376
Dennis, Brigadier Meade, 189, 205p, 223, 245, 286, 291, 340
Dennis, Lance Bombardier C.A., 368, 386
Desert Air Force (DAF), 154–155, 160, 183, 189–191, 202, 207–208, 219, 231–232, 246, 248–249, 286–287, 291, 293, 300, 302, 310
    definition of, 335, 390
Deutsches Afrika Korps (DAK)
    ammunition lethality, 207
    anti-aircraft/anti-tank 88-mm GP gun, characteristics, 121, 166, 206
    armoured assault tactics, 161–162, 198–200, 205–206
    artillery, 121, 135, 164, 203–204
    artillery organisational structures, 44
    artillery tactics, 205–206
    characteristics of guns, 357
    definition of, 323
    intelligence, 200–202
    PAK 38 50-mm gun, characteristics, 121, 166, 300–301
    use of captured British equipment, 200, 208, 290
Deutsches Afrika Korps (DAK) formations and units
    5th Light Division, 109, 111
    15th Panzer Division, 156, 203, 256, 262, 266, 291–293, 314, 343
    21st Panzer Division, 146, 154, 159, 162, 164, 168–169, 203, 268, 271, 291, 298, 306, 314

90th Light Division, 159, 169, 189–190, 203–205, 217, 263–264, 266, 270, 291, 298, 306, 314
164th Infantry/Light Division, 203–205, 217, 249, 254, 292, 306
Ramcke Parachute Brigade, 203, 217
Kiehl Group, 217
33rd Reconnaissance Battalion, 217
159th Battle Group, 263
361st Battle Group, 263
Intelligence Company U621, 138, 189, 200–201
Deverell, Field Marshal Sir Cyril, 32
di Stefanis, General Guiseppe, 206
Dibb, Brigadier, 75, 83
discipline
    road and track, 21, 246
    troops, 97–98, 112, 123, 211, 219
Dixon, Gunner L.V.C., 375
doctrine development
    artillery, interwar period, 25–26
    command and control, 296–298
    land-air warfare, 302
Donaldson, Lance Sergeant J.B., 127
Dooley, Gunner, 257
Dorman-Smith, Major General Eric, 170–171, 175–177, 314, 336
Douglas, Captain R.A., 226, 340
Douglas, Lieutenant Colonel J.S., 120
Downie, Lance Sergeant J.E., 377
Doyle, Gunner H.P., 377
Drummond, Gunner N., 372
Drysdale, Gunner C.E., 376
Duggan, Gunner C., 372
Dumaresq, Captain M.C.W., 86
duToit, Colonel C.L.deW., 224
Dwyer, Captain F., 372
Dwyer, Captain K.F., 79–80, 367

E

Eastick, Brigadier Sir Thomas, ix, 50, 100–101, 145, 150, 152, 154, 216p, 241–242, 278, 281, 310, 337, 341, 368, 384–385
Edwards, Captain, 229
Edwards, Gunner E., 265, 373
Edwards, Major General J.B., 1
El Aghelia, defence of, 109–110
El Aghelia to Tobruk withdrawal, 109–115
    air support, 113
    artillery plan, 110–115
    lessons, 114–115
El Alamein, First Battle of, 170–172, 175–177, 181–194
    ammunition expenditure, 185, 191
    artillery, method of employment, 184
    casualties, 186, 191
El Alamein, Second Battle of
    ammunition used, 279–280
    anti-aircraft defence, 230
    anti-tank defence, 230
    artillery communications planning, 239
    artillery preparations for, 211–220
    artillery resources, 225–227, 229–230
    casualties, 277–279, 305–306
    counter-bombardment planning, 224, 231–235, 282, 287
    equipment deficiencies, 283
    final preparations, 243–249
    fire planning, 245, 281–282, 284
    flash spotting support, 236–238
    lessons learned, 281–283
    medium artillery effectiveness, 287
    meteorological support, 230
    opening salvos, 248, 251–252
    Operation Order, 361–365
    Phase A, initial assault, 253, 256–261
    Phase B, Trig 29 & first thrust north, 253, 261–265
    Phase C, second northward thrust, 253, 266–269
    Phase D, Supercharge, 253, 270–274, see also Supercharge, Operation
    sound ranging support, 235–238, 255
    survey support, 228–230, 233, 236, 287
Elder, Captain J.S., 182, 375
Elliott, Brigadier G.M., 74, 224, 330
Ellis, Major L.F., 38
Elton, Lieutenant Colonel H.C., 191, 225–226,

308, 340
Everett, Captain K.W., 244, 265, 374
Exercise Bumper, 34, 44–47
Eyre, D.G.E., 379

F

Fairthorne, Bombardier F.F., 375
Farmaner, Gunner H., 373
Fawcett, H., 380
Feitel, Major M., 130, 134, 141, 214
Fenton, Driver B., 182
Feuerheerdt, Gunner A.B., 373
Fidden, Gunner W.B., 374
Fiddler, Gunner W.B., 374
Fiddler, W.J., 378
Field, L.N., 378
field artillery against tanks, 146–148, 165, 184, 284, 288
field regiment
    attached personnel, 351
    command and control, 56
    communications, 62–63
    employment of, 63
    key appointments, 56–58
    Light Aid Detachment, 57
    method of operation, 55–66
    OP officers, 61–62
    organisational structures, 39–40, 55–56, 63
    paperwork, 64–65
    personal kit, 65
    rations and water, 65–66
    strength, 56, 351
    survey, 63
    vehicle identification signs, 347–348
    vehicles, 62, 348–349
    weapons, anti-aircraft defence, 62, 349
    weapons, personal, 62, 349
    wireless set distribution, 349–351
Fielding, Captain A.W., 185, 373
Fisher, Gunner A.P., 373
Fisher, K., 379
Fisher, R.C.McG., 379
Fitzhardinge, Lieutenant Colonel Roger, xiii, 181, 242, 316, 385

flash spotting, 8, 10, 38, 40, 83, 123, 127, 129–133, 167, 178, 191, 200, 202, 208, 218, 228–229, 234–238, 254, 281, 285, 287, 289, 391
    effectiveness of, 41
    German use of, 151, 172, 187, 192, 203
    layout of, 128
flashless propellant, 60, 130, 187, 289
Fleming, K.D., 378
Forbes, Captain W., 134
Ford, Signaller Keith, 125
Forster, Gunner T.A., 375
Forwood, Captain Archie, 242, 365
Fowler, Sergeant R.W., 368
Francis, E.G.F., 377
Francis, Lieutenant Colonel, xiv
Frank, L.V., 378
Freeman, Lance Sergeant F.R., 265, 278, 369
Freeman, Sergeant N., 372
Freitel, Captain M., 130, 134, 141, 214
Freyberg, Major General Bernard, 193, 302
Frowen, Colonel J.H., 74–76, 79, 84
Fryett, Major, 129
Fuller, General J.C.F., 13, 29, 35, 175, 204, 339
Fulton, Gunner W.H., 375
Fysh, Gunner H.D., 375

G

Gabriel, Gunner G.T., 375
Gambara, General, 160
Gambier-Parry, Major General Michael, 110, 112, 332
Garrard, Sergeant Vernon, 184
Garwood, Gunner H.C., 376
Gazala, battle of, 160, 168, 227, 321–322, 334, 340
Geddes, Captain A.W.R., 101
Gentle, Gunner C.H., 373
George, Gunner S.T., 369
German, pre-war training, 298
Gibbs, F.C., 379
Gilbert, Sergeant J.A., 373
Gilchrist, Gunner A., 377
Gillespie, Gunner R.W., 369
Glatte, Gunner H.G., 375

Gleeson, Gunner F.L.G., 376
Godfrey, Brigadier A.H.L., 110, 116, 270
Goodhart, David, xix, 152, 154, 185, 321–322, 385
Goodwin, Brigadier Shirley, 103–104, 119–120, 125–126, 129, 135–136, 141, 220, 272, 278, 310, 343, 369, 385–386
Gordon, Gunner R.A., 373
Gort, General Lord, 37
Gott, Lieutenant General 'Strafer', 159, 170, 176–177
Gowing, Gunner B., 2
Graziani, Marshal Rodolfo, 69, 108
Greenwood, Sergeant J.H., 372
Gregory, Gunner H.S., 377
Gregory, Gunner J., 373
Greville, W.P., 378
Griffiths, Bombardier D.L., 374
Griffiths, Captain W., 78
Grimston, Captain, 187
Grimwade, Harold, 8
Guderian, General Heinz, 32, 204
Gudmundsson, Bruce, xxv, 204, 208
gun aiming post, 248, 391
gun pits, construction of, 60, 124, 187–188
Gunton, Gunner W.A.L., 374

H

Haig, Field Marshal Sir Douglas, 8, 47
Hale, L.R., 379
Hamer, Lieutenant Colonel Frank, xiv, 234, 234p, 235, 308
Hamilton, Captain H.P., 126
Hancock, Sapper, 86
Hanna, Gunner L.W., 375
Hanna, Gunner R.B., 373
Hanson, Captain, A.G., 367
Harders, G.G., 377
Harris, J.P., 30
Harris, R.J., 379
Harrison, Bombardier L.J., 376
Harrison, V.C., 379
Harvey, Gunner E., 377
Hay, Sergeant A.R., 374
Hay, Sergeant I., 368–369
Hayward, Mick, 54
Hayward, Sergeant J.M., 375
Haywood, Lieutenant E.V., 372
Hazell, Sergeant R.H., 373
Heathcote, Gunner M.E., 376
Helyer, Gunner L.G., 377
Hemming, Lieutenant Colonel H., 322
Henty, Captain W.M., 214
Hercus, Bombardier V.W., 127
Herring, Lieutenant General Sir Edmund, ix, xx, 50, 52–53, 70–71, 72p, 73, 75–77, 81–84, 86, 92, 295, 305, 328, 330, 367, 383
Herz, Leutnant H., 201–202
Hetherington, John, 141
Hill, Gunner T.P., 183, 368
Hill, Sergeant C., 374
Hills, Gunner A.R., 375
Hiscock, Captain W.R.G., 79, 84
Hobart, General, 29, 299
Hobbs, Colonel Athol, 54, 383
Hobbs, Major General Sir John Talbot, 8, 54
Holder, Sergeant A.D., 373
Holding, Lance Sergeant W.E., 278, 369
Hollit, Gunner S.P.R., 374
Holmes, Captain D.L., 126
Holmes, Lieutenant General, 170
Holt, Captain D., 103
honours and awards, 367–369
Hooper, Gunner G.L.G., 373
Horrocks, Lieutenant General Sir Brian, 281, 292
Horsell, Lance Bombardier W.R., 374
Horton, Gunner K.G., 376
Hoskin, Gunner R.W., 369
Hoskin, Gunner W.M., 373
Houston, Temporary Lieutenant Colonel G.D., 104, 119–120, 272, 278, 310, 343, 369
Howard, Gunner G.H., 373
Howard, Major John, 328
Huggett, Captain G., 152
Hughes, Prime Minister Billy, 51
Hunt, Lieutenant Colonel H.S., 223
Hunter, Sergeant R.W., 374

Huntingdon, Gunner G., 373
Hussey, Sergeant E.J., 375
Hutchinson, Gunner K., 375

I

Indian formations and units
    4th Indian Division, 116, 127, 145–146, 148, 155, 164–165, 200, 205, 217, 236, 337
    5th Indian Division, 120, 189
    10th Indian Brigade, 168
    15th Indian Brigade, 189
Ingram, Gunner D.J., 373
Irwin, Major Jim, 100–101, 213
Italian artillery units
    3rd Mobile Artillery Regiment, 206
    8th Artillery Regiment, 206
    21st Artillery Regiment, 206
    26th Artillery Regiment, 206
    46th Artillery Regiment, 206
    55th Artillery Regiment, 206
    132nd Artillery Regiment, 206
    185th Artillery Regiment, 206
    205th Artillery Regiment, 206
    357th Light Artillery Regiment, 206
    49th Artillery Battalion, 206
    133rd Littorio Semovente Battalion (SP), 206
    147th Artillery Battalion, 206
    161st SP Artillery Battalion, 206
Italian artillery weapons characteristics, 358

J

Jaboor, Major R.F., 84, 330
Jackson, Gunner A.C., 374, 376
Jenkins, T.S., 380
Jock columns, 148–150, 183, 190, 298
Johnson, D., 380
Johnson, Gunner A.D., 374
Johnson, Gunner H.D., 374
Johnson, Gunner R.H.I., 376
Johnson, Major R.L., 149, 244, 363
Johnston, Gunner George, 8
Johnston, Major R., 103, 118, 188, 242, 245, 256

Jones, Captain A.G., 244
Jones, Gunner J.R., 374
Jones, Major R.M., 104, 244, 369
Jones, R.J., 377
Jordan, W.A., 378

K

Kaiser, Major H.C., 213, 369
Kellet, Lieutenant Colonel E.O., 118, 129–130, 132, 229
Kellett, Major L., 129–130, 132, 229
Kelly, B.P., 378
Kelly, Lieutenant Colonel Leo, 53
Kenna, Lance Bombardier J.R., 377
Kennedy, H.S.A., 378
Kennedy, Lance Sergeant P.StJ., 373
Kennedy, Major, 25, 344
Kerry, H.F., 377
Kimpton, J.P., 379
King, Bombardier G.D.C., 368
King, Sergeant C.D.C., 373
Kinghorn, Gunner Alan, 282, 368
Kirkman, Brigadier Sidney, 44–45, 47, 190, 208, 223, 232–233, 249, 253p, 279, 281–282, 284–291, 299, 307, 341
Klein, Lieutenant Colonel Bruce, 53, 102, 127, 129, 132, 134, 139, 286, 328, 334
Klopper, Major General, 142
Knight, Gunner J.R., 374
Knight, Gunner N.R., 375
Krause, Generalleutnant, 191, 192p, 203, 207, 287, 289–291
Krumbeck, Gunner M.S., 372

L

Lackmann, Bombardier C.J., 377
ladder ranging, 61, 172
Lampe, Lance Bombardier D.P., 127
Latham, Brigadier, 71, 114
Lavarack, Lieutenant General John, 18, 109, 113, 115, 141, 160
Laverty, H.G., 379
Lee, Lieutenant Colonel E.A., 73
Lees, Gunner H.J., 375

Leese, Lieutenant General Sir Oliver., 189, 280, 310, 345
left out of battle (LOB), 211, 247, 392
Lewin, Lance Bombardier R.A., 377
Lewis, Essington, 21
Lewis, Gunner T.A., 258, 373
Lewis, R., 377
Liddell Hart, Basil, 29–30, 32, 43
Ligertwood, Captain W.L., 185, 244, 258, 278, 368, 373
Light Aid Detachments (EME)
    58th LAD, 354
    61st LAD, 104, 353
    63rd LAD, 101, 353
    64th LAD, 102, 353
    67th LAD, 354
    71st LAD, 353
    72nd LAD, 354
    76th LAD, 354
    77th LAD, 354
    78th LAD, 354
    82nd LAD, 353
Lightfoot, Operation, 177, 219, 253
Lillingstone, L.R., 377
Lindsay, Gunner W.H., 374
Lingerwood, Sergeant W., 374
Little, Gunner R.A., 376
Lloyd, Gunner W.J., 369, 380
Lloyd, H.W., 380
Lloyd, Major General Charles, 99, 140, 177, 384
Lloyd, W.J., 380
Loder-Symonds, Major R.G., 121–122, 333
Long, Gavin, xxiv, 71, 76, 81, 384
Lonne, K.F., 380
Lord, Gunner M.A.A., 374
Loughrey, Lieutenant Colonel, 269
Loveband, Captain Harry, 153, 242–243
Luck, Gunner D.E., 374
Luftwaffe, 33, 114, 132, 135, 137–138, 152, 177, 180, 192, 202–203, 219, 232, 246, 291, 298, 302, 323
    25th Anti-Aircraft Regiment, 203
    33rd Anti-Aircraft Regiment, 203
    102nd Anti-Aircraft Regiment, 203
    135th Anti-Aircraft Regiment, 203
    in Europe, xxv
    Ju87 Stuka, 113, 184, 186, 226, 267–271
    Ju88 Junkers, 113
Luke, R.T., 379
Lungerhausen, Generalmajor Karl-Hans, 203–205, 249
Lustreforce, 96, 328
Lutz, Lieutenant Tom, 242
Lyle, R., 378

M

Mackay, Captain Ken, 103, 154, 332
Mackay, Gunner J., 377
Mackay, Lieutenant General Sir Iven, xxiii, 70, 71p, 75, 79, 81–84, 86, 88, 92, 97, 295, 305, 383
Maddison, Bombardier L.W., 369
Mahoney, Gunner K.W., 373
Main, A., 377
Main, E.K., 378
Maine, Gunner, 267
Mair, Lieutenant, 86
Major, Gunner S., 374
Maltman, H.J., 379
Manning, Gunner A., 185, 368
Martel, General, 29
Martin, W.J., 379
Mason, Lieutenant Alan, 244
Mason, Sergeant L.A., 377
Matthews, E.L., 378
Matthews, L.A., 377
Matthews, Lieutenant Colonel A.G., 120
Maxse, General, 29
Maxwell, Aylmer, 161
Maxwell, Major General Amer, 283, 286
McCann, Gunner D.J., 376
McCashney, Gunner J.N., 372
McDermott, Captain D.H., 126
McIlrick, Sergeant A., 186
McKay, Gunner G.W., 375
McKeddie, Captain J., 124
McKenzie, C.A.L., 379
McKenzie, Charles G., xiii, 321
McKercher, Sergeant R.W., 375

McKinna, Sergeant 'Darky', 150
McKinnon, W.C.B., 379
McQueen, Sergeant F.J., 369
McRobert, A.G., 377
McVeigh, E.W., 379
Meloury, W.J., 379
Melville, J.F., 378
Menzies, Lieutenant R.W.J., 256, 258, 278, 368
Menzies, Prime Minister R.G., 99
Meredith, Lieutenant Colonel G.P.W., 51
Messervy, Major General Frank, 146, 165, 205p
Miles, Brigadier, 224
Milford, Colonel Ted, 50
Milledge, Major R.A., 104, 124, 214, 337
Millen, Gunner W.R., 374
Miller, Gunner E.M., 374
Milne, Field Marshal, 29
Milner, Gunner I., 375
Minson, Lance Sergeant C.S., 368
Mitchell, Gunner L., 374
Monash, Major General John, 7–8, 10–11, 97, 193
Monkhouse, Bombardier W.D., 373
Monro, V., 378
Montgomery, Field Marshal Sir Bernard, xi, 44–45, 47, 168, 177, 180, 190–191, 194, 201–202, 207–208, 215, 217, 219, 232, 241, 251, 253p, 263, 267, 280, 287, 292, 298, 302, 307–309, 313–314, 336–337, 341
Mooney, Gunner E., 372
Moore, Gunner S.G., 376
Morely-Mower, Lieutenant Geoff, 132
Morgan-Taylor, F.B., 379
Moroney, J.P., 380
Morrow, W.A., 377
Morse, Sergeant W., 78
Morshead, Lieutenant General Sir Leslie, xx, xxvi, 53, 83, 93, 95–100, 104, 110–118, 120, 122, 124–125, 128–129, 132–133, 135, 138–142, 145, 150, 175, 177, 179–181, 188–190, 193–194, 211, 214, 217, 225, 230, 245, 251–252, 263, 266–267, 269, 280–282, 285, 291, 295, 297, 305–307, 314, 332, 337–338, 342, 344–345, 384

mortars
    3-inch, 18, 285
    80-mm, 285
    81-mm, 75, 121, 204, 206
    4.2-inch, 228, 315–316
    locating of, 46
    practices, 41
Morton, Bombardier, H.L., 367
Morton, Lieutenant Charles, 375, 385
Mountjoy, Gunner R., 180, 374
Mowbray, C.D., 378
'Mr Clarke's Guns', 115
Muller, J.F., 379
Mummery, Lieutenant S.W., 373
Munro, Lieutenant Colonel E., 112
    Munro, Major Norman, 242
Murname, Gunner, 260
Murphy, Lieutenant 'Spud', 53
Murray, Brigadier J.J., 93, 116
Murray, Gunner A.R., 373
Murray, Lieutenant, 272

N

Nairn, V.E.A., 379
Nankervis, Gunner J.L., 376
Nash, Sergeant B.V., 377
naval gunfire support, 81, 84, 132, 307, 318
Navarini, General Enea, 160, 206
Neame, Lieutenant General Phillip, 109–116
Nebbia, General Edouardo, 206
Nelligan, Captain P.W., 244, 256, 264, 278, 342, 369
Nethercote, Lieutenant C.R., 79, 372
Neumann-Silkow, Generalmajor, 159
New Zealand formations and units
    2nd New Zealand Division, 170, 175, 217, 224, 229–230, 239, 244, 272–273, 291, 293, 296, 306, 361, 371
    4th Field Regiment, 229
    5th Field Regiment, 229
    6th Field Regiment, 156, 229
    16th Field Regiment, 316
Newell, Gunner A.E., 372
Newham, Gunner S.C., 373

Newham, Gunner T.C., 373
Newland, Lieutenant, 242
Nicholson, Gunner J.N.W., 375
night laager routine, 152–154, 300
Norrie, Lieutenant General W., 181, 337
Norris, Gunner R.A., 374
North, J.H., 380
Norton, E.C., 379
Nugent, H.W., 378

O

O'Brien, Gunner W.J., 374
O'Brien, Lieutenant Colonel John, 101, 322
O'Brien, Major George, 53, 72, 367
O'Connor, Lieutenant General Richard, xxiii, 69–72, 74–75, 78, 82–83, 87–88, 90–92, 107–109, 112–113, 227, 302
Ogden, Major R.H., 308
Ogilvie, Gunner S., 377
O'Halloran, Gunner J.C., 374
Oldfield, Gunner A., 373
Olney, Gunner L.B., 377
Olsen, C., 380
OP tanks, 168, see also bren gun carriers for observation post officers
Orr, Gunner H.C., 377
Osborne, C.C., 380
O'Shea, Gunner J.L., 373
O'Sullivan, Gunner J.H., 372
Overlord, Operation, 309–310, 337

P

Palmer, Bombardier G.H., 376
panoramic sketching, 125
Panzerarmee Afrika (PAA) artillery
    anti-tank artillery, 166
    artillery resources, 292
    definition of, 323
    effectiveness of, 285, 288–293
Parham, Brigadier John, 40, 41p, 45, 328
Parker, Lance Bombardier H.C., 373
Parker, W.H., 377
Parr, Gunner K.W., 258, 373
Payne, A.M., 378

Peacock, Gunner H.G., 377
Pearce, Gunner F.E.G., 369, 378
Pearse, Sergeant A.E., 70p, 89, 367, 372
Pemberton, Brigadier A.L., 41, 149, 284, 299, 323
Percival, General, 156
Peters, Major C.N., 72, 80, 328, 330
Phillip, Governor Arthur, 1
Phillips, Lance Sergeant G.W., 373
Phillipson, Lieutenant E., 154
Phillps, J.C., 379
Piennaar, Major General Daniel, 193
Pilgrim, Gunner B.C., 378
Pilgrim, Gunner R.F., 373
pistol gun, 171, see also sniping guns
Pitt, Lieutenant, 267
Pope, Acting Bombardier R.H., 373
Poyntz, Lance Bombardier J.S., 373
predicted fire, 9–10, 81, 83, 131, 139, 155, 172, 228, 243, 245, 282, 285, 287, 293, 392
    night shoot, 155
Price, P.D., 377
Prouse, Gunner J.A., 375

Q

Quilliam, Gunner J., 150
Quinn, H.B., 378

R

Ralph, Major M.R., 149, 154, 214
Ramchke, Generalmajor Hermann, 203, 339
Ramsay, Major General Sir Alan, ix, xiii, xx, xxvi, 54, 100, 150–152, 155, 177–178, 181, 186–188, 201, 208–209, 212p, 213–214, 224, 227, 230–231, 245, 249, 252p, 255, 269, 279, 281–285, 292, 296, 305, 310, 332, 368, 384–385
Ramsden, Lieutenant General W.H.C., 176, 181, 189, 193, 297
Ramsford, Bombardier P.S., 127
Reid, D., 378
resupply in the desert, 152–154
Reynolds, Gunner J.J., 375
Richardson, Lieutenant R.E., 368, 376–377

Richardson, Major Frank, 72, 330
Richardson, Sergeant B.B., 377
Richie, Lieutenant General Neil, 166, 168, 170, 331
Rigby, Gunner W., 377
Roberts, Captain T.L., 151, 182, 368, 376
Roberts, Gunner R.A., 377
Robertson, Brigadier, 85–86, 99
Robinson, Gunner J.V., 374
Rodriguez, Colonel Tim, xiii, 178, 192, 218, 242, 244, 260, 278, 341, 368, 385
Rogers, Lance Bombardier P.K., 374–375
Rogers, Lieutenant Colonel A.L., 316
Rogers, Major Alf, 242
Rommel, Feldmarschal Erwin, xi, xxiii, xxv, 25, 92, 109–115, 118–122, 138, 140–142, 145–148, 150, 152, 154–155, 159, 161–162, 165–166, 168–171, 175–177, 180, 185, 188–191, 193, 197, 200–201, 203–204, 207, 217, 223, 225, 227, 263–264, 268–270, 283, 285, 290–293, 298–299, 302, 306–307, 309–311, 313–314, 323, 332, 334–337, 339, 344, 390
Rosenthal, Major General Sir Charles, 8
Ross, W.L., 379
Rowlands, Lieutenant F.W., 376
Royal Air Force, 9, 13, 40, 63, 78, 81, 110, 113, 122, 131–132, 169, 171, 177, 249, 257, 270, 292, 302, 316, 328, 335
    No. 2 Photo Reconnaissance Squadron, 335
    No. 6 Squadron, 82
    No. 208 Squadron, 83, 87, 335
    No. 451 Squadron, 132
    Artillery Reconnaissance Squadrons, 33
    Lysander Army Cooperation Squadrons, 41, 82, 87, 134
    meteorological support, 128–129, 172, 342
    Photo Reconnaissance, 234
Royal Artillery formations and units
    5th Army Group Royal Artillery, 223, 225, 232, 308, 340
    31st Field Regiment, 147–148
    51st (Cumberland) Army Field Regiment, 74, 76, 79, 84, 110, 113–114, 119–120, 129, 134
    126th Field Regiment, 219, 229, 254, 268
    127th Field Regiment, 219, 229, 254, 268
    128th Field Regiment, 219, 229, 254, 268
    146th Field Regiment, 239, 247, 269, 281–282
    7th Medium Regiment, xi, xiv, xx, 69, 74–78, 81, 84, 111, 114, 118, 121, 169, 181, 189, 191–192, 224–226, 228, 230, 236, 243, 245, 262, 268, 281, 285, 288, 308, 321, 330, 344
    32nd Medium Regiment, 308
    64th Medium Regiment, xvii, 84, 169, 225, 227–228, 230, 232, 239, 243, 245, 254–255, 262, 266, 269, 277, 286, 288, 308, 321, 342
    68th Medium Regiment, 74, 76, 84
    69th Medium Regiment, 225, 228, 230, 232, 236, 308
    4th Survey Regiment, xiv, xvii, 127, 129, 131, 172, 191–192, 202, 228, 230, 233–235, 237, 277, 292, 308, 310, 322, 342
    94th Observation Regiment, xx, 309
    14th Light Anti-Aircraft Regiment, 120
    Royal Horse Artillery
        1st Regiment, 74, 83, 91, 113–114, 118, 120–122, 129–130, 147, 156
        2nd Regiment, 239, 269
        3rd Regiment, 318
        4th Regiment, 74, 76, 79–80, 84, 269, 330
        7th (Para) Regiment, 318
        103rd Regiment, 114
        104th Regiment, 74, 76, 82, 85–86, 111–112, 114, 120, 141
        107th Regiment, 118, 120, 122, 139, 169, 225, 340
        144th Regiment, 169, 334
        3rd Anti-Tank Regiment, 74, 84–85, 120, 140
        106th Anti-Tank Regiment, 74, 91
Royal Australian Air Force, 113, 160, 335
    No. 3 Squadron, 82, 335
    No. 450 Squadron, 335
    No. 451 Squadron, 132, 167, 179, 187

No. 459 Squadron, 335
Royal Australian Artillery
    casualties, 371–380
    honours and awards, 367–369
    post-World War II, 314–319
Royal Australian Artillery units
    1st Field Regiment, 317
    2/1st Field Regiment, xiv, xx, 51–53, 70, 73, 76, 78–80, 82, 84, 87, 89, 92, 310, 329, 331
        casualties, 372
    2/2nd Field Regiment, casualties, 372
    2/3rd Field Regiment, xiv, 53–55, 70, 84–85, 87, 310, 329, 331–332
        casualties, 372
    2/3rd Light Anti-Aircraft Regiment, 96, 120
    2/4th Field Regiment, 54
    2/4th Light Anti-Aircraft Regiment, 186, 353
    2/7th Field Regiment, ix, xiii, xx, 96, 100–104, 124, 145, 149–156, 169, 178–181, 183–186, 188, 191–193, 213–216, 218–219, 229, 232, 236, 239, 243–244, 254–258, 260–266, 268–274, 277–282, 285, 287, 290, 310, 316, 321, 323, 332, 335, 337–340, 342, 353, 361–363, 365, 385
        casualties, 372–374
        final preparations for El Alamein battle, 245–249
        honours and awards, 368
        personnel, 241–242, 244
    2/8th Field Regiment, ix, xiii, xx, 96, 100–103, 145, 149, 151–154, 156, 169, 178, 180–184, 186, 188, 191, 213–214, 216, 219, 226, 229–230, 236, 239, 241, 245–247, 254–256, 260, 262–263, 266–269, 271, 273–274, 277–279, 310, 316, 321, 332, 335–337, 340–341, 343, 353, 365, 385
        casualties, 374–377
        honours and awards, 368
    2/12th Field Regiment, ix, xiii, xx, 100, 103–104, 116, 118–121, 123–124, 126, 129–131, 135, 137–138, 141–142, 145, 149–150, 155, 178–179, 181, 185–188, 191, 213–214, 219–220, 229, 239, 244–247, 254–257, 259–265, 267–274, 277–279, 310, 321, 332–333, 343, 348, 353, 365, 385
        casualties, 377–380
        honours and awards, 369
    3rd Field Regiment, 316
    4th Field Regiment, 317
    6th Field Regiment, 316
    10th Field Regiment, 40
    12th Field Regiment, 317
    13th Field Regiment, 316
    23rd Field Regiment, 316
    36th Heavy Artillery Group, 3
    2/3rd Anti-Tank Regiment, 95–96, 112, 120, 140, 186, 334, 339, 353
    3rd Anti-Tank Regiment, 316
    4th Australian Light Anti-Aircraft Regiment, 353
    Artillery Training Regiment, 220
    A Battery, 1–2
Rumberg, C.R., 378
Rungie, Major R.H., 244, 316, 385
Russell, Bombardier P., 79, 372
Ryan, D.J., 377
Ryan, Gunner J.M., 278, 369, 374
Ryan, Gunner J.T., 372

S

Salter, Major A.A., 103, 214, 316
Saunders, Gunner A.T., 376
Saunders, Gunner R.J., 378
Savige, Brigadier Stan, 76, 79, 89, 331
Scarlett, Captain G.Y.D, 80, 367
Schnaars, Gunner H.J., 374
Schools of Artillery
    Al Maza, 100, 156, 180, 185
    Larkhill, 14, 55, 309, 342
    South Head, 50
Schrader, Captain C.L., 150
schwerpunkt, 146, 165, 198, 200, 298, 334
Scobie, Lieutenant General, 155
Scorpion (enigma code), 334

Scull, Lance Bombardier C.C., 373
Sea Lion, Operation, 44
Seabert, Captain L.R., 214
Searle, A.E., 378
Seater, A.J., 379
Seebohm, Hauptman, 201
Seely, Lieutenant Colonel W.E., 118, 120, 139, 169
Sewell, R.A.E., 380
Sewell, R.F., 380
Sharp, Gunner A.L., 375
Shave, Lieutenant L.K., 114
Sheehan, Lance Sergeant W.E., 375
Sheenan, Lance Sergeant W.K., 374
Shmith, Bombardier R.A., 377
Shovelton, R.J., 377
Sibree, Captain E.W., 214, 281
Siggs, Brigadier, 147
Skeels, Lance Sergeant F.J., 373
Slender, Operation, 219
slidex (code), 202
Smart, Warrant Officer II C.J., 180, 369
Smith, C., 377
Smith, Captain, A.F.A., 262
Smith, D.W., 374
Smith, Gunner C.W., 372
Smith, Gunner W.K., 372
Smith, Howard, 220
Smith, Lieutenant N.F., 373
Smith, Lieutenant T.D., 375
Smith, Major J.C., 120
Smith, R.T., 377
Smith, Warrant Officer I Fred, 104
smokescreens, 5, 7, 28, 88, 171, 184–186, 267, 273, 290
Sneath, Gunner H.H., 373
sniping guns, 130, 150–151, 187, 206, see also pistol gun
sound ranging, 34–35, 128–129, 132, 235–238, 255
Souter, Gunner W., 373
South African formations and units, 181, 186, 189–190, 225, 229
    1st South African Division, 101, 164, 217, 224, 229–230, 236, 244, 371
    4th Field Regiment, 229, 236
    7th Field Regiment, 224
    15th Field Regiment, 229
    75th Field Regiment, 229
Spreadborough, Sergeant E.F., 127
Squires, Lance Corporal, 79
Stammers, Gunner R.R., 376
Starkey, E.A., 378
Steele, Lance Bombardier G.L., 278, 369
Steer, Lieutenant, J.L., 374
Stephen, Gunner A.C., 375
Stephen, J.M., 378–379
Stevens, Captain Bill, xiii, 213–214, 242
Stevens, Major F.D., 214, 278, 368
Stewart, Captain G., 278, 369
Stigant, A.E., 377
stonk, x, 27, 61, 185, 244, 262, 285, 393
Streich, Generalmajor, 159
strosslinie device, 200
Strutt, Brigadier Horace, 55, 84, 85p, 383
Stumme, General, 290
Sturdee, Colonel Vernon, 18
Supercharge, Operation, xvii, 236, 253, 267, 270–273, 280, 342
survey, 9, 40, 63, 127, 164, 228–230, 233, 236, 287, 393
    and counter-battery fire, 131, 281
    decrease in importance, 46
    and predicted fire, 83
    role of survey officer, 56
    shooting for, 26
Syer, Lieutenant Colonel R.L., 227
Syria, artillery training in, 177–179

T

Talbot, Gunner R.E., 375
Tatchell, Lieutenant, 86
Taylor, T.A., 379
Tedder, Air Marshal, 189, 291
Teirney, W.J., 379
Tel el Eisa, attack on, 181–183, 186
Telic, Operation, 318
Tellera, General, 92

Theron, Brigadier F., 224, 340
Thomas, Bombardier G.A., 372
Thurley, G.G., 380
Tinsley, Brigadier Walter, 102, 145, 150–151, 216p, 278, 310, 332, 368, 385
Tobruk, attack on Italian position, 82–88
    artillery plan, 82–86
    artillery weapons captured, 88
    lessons, 87
Tobruk, defence of, 115–142
    ammunition expenditure, 136
    ammunition requirements, 124
    anti-tank gunnery, 140
    artillery order of battle, 229
    casualties, 137
    counter-bombardment, 127, 129–130, 135
    DAK artillery, 121, 135
    employment of artillery, 117–142
    flash spotting, 128–133
    German air support, 138–139
    Italian artillery, 135
    lessons, 139–140
    living conditions, 137
    meteorology, 128
    rations, 137
    sound ranging, 128–129, 132
    survey, 127
    use of captured Italian artillery, 125–126, 136
    water, 136–137
Todhunter, Lieutenant Colonel E.J., 112, 114, 332
Torch, Operation, 177, 314
Tovell, Brigadier, 116
Trengrove, Lance Bombardier A.E., 369
Trenwith, Major R.G., 272, 379
Trott, Sergeant A.L., 373
Trounson, J., 379
Tudor, General Henry, 8
Turner, Gunner A.B., 375–376
Turner, Lieutenant P., 244, 262–263, 369
Tutton, F., 334
Tutton, J.K., 378

U

Ultra special intelligence, xxiii, 201, 217
Uniacke, Lieutenant General Herbert, 8, 47
Upcher, Gunner P.R., 376

V

Vickery, Captain Norman, 78, 89, 90p, 367
Vidal, Bombardier B.R., 372
Vinden, Gunner W.A., 372
von Barst, Generalmajor Gustav, 203, 292
von Radow, Generalmajor Heinz, 203
von Sponeck, Generalmajor Theodore Graf, 203–205
von Thoma, General, 288, 290

W

Wagstaff, Captain John, 260
Wall, Gunner S.R., 373
Wallworth, Sergeant T.H., 375
Ward, Gunner S.A., 265, 268, 278, 368
Ward, R.C., 378
Ware, Gunner R.T., 376
Ware, R.E., 378
Ware, Signaller R., 182
Warren, A.F., 378
Watson, Gunner D.H.J., 373
Wavell, General Sir Archibald, 28, 96, 109, 111–113, 115–116, 141, 146, 148, 384
Weakly, Warrant Officer I, 101
weapons, Allied artillery
    25-pounder gun howitzer
        ammunition, 59–60
        description of, 15–16, 58–60, 358
        the Priest, 35
        production of, 18
    4.5-inch gun, characteristics, 4–5, 17–18, 358
    5.5-inch gun, characteristics, 17, 121, 226, 229, 288, 358
Weir, Brigadier C.E., 224
Weller, Gunner G., 373
Wells, Lieutenant General Sir Henry, 97p, 177, 181, 336, 384

Whetton, Lieutenant Colonel J., 120, 130, 308
Whinfield, F.R., 379
Whitelaw, Major General John, 51
Whyte, Lieutenant Ken, 242
Wicks, Gunner W.S., 377
Wilcox, Gunner V.T., 376
Williams, Lieutenant Colonel S., 120
Williams, Major General Robert E., 339
Williams, Major Hylton E., 100, 213–214, 332, 339
Williams, Major (US Army), 187
Williams, Warrant Officer II R.D., 377
Wilmot, Chester, 141
Wilson, Bombardier K.G., 179
Wilson, General 'Jumbo', 96
Wilson, Gunner T.F., 373
Wilson, Lance Sergeant A.T.R., 374
Windeyer, Brigadier, 263, 267
Wittus, Lieutenant J.O.F., 131
Wood, F.D., 378
Woods, Gunner S.W., 374
Woodstock, K.V., 377
Wootten, Lieutenant Colonel G., 96, 116
Wragge, Gunner S.M., 375
Wright, Gunner C.J., 377
Wynter, Major General H.D., 95–96

Y

Yates, Brigadier, 308, 344
Yates, Lieutenant Colonel M. 'Stag', 86–87, 192, 223
Young, Captain V.L., 125

www.ingramcontent.com/pod-product-compliance
Lightning Source LLC
Chambersburg PA
CBHW082058230426
43670CB00017B/2880